D0945238

THE ECONOMIC EFFECTS OF REGULATION

The Trunk-Line Railroad Cartels and the Interstate Commerce Commission Before 1900

THE ECONOMIC EFFECTS OF REGULATION

The Trunk-Line Railroad Cartels and the Interstate Commerce Commission Before 1900

PAUL W. MACAVOY

THE M.I.T. PRESS

Massachusetts Institute of Technology
Cambridge, Massachusetts

Library of Congress Catalog Card Number 65-23542
Printed in the United States of America

Preface

To avoid all suggestions of mislabeling the product — an important consideration given the price — an advertisement might be provided at the outset as to what this study includes, and what it does not.

It is an attempt at analysis of the effects of cartel agreements among the major railroads east of the Mississippi River and north of the Ohio River before 1900. The effects take the form of changes in the time pattern of rates, tonnage, and profits following from formation or dissolution of such agreements. These effects are identified by the economic theories of competition, oligopoly, or monopoly.

The cartel agreement of greatest interest is that in operation at the time of the introduction of Interstate Commerce Commission regulation in 1887. The new Commission, from an economic point of view, placed controls over the railroads' business conduct. Particularly, it sought to prevent rate discrimination, to require certain services to be offered to all shippers, and to require that rates or services be offered as publicized. These rules had some effect upon the actual rates set by the railroads — an effect estimated from a comparison of prices and outputs under regulation with those before regulation.

The regulation of the structure of rates, it is argued, tended to make more effective the cartel's control of the level of long-distance rates. The rates on eastbound transport between the Mississippi River Valley and the East Coast, in the fifteen years preceding regulation in 1887, fluctuated in consequence of frequent changes in the extent of cartel control. After 1875, there seems to have been persistent "cheating" on the rates set in conference, so that the level of rates declined markedly as one agreement after another broke down (particularly during 1876 to 1879 and 1883 to 1887). The effect of regulation, as assessed by comparing this period with 1887 to 1893, seems to have been to establish the cartel rates as the actual rates. Rates were "stabilized," and "increased," for long-distance transport of bulk commodities. Rates were "destabilized" when regulation was weakened by the Supreme Court in the later 1890's.

This study is not history, nor is it econometrics. The first would require discussion of the effects on the railroads of changes in the

United States economy and of the personalities of the officers of the cartel and member lines — discussion that is lacking for reasons including that of limits on information and on financial resources. The second requires data not found here; in the presence of incomplete and inexact statistics, there is reliance on "qualitative" assessment and on simple regression analyses as means for summarizing, rather than on econometric techniques as means for testing specific hypotheses. Only in a world of full disclosure of rates, rebates, and costs — and the reasons for each — would history or econometrics have been possible.

Whatever the results, any gains are due to large expenditures of time and funds by individuals other than the author. The officers of the Pennsylvania Railroad, the New York Central Railroad, and the Erie Railroad made available records in various warehouses throughout the country; the records of the Baltimore and Ohio Railroad for the period before 1880 were made available, with gracious assistance, by the Maryland Historical Society (unfortunately those for later periods are suspected to be in an abandoned, but sealed, warehouse of this railroad). A Ford Foundation Faculty Research Fellowship in 1961–1962 made possible the search for and collection of these railroad materials. Large monetary commitments were made from Faculty Research Funds of the Graduate School of Business, University of Chicago, for card reproduction of the statistics on daily grain and lard prices, on tonnage shipments from Chicago and into New York, and on weekly prices of railroad common stocks for 1870 to 1900. A Law and Economics Fellowship in the Law School of the University of Chicago in 1961–1962 provided stimulation from interested colleagues during the search for reasons for cartel instability. A grant from the Sloan Research Fund of the Alfred P. Sloan School of Management, Massachusetts Institute of Technology, subsidized the trend and annual regression analyses, and the valuable research assistance of Frederic Grzybowski. Counsel or criticism have been generously provided by Aaron Director, George Hilton, Brian Leslie Johns, Daniel Orr, Joel Segall, and George J. Stigler. Sidney Alexander, Robert Bishop, James Ferguson, and Robert Fogel have completed thorough, helpful critiques of the manuscript. Would that recognition for their accomplishments could be bestowed without accompanying responsibility for the errors; would that it was what they wished.

PAUL W. MACAVOY

Contents

1 Markets for Transport in the Midwest and Sources of Transport Services in the Early 1870's 1

Steamers and Railroads from Chicago 6

Summary: Market Definition and Sources of Transport Services 10

2 Prerequisites and Effects of a Successful Cartel 13

Requirements for Loyalty to a Cartel 14

The Results of a Successful Cartel 21

Measuring Reasons and Results 22

3 The Beginning of Cartel Control of Trunk-Line Railroad Rates, 1871 to 1879 25

The Effects of Collusive Rates, 1871 to 1874 30

The Effect of Collusive Rates Upon Grain Markets, 1871 to 1874 32

The Entry of the Baltimore and Ohio into Chicago and the Collapse of the Agreement, 1874 to 1876 39

The New Trunk-line Rate Agreements, 1876 to 1879 50

The Consequences for the Shippers of Cartel Instability, 1875 to 1879 63

The Consequences for the Railroads of Cartel Instability, 1875 to 1879 71

4 Elaborate Trunk-Line Cartels and the Control of Rates, 1880 to 1886 79

Harmony in 1880 and Rate Wars in 1881 80

The Pooling of Revenues and Joint Control of Rates During 1882 and 1883 91

Independent Railroads and a Breakdown in the Rate Agreement, 1884–1885 95

Last Attempts to Set and Maintain Rates, 1885–1886 103

The Resulting Structure of Through and Local Rates 106

5 Trunk-Line Railroad Rates and the Interstate Commerce Commission, 1887 to 1899 110

The Act to Regulate Commerce of 1887 111

The Interstate Commerce Commission's Standards for "Just and Reasonable" Rates 113

The Interstate Commerce Commission's First Attempts at Enforcement 115

The Trunk Lines' Initial Adjustments to Interstate Commerce Commission Regulation 119

Cartel Control of Long-Distance Rates, 1887 to 1890 125

Strengthened Regulation and Strict Cartel Control of Rates, 1891 to 1893 144

Regulation and Cartellization: The Weakening of Regulation and the Disruption of Rates in 1894 153

Regulation and Cartellization: Evasion of Regulation and the Rate Demoralizations of 1894–1895 164

Regulation and Cartellization: The Strengthening of Cartellization and Effective Control of Rates in 1896 176

The Destruction of Regulation and Cartellization, 1897 to 1899 183

6 The Effects of Regulation and Cartelization, 1887 to 1899 193

Appendix to Chapter II: Price Strategies to Promote Cartel Stability, by Daniel Orr and Paul W. MacAvoy 205

Statistical Appendices 221

Selected Bibliography 267

Index 273

". . . it is a wise shipper who can guess the rates of freight east for two consecutive days."

(The *Chicago Daily Commercial Bulletin,* March 4, 1895.)

1

Markets for Transport in the Midwest and Sources of Transport Services in the Early 1870's[1]

In most years of the first decade after the Civil War, farmers in seven states in the Midwest were responsible for approximately one-half of the nation's output of corn, wheat, and oats.[2] Illinois farms were the leaders in the production of each of these grains; farmers to the North provided large amounts of the hard varieties of wheat, while those to the South and East produced most of the remaining corn. With the presence of abundant feed, these producers established locations for fattening livestock and producing meat products: the seven states contained over 30 per cent of the cattle and over 47 per cent of the hogs in the country in 1870.[3] These individuals did not constitute a large proportion of the food-consuming population, nor did they produce their own machinery or merchandise; given this state of affairs, they were "in the market" for transport services to carry crops and livestock to major produce exchanges.

Chicago and St. Louis were the two large midwest produce exchange centers. More than 13 million bushels of the harder varieties of wheat were

[1] The following is in part a summary of H. F. Williamson, ed., *The Growth of the American Economy* (New York, 1951), Chapters 20, 22, 26; G. R. Johnson *et al.*, *History of Domestic Trade and Foreign Commerce* (Washington, 1915); and intensive studies such as H. C. Hill, "The Development of Chicago as a Center of the Meat Packing Industry," *Mississippi Valley Historical Review*, *X* (1923), 253–273, C. H. Taylor, *History of the Board of Trade of Chicago* (Chicago, 1917), and V. S. Clark, *History of Manufactures in the United States* (New York, 1929), Vol. II.

Cf. also G. R. Taylor, *The Transportation Revolution 1815–1860* (New York: Rinehart and Company, 1951), Chapters V, VI, VIII.

[2] As shown in United States Commissioner of Agriculture, *Annual Report*, *1870* (Washington, 1870) and *Annual Report*, *1871* (Washington, 1871), farms in the states of Wisconsin, Minnesota, Illinois, Iowa, Missouri, Ohio, and Indiana produced 55.9 per cent of the corn in successive years, 54.5 and 53.0 per cent of the wheat, and 40.9 and 49.8 per cent of the national output of oats.

[3] *Ibid.*

received in Chicago in 1871 from Wisconsin, Minnesota, and central Illinois (as shown in Table 1.1). Approximately 45 million bushels of corn

TABLE 1.1

Receipts in Chicago, 1871

State	*Wheat*	*Corn* (*Millions of Bushels*)	*Oats*	*Cattle* (*Thousands*)	*Hogs* (*Thousands*)
From Wisconsin, Minnesota, and Northern Iowa[a]	7.0	2.7	3.3	78.6	532.2
From Central-South Iowa[b]	2.9	15.0	5.4	261.0	1,063.0
From Missouri and Southern Illinois[c]	0.1	7.1	1.2	105.3	260.0
From South-Central Illinois[d]	4.0	10.4	3.1	79.4	440.5

[a] Transported by the Chicago and Northwestern Railway.
[b] Transported by the Chicago, Burlington and Quincy Railroad and the Chicago, Rock Island and Pacific Railroad.
[c] Transported by the Chicago and Alton Railroad.
[d] Transported by the Illinois Central Railroad; the totals may include shipments from points in Tennessee and Mississippi as well.
Source: *The Annual Report*, the Chicago Board of Trade (Chicago, 1872).

were received there; of that for which the origin was known, the majority came from southern Illinois and central Iowa. More than 2 million hogs and 0.5 million cattle were received in Chicago (one-half from Iowa, as shown in Table 1.1). With transport lines from points as diverse as those in Wisconsin and southern Iowa, and with the largest volumes actually received in a midwest city, Chicago was the center of commodity trading.

TABLE 1.2

Receipts in St. Louis, 1871

State	*Wheat*	*Corn* (*Millions of Bushels*)	*Oats*	*Cattle* (*Thousands*)	*Hogs* (*Thousands*)
From Central-Southern Missouri[a]	4.0	1.6	2.0	184.1	452.6
From Arkansas-Texas[b]	0.01	0.001	0.001	1.2	2.8
From Northern Missouri–Iowa[c]	1.1	0.3	1.1	4.5	13.8

[a] Transported by the Missouri Pacific Railroad; the St. Louis, Kansas City and Northern Railway; the Atlantic and Pacific Railroad; and Missouri River Boats.
[b] Transported by the St. Louis and Iron Mountain Railroad and Arkansas River or White River Packets.
[c] Transported by Upper Mississippi River Packets.
Source: The *Annual Report* of the St. Louis Union Merchants' Exchange (St. Louis, 1871).

St. Louis was a smaller exchange center serving agricultural regions farther South and West. Shipments of 1 million bushels of wheat, and approximately the same amount of oats, were made from northern Missouri and southern Iowa into St. Louis on railroads adjoining those transporting to Chicago in 1871 (as shown in Table 1.2). Four million bushels of wheat, and 1.6 million bushels of corn, and 0.2 million cattle came from south and central Missouri that year. The city exchanges were a smaller version of those in Chicago: a full range of trading was offered in spot and futures transactions in wheat, corn, and oats; spot sales were made in cattle, hogs, sheep, and processed meats from these animals; but the volume of transactions was less.[4]

Chicago and St. Louis were collection centers for agricultural products but not important locations of final consumption. Shipments out of these two cities in 1870 included more than 15 million bushels of wheat, more than 20 million bushels of corn, and more than 11 million bushels of oats.[5] The final destination was one or the other of the Atlantic seacoast cities. The daily reports of activity in the Chicago Board of Trade note "purchases by Boston, New York, New Haven and city buyers" (for example, in the *Chicago Daily Commercial Bulletin* for October 3, 1870) and include a record of charters for steamer shipments of grain "on through rates to New York" via the railroads from Oswego, Port Sarnia, and Buffalo. More than 30 million bushels of wheat, 18 million bushels of corn, and 15 million bushels of oats were received in New York, Boston, Baltimore, and Philadelphia in that year (as shown in Table 1.3).[6]

The large receipts in New York City, and the smaller receipts elsewhere, were in keeping with consumption needs, but were prompted by growing international trade in grain and produce. As trade expanded during the 1870's, receipts in all the Atlantic ports together were greatly increased

[4] There were other market centers in Peoria and Milwaukee which adjoined the Chicago and St. Louis markets. These cities had facilities for limited exchange and storage; for example, both provided exchange in spot grain but no futures markets, and the Chicago grain storage capacity of 11.3 million bushels in 1870 was twice that of Milwaukee and Peoria combined.

[5] It is apparent from the eastern seaboard receipts, as compared to Chicago–St. Louis shipments, that not all of the grain received in the East came originally from Chicago and St. Louis. Additional sources of grain were found in Ohio, Indiana, and in the home states of the eastern cities themselves.

[6] There was considerable movement of animals from Chicago to St. Louis to the eastern population centers as well. Of the 356,026 cattle received in New York, 205,255 were from Illinois cattle markets [*Report of the Commissioner of Agriculture for the Year 1870* (Washington: Government Printing Office, 1871), p. 49]. More than 33 thousand of the 112 thousand dressed hogs received in New York were from Chicago packing houses and were delivered by one railroad. As a matter of fact, "The falling off in the number of hogs (received in New York) is attributed to the increased number of dressed hogs shipped in the winter from the west." (*Ibid.*, pp. 50–51.)

TABLE 1.3

Receipts of Grain on the Atlantic Seacoast, 1870

	Wheat (*Millions of Bushels*)	*Corn* (*Millions of Bushels*)	*Oats* (*Millions of Bushels*)
New York	23.9	9.2	9.6
Boston	0.2	2.4	2.1
Baltimore	3.0	3.8	1.2
Philadelphia	3.2	3.0	2.3
Total	30.3	18.4	15.2

Source: *Annual Report*, the New York Produce Exchange (New York, 1870).

and were accompanied by comparable increases in exports to Great Britain and Europe (as shown in Table 1.4). At the same time, shipments

TABLE 1.4

Wheat Receipts and Exports in Eastern Cities

Year	*Receipts of Wheat at Seven Atlantic Ports* (*Millions of Bushels*)	*U.S. Exports of Wheat to Great Britain*[a] (*Millions of Bushels*)	*Receipts of Corn at Seven Atlantic Ports* (*Millions of Bushels*)	*U.S. Exports of Corn to Great Britain* (*Millions of Bushels*)
1868	—	12	—	9
1869	—	13	—	4
1870	—	27	—	0
1871	43	22	53	5
1872	28	19	77	25
1873	52	31 (39)	54	29 (38)
1874	63	51 (71)	54	26 (34)
1875	54	42 (53)	51	23 (28)
1876	43	42 (55)	88	42 (49)
1877	47	31 (40)	88	55 (70)
1878	112	54 (72)	104	65 (85)
1879	164	57 (122)	105	64 (86)

Sources: [a] The U.S. Bureau of Statistics, as shown in the *Annual Report*, the Chicago Board of Trade, 1879 (p. 43). Figures in parentheses are total exports to all foreign countries, as shown in the same source.

from Chicago grew in volume — particularly shipments of wheat and corn, which were export grains, but not of oats, which were destined for home consumption (cf. Table 1.5).

Because of this international trade, Chicago was a first collection center

TABLE 1.5

Shipments of Grain from Chicago

Year	Shipments of Wheat (Millions of Bushels)	Shipments of Corn (Millions of Bushels)	Shipments of Oats (Millions of Bushels)
1870	16	17	8
1871	12	36	15
1872	12	47	12
1873	24	36	15
1874	27	32	10
1875	23	26	10
1876	14	45	11
1877	14	46	12
1878	24	59	16
1879	31	61	13
1880	22	93	20

Source: *Annual Reports* of the Chicago Board of Trade.

for corn and northern wheat, while New York or Philadelphia or Baltimore or Boston was the final collection center for sale in Great Britain. Farmers or (more often) commission merchants in the Chicago exchange markets were "in the market" for transport services to the eastern seaboard. It seems possible to argue that a commission merchant must have been able to obtain comparable transport to a number of eastern cities — there must have been rates at which transport services from Chicago to two or more of the seaboard cities were effective alternatives, given that such services were actually provided in the early 1870's. With "the transport market" consisting of all transporters with the same service to offer any one shipper, and all shippers seeking the same service from any one transporter (to any one of the locations), "one market" for transport services included railroads and other transporters able to carry from Chicago to any one of the larger north Atlantic seaboard cities.[7]

[7] This holds only within a "comparable" time period, of course, so that wagon transport need not be considered part of this transport market. It is possible that "St. Louis to the eastern seaboard" was an alternative service in this market: the commission merchant could have considered purchasing "similar corn" in Iowa for shipment via St. Louis to Baltimore, rather than in Illinois for shipment via Chicago to New York. Unfortunately there is no information as to actual transactions to support this extension of the defined market. There were shipments from St. Louis to the East, as there were from Chicago to the East, but there is no information that substantial amounts consisted of northern grains diverted from Chicago (as compared to southern wheat and Missouri corn) or consisted of St. Louis grains purchased instead of Chicago grains. Consequently, attention will be centered on the more limited transport market consisting of the Chicago–eastern seaboard traffic.

Steamers and Railroads from Chicago

There were well-established "lines of supply" between Chicago and the larger cities on the eastern seaboard by 1870. Great Lakes steamers and schooners provided chartered transport to Lake Erie ports in the warmer seasons; from there a number of railroads and the Erie Canal could be used for transport to Montreal, Boston, New York City, Philadelphia, or Baltimore. There were two extensive railroad systems that provided direct transport from Boston, New York, Baltimore, or Philadelphia to the Chicago region. And there were lesser railroads able to provide transport from one or two of the eastern seaports to some point with connections to Chicago over the two larger systems.

The Great Lakes steamers and schooners provided most of the service between Chicago and the northern portions of the states of Ohio, Pennsylvania, and New York in the early 1870's. Lake vessels carried almost all of the grain to the East: in 1871, 12.1 of 12.9 million bushels of wheat shipped from Chicago were transported by lake, while 34.2 of 36.7 million bushels of corn and 12.2 of 12.3 million bushels of oats were shipped from Chicago in the same manner (as reported in the 1871 *Annual Report* of the Chicago Board of Trade).

Lake vessels provided the means for receiving bulk commodities in Chicago, as well: more than 984 million board feet of lumber, out of a total 1,039 million board feet received in Chicago, were transported over the Great Lakes; approximately 515 thousand tons of coal out of the 1,081 thousand tons received in Chicago were transported by steamer or schooner. There was an established pattern of transport of grain and provisions eastward from Chicago[8] to Lake Erie ports and of westward shipments of lumber from Michigan or Canadian forests or of coal from Pennsylvania.[9]

[8] The ship companies were not an extensive source of transport East for provisions, however. There is no mention of cattle shipments by lake during the early 1870's, nor was a very large proportion of beef or processed pork shipped from lake port to lake port. During 1871, only 5 of 89 thousand barrels of beef were shipped from Chicago by lake and 34 of 149 thousand barrels of pork were listed on lake shipments. In total, 155 thousand pounds of provisions and cured meats of 163 million pounds, were shipped from Chicago via Lake Michigan. The processed meats, along with lard, were mostly transported by the railroads between Chicago and the eastern cities.

[9] This trade involved a large number of vessels. The Chicago Board of Trade lists vessels laid up in Chicago with cargoes of corn, during the winter of 1871–1872, as including 17 steamers with a total capacity of 0.4 million bushels and 104 sailing vessels with a total capacity of 2.4 million bushels. Chicago vessels were not the only sources of transport. The official listing of registered merchant vessels by customs districts, for the northern Lakes, indicates that there were 1,548 sailing vessels, 641 steam vessels, and 3,154 unrigged vessels on the Lakes as of June 30, 1870.

There were four railroads engaged in disrupting this established pattern, however; each provided transport service between Chicago, or some city connecting to Chicago, and at least one city on the East Coast. They had announced service at "published rates" on all types of freight (grouped according to four or more classes, with the first class at the highest rate). Moreover, service was regularly provided after 1870, from all indications, on the basis of a week in transit for most of the meat products, cattle, and grain, and almost all of the merchandise.[10]

The two railroads with independent and complete service were the New York Central system and the Pennsylvania system. The "lesser" railroads were the Erie, and the Baltimore and Ohio.

The New York Central Railroad system provided transport between New York City or Boston and Chicago along the perimeter of the Great Lakes. A through line from Chicago to Buffalo was obtained by the Vanderbilt interests via the merger late in 1869 of independent local lines between (1) Buffalo and Erie, Pennsylvania, (2) Erie and Cleveland, Ohio, (3) Cleveland and Toledo, Ohio, and (4) Toledo and Chicago. The "Lake Shore and Michigan Southern Railroad" that resulted put in uninterrupted service between these important cities. At the same time the formal union of the New York Central Railroad Company running from Buffalo to Albany, and the Hudson River Railroad Company between Albany and New York City, provided a Vanderbilt line from the Great Lakes to the seaboard. The Lake Shore and New York Central together provided the service from Chicago — service that had scarcely begun by 1870, given that grain dealers shipped no more than 1 million bushels of corn, 0.75 million bushels of oats, and 0.06 million bushels of wheat on this line. The line did transport some part of the trade in provisions even in the early 1870's: shipments from Chicago in 1871 included 42.5 million pounds of cured meats, 22.2 million pounds of lard, 4.5 million pounds of fresh provisions, and 2.4 million pounds of tallow and grease.[11]

The Pennsylvania Railroad offered more extensive transport service than did the New York Central lines. Its central artery ran the length of south

[10] Cf., e.g., T. B. Veblan, "The Price of Wheat Since 1867," *Journal of Political Economy*, *I* (1892), 90–95, where are outlined the improvements in handling and bulk transporting that made this service possible.

[11] There were also large shipments from Chicago to Buffalo by way of Detroit and the north shore of Lake Erie on the Michigan Central Railroad operating in cooperation with the Great Western Railroad of Canada. The Michigan Central was in the process of becoming a "Vanderbilt line"; it carried more than 1 million bushels of corn, 0.03 million bushels of wheat, and 1.2 million bushels of oats. It was not as important a transporter of meat products as the Lake Shore, however (it carried 18.2 million pounds of cured meats and 5.9 million pounds of lard in 1871). This tonnage of the two Central roads constituted from 40 to 50 per cent of the Chicago shipments of meat and provisions, and lard, as shown in Table 1.6.

Pennsylvania and connected with two affiliates traversing Ohio and Indiana which led into Chicago. The first affiliate, the Pittsburgh, Cincinnati and St. Louis, was a pre-1870 acquisition running from Pittsburgh to Columbus, Ohio. The second affiliate, the Columbus, Chicago and Indiana Central, provided a connection between Columbus and Indianapolis and also ran directly to Chicago. The Pennsylvania began operation of the C.C. & I.C. in February, 1869 and from that date provided transport for agricultural products accumulated at Chicago, as well as at collection points in central Indiana and Ohio, to the seacoast at Philadelphia. At the beginning, shipments were small: 0.05 million bushels of wheat, 0.09 million bushels of corn, and 0.05 million bushels of oats were transported in 1871; also, the line moved only 26.9 million pounds of cured meats and other provisions and 4.5 million pounds of lard.[12]

Control of the Pittsburgh, Fort Wayne and Chicago Railroad provided the Pennsylvania system with a second, and more direct, route from the East Coast to Chicago. This line offered through service from the seaport at Philadelphia to Chicago, along a shorter more northerly route, soon after the Pennsylvania's purchase of the Fort Wayne stock in February, 1869. In 1871 the Pennsylvania leased the United Railroad and Canal Company of New Jersey, which had access to terminal and storage facilities in New York City, and offered "through" transport service from Chicago to New York. Service began on a modest scale: the Fort Wayne received 0.06 million bushels of wheat, 0.3 million bushels of corn, 8 million bushels of oats at Chicago, as well as 63.5 million pounds of provisions and cured meats, and 26.3 million pounds of lard; and only a part of the shipments went "through" given that approximately 4.7 per cent of this amount of wheat, 24.1 per cent of this volume of corn, and 8.3

[12] Control of the C.C. & I.C. did provide the basis for the extension of transport services to shippers of goods South and West of Chicago, however. The line connected at Indianapolis with the Indianapolis and Terre Haute Railroad, and the latter road provided direct access to many of the smaller gathering points in southern Indiana. Unfortunately the Indianapolis and Terre Haute road could receive and ship to points farther West only over the independent St. Louis, Alton and Terre Haute Railroad — the only line providing access to the St. Louis area from southern Indiana. Rather than deal with this single source of transport, the Pennsylvania Company constructed the St. Louis, Vandalia and Terre Haute Railway during 1870; the completed line provided direct transport from Philadelphia to St. Louis. With the lease of the Little Miami Railroad and of a number of smaller lines, the Pennsylvania also added direct transport from Philadelphia to Cincinnati and Louisville. This completed coverage of most of the larger market centers in the Ohio River–Wabash River–Upper Mississippi River region. The beginnings, once again, were modest: from St. Louis, at least, the Vandalia line carried only 0.2 million bushels of corn, 0.05 million bushels of wheat, and 0.03 million bushels of oats in 1871; the line also carried 30 thousand head of cattle and 2.7 million pounds of lard.

per cent of these oats arrived in New York over the New Jersey (United) Railroad.[13]

The Erie Railroad provided no direct service from Chicago. In the immediate post-Civil War period, the Erie was entirely an upstate New York railroad. It provided transportation from New York City to Buffalo and Dunkirk, and transferred freight at these points to other railroads' cars (since the Erie operated on a 6-foot gauge track, rather than the standard 4 feet 8½ inches of the Pennsylvania, New York Central, or Baltimore and Ohio). Attempts were made between 1868 and 1870 to expand transport into the Midwest: an agreement was sought with the Atlantic and Great Western Railroad (a "wide gauge" railroad) for through service to Cincinnati and St. Louis; arrangements were proposed with both the C.C. & I.C. and the Fort Wayne by which one or the other would provide direct transport from the Atlantic and Great Western to Chicago. The agreement with the Atlantic and Great Western was confirmed in 1868. This line provided the Erie with an integrated route from New York to Cincinnati, if not with the sought-after route to Chicago and the upper Midwest. The lease of the C.C. & I.C. was not obtained by the Erie, but rather was the prize of the Pennsylvania Railroad.[14] The Erie, and Jay Gould, did obtain a substantial portion of the voting stock of the Pittsburgh, Fort Wayne and Chicago but was prevented from using the control for setting up "through" transport service.[15]

The failure of the Erie to acquire connecting lines resulted in more limited service than that of the New York Central lines. Mr. Gould's railway reached down into the Ohio River Valley, but it did not have direct access to the large trade center in Illinois. It had to receive rail shipments from the New York Central or Pennsylvania. As noted by the *Railroad Gazette*, "[The Erie does not have] relations with any connecting railroad accept [sic] the Atlantic and Great Western. It has not been able to load cars in

[13] The remaining volumes eastbound had been delivered at storage facilities in Philadelphia or Baltimore and delivered in New York over the Pennsylvania steamship line, the Empire Transportation Company.

[14] Cf. G. H. Burgess and M. C. Kennedy, *Centennial History of the Pennsylvania* (Philadelphia: Pennsylvania Railroad Company, 1949), pp. 195–200.

[15] One reason given in the corporate histories was that "The Fort Wayne management and directors did not want their property to come under Erie domination and . . . obtained an amendment to the Charter providing that the directors might be classified so that only one-fourth of the board came up for election each year [rather than the entire number]. Since there were 13 directors, and only six were classified for reelection in 1869 and 1870, it would have required two years for the Erie to have gained a majority of the Board and its attempt was not persisted in." *Ibid.*, p. 199. This gives the responsibility for failure to the short-term separation of ownership and control; it provides little information on either the source or extent of the Erie's impatience, or the degree of management control in the interim.

Milwaukee, Chicago, Detroit, Toledo and the interior towns of the West for New York. The consequence has been that they have had to depend more than any other line on the traffic between New York City and the Great Lakes."[16]

Another railroad with limited service from the upper Midwest was the Baltimore and Ohio. This road, founded in 1827, had not greatly extended beyond Maryland and southern Pennsylvania before the Civil War; it carried traffic from Pittsburgh and West Virginia to Baltimore, where its Locust Point terminal was the major port facility. In the late 1860's, however, the railroad gained control of the Central Ohio Railroad so that it became possible for the Baltimore road to transport from Columbus, or from Sandusky on Lake Erie, to the eastern seaboard.[17] The railroad was able to provide independent transport service from the Great Lakes, but it did not go as far as the large grain collection centers in upper Illinois and Wisconsin. It could not receive railroad traffic from Chicago except from the Pennsylvania or the New York Central.

Summary. Market Definition and Sources of Transport Services

On the whole, it would seem that Chicago served as a gathering center for newly produced grain from the upper Midwest and that, over the 1870's, an increasing percentage of foodstuffs for the East was for transshipment to European markets. Farmers and commission merchants were in the position of seeking transport service to the Atlantic Coast. It was possible to consider sale in Liverpool on the basis of shipment via Philadelphia or Baltimore or New York or Boston. As a consequence, the

[16] The *Railroad Gazette* (January 6, 1872), 422.

[17] Access to Sandusky was obtained from a 17-year lease of the Sandusky, Mansfield and Newark Railroad to the Central Ohio in July, 1869. This road connected the Central Ohio at Newark, Ohio to the lake port. The railroad also obtained direct access to St. Louis by completing an agreement for a through line from the end of its Parkersburg branch in West Virginia. This line was composed of the Marietta and Cincinnati Railroad, and the Ohio and Mississippi Railroad. The first "worked closely" with the Baltimore and Ohio, particularly since its president, Mr. John King, was the first vice-president of the Baltimore and Ohio; the second railroad operated a shipping company jointly with the Baltimore and Ohio (termed the "Continental Fast Freight Line") that provided through transport service from St. Louis to Baltimore. This connection was the means for shipping small volumes of grain, including 0.6 million bushels of corn, 0.02 million bushels of oats, and 0.2 million bushels of wheat. Also, the line shipped 11,500 cattle, 28,754 hogs, and 6.1 million pounds of lard (this last being the largest amount from that city in 1871).

market for transport services included, as suppliers, steamers and railroads providing continuous service to any one of these cities.

The transporters in this "service market" included a large number of lake transporters and, to some extent or other, four railroad transport systems. The lake transporters carried most of the wheat and corn and the majority of the oats from Chicago to the East at the beginning of the 1870's.[18] The railroads were responsible for most of the shipments of meat, provisions, and lard from Chicago (as shown in Table 1.6).

The New York Central received grain and provisions over two lines in Chicago for through shipment over its own road to its terminals in New York or in Boston. The Pennsylvania Railroad shipped out of Chicago over its own lines to New York or to Philadelphia or (at approximately this time) to Baltimore. In contrast, the Erie Railway did not have a direct connection to Chicago; rather, it was confined to accepting shipments from Lake Erie or from producers in the Ohio River Valley. The Baltimore and Ohio also lacked independent facilities for all-rail shipment from Chicago and had transfer facilities from only one lake port.

TABLE 1.6

Shipments by Rail and Lake from the Midwest, 1871

Route: From Chicago	*Wheat* (*Millions of Bushels*)	*Corn*	*Oats*	*Meat and Provisions* (*Millions of Pounds*)	*Lard* (*Millions of Pounds*)
Lake Steamer	12.1	34.2	8.7	—	—
New York Central	0.4	2.0	2.2	64	27
Erie Railroad	—	—	—	—	—
Pennsylvania Railroad	0.1	0.3	1.0	89	30
Baltimore and Ohio	—	—	—	—	—
Total Shipments from Chicago	12.9	36.7	12.1	163	61

Source: The preceding tables and text of this chapter. The totals include some winter shipments to points in central Illinois, Wisconsin *et al.* from storage facilities in Chicago, so that they are greater than the sum of the figures shown for the eastbound roads.

It was possible for a shipper to choose between two railroads and many lake transporters from Chicago. The two railroads early sought to limit the results of choice by making agreements on rates to be quoted and services to be offered. By the middle 1870's, the agreements took the form

[18] The railroads' share of the grain traffic greatly increased in the later 1870's, as will be noted. But, at the time of initiation of service, their total share was small.

of elaborate organizations for setting rates — organizations that can be termed "cartels."

Before reviewing the actual effects of such cartels, some attention can be given to examining the conditions necessary for them to operate successfully in any market.

2

Prerequisites and Effects of a Successful Cartel

The trunk-line railroads attempted to control rates on shipments of agricultural commodities and merchandise at first by means of simple oral agreement. There were occasional meetings of the chief freight agents of the Pennsylvania and the New York Central Railroads in the early 1870's from which followed uniform timetables, a common system of track signals, and the same set of published rates for each railroad. These meetings were supplemented by those of an administrative committee, however, after the middle 1870's — a committee having "the duty . . . to regulate through rates for the transportation of passengers and freight over the several lines between competitive points East and West."[1]

The committee was, in effect, the cartel organization for making changes in rates and fares and for making certain that the agreed rates were actual rates charged. When this organization was put into operation, it was recognized that oral agreement was not sufficient to set actual fares and charges — too much was gained by an individual railroad from agreeing and then undercutting the announced rate so as to obtain a larger share of the tonnage. When this organization was abandoned, the individual railroads were admitting that group control of rates was impossible.

The first such organization was in operation only a few months before dissolution in January, 1875. The presidents of the Erie and the Baltimore and Ohio joined those of the two larger railroads in establishing a second administrative body in April, 1879. This group was to have held rates up to the level of the agreed rates by removing from "the rate cutter" the tonnage gained at the expense of the other railroads; committee meetings were discontinued for short periods during 1881 and 1882 and were abandoned altogether in January, 1885, after an extended period of "demoralized conditions of eastbound freight rates and of the (tonnage) division."[2] The third working organization was formed late in 1886. It reformed the structure of agreed rates (in 1887, in accord with the new

[1] The agreement of the Western Railroad Bureau, Section (1) as reproduced in the *Railroad Gazette* (October 13, 1874), 403.

[2] The *Railroad Gazette* (January 23, 1885), 56.

Interstate Commerce Commission regulation) and put into effect new means for controlling relative shares of tonnage. Such reform seems to have been in effect until the summer of 1895, when once again rates were said to have been demoralized, the association dissolved, and each railroad announced its own rates. The fourth administrative body, organized almost immediately after the demise of the third, was given a strong mandate to "secure to each company equitable proportions of the competitive traffic covered by the agreement," and also to make "such changes in rates, fare, charges, and rules that may be reasonable and just and necessary for governing the traffic covered by the agreement." [3] There was an elaborate system of penalties for deviations from the announced rates and fares as well. The organization was dissolved in 1898.

From the first meeting in 1874 to the last dissolution in 1898, the trunk-line cartel went through four major, and many more minor, reorganizations. Each major reorganization followed in the path of dissolution, and each was a new attempt to provide means for detecting deviations from agreed rates and to introduce penalties against deviations. Such detection and enforcement were to guarantee loyalty to the agreement; it remains to investigate why there was immediate failure in some instances and relative success in others.

Gaining allegiance to a cartel agreement involves no more than convincing each separate railroad to charge the agreed rate. There are many convincing reasons: it is profitable for each railroad to be loyal; it is more profitable to be disloyal, but this is not recognized or this is ignored in order to maintain the agreement for its own sake. It is assumed that the agreement might be maintained for any one of these reasons, but that it is not as likely to be maintained when profits can be made by "cheating." That is, loyalty can follow from ignorance of its cost or for its own sake, but it is less likely than loyalty for profit.

Given such an assumption it follows that, when conditions in the market make it possible for an individual firm to make more money by being loyal to a cartel agreement than by being disloyal, the agreement is not likely to break down. The simple theory that follows specifies the market conditions for profitable loyalty. If these particular conditions can be guaranteed, then the cartel can expect success.

Requirements for Loyalty to a Cartel

In the process of setting up an agreement, a cartel might be assumed to set prices so as to maximize the total profits of the members. At least, more

[3] *The Article of Organization of the Joint Traffic Association* (November 19, 1895), Sections 7, 8.

profits are more satisfactory than less in most instances.[4] The members of the association might encounter marginal costs MC and demand for output D, and might be instructed to each set price similar to R so that marginal costs equal the marginal addition to revenue (as in Figure 2.1).

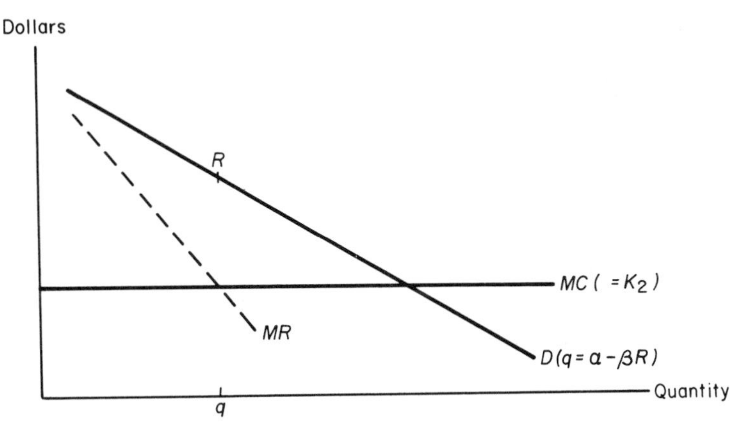

Figure 2.1. The price for maximum cartel profits.

When cost and demand conditions can be approximated by straight lines and each company receives an equal market share, then each company encounters demand for output q (at official agreed price R_0) $q = \alpha - \beta R_0$, and total costs $TC = K_1 + K_2 q$. The demand for output from the association of $(n + 1)$ companies is $Q = (n + 1)\alpha - (n + 1)\beta R_0$, and the costs of transport are $(n + 1)(K_1 + K_2 q)$. To maximize profits, each member of the association sets price so that $\pi = (n + 1)[(R_0 - K_2)(\alpha - \beta R_0) - K_1]$ is as large as possible,[5] and this requires that official price $R_0 = (\alpha + \beta K_2)/2\beta$.[6]

[4] The cost curve for setting prices for maximum group profits should be derived from individual firms' marginal costs by using the capacity first of the lowest cost line. If one firm has lower marginal costs for all quantities of output, then the *group MC* curve would be the same as this one firm's *MC* curve because all production would take place there. In a second instance, if one firm's marginal costs are lower only over the range 0 to 25 per cent of total production, then this firm should produce 25 per cent plus an equal share of the remaining (so that costs at the margin are the same for each line). This rule for lowest costs may not be adhered to; for one, output shares may follow from bargaining strength, so that a larger firm receives too large an output share. Without this rule, profits are not maximized as a deliberate policy.

[5] That is,

$$\pi = (n + 1)(R_0 q - TC) = (n + 1)[R_0(\alpha - \beta R_0) - K_1 - K_2 q]$$
$$\pi = (n + 1)[R_0(\alpha - \beta R_0) - K_1 - K_2(\alpha - \beta R_0)]$$
$$\pi = (n + 1)[(R_0 - K_2)(\alpha - \beta R_0) - K_1]$$

[6] For a maximum,

$$d\pi/dR_0 = 0 = (n + 1)[(R_0 - K_2)(-\beta) + (\alpha - \beta R_0)]$$

and

$$R_0 = (\alpha + \beta K_2)/2\beta$$

Maximum profits $\pi = [(\alpha - \beta K_2)^2/4\beta] - K_1$,[7] for each member, and $(n+1)[(\alpha - \beta K_2)^2/4\beta] - (n+1)K_1$ for the association as a whole.

It is possible, under some circumstances, for an individual producer to offer output for less than the association price R_0 and to profit from doing so. A producer's alloted share of output at official price R_0 is indicated by q_0 on demand curve D in Figure 2.2 (from Figure 2.1); the firm's actual

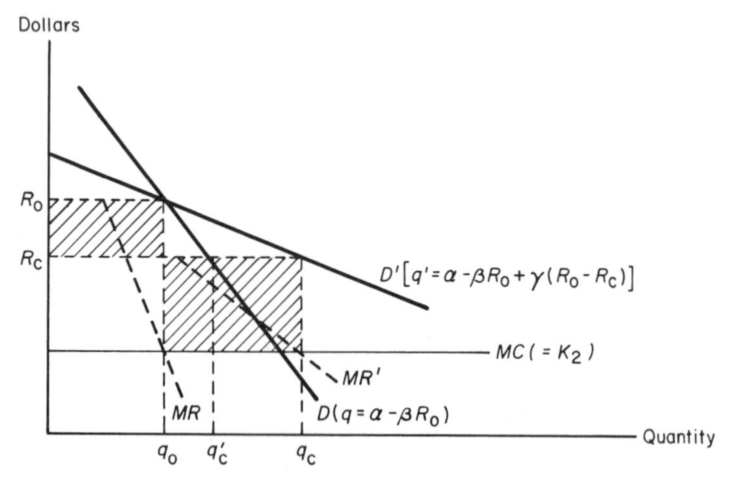

Figure 2.2. Profits from not adhering to the cartel price.

share at some lower price, given that the other producers quote the official price R_0, is shown by quantities on demand D' as including the allotted share and additional output from transfers of patronage at the lower relative price.[8] This single "cheater" receives larger profits from price-cutting during the period in which the association maintains the price offered by all other firms. [The additional profits from R_c are equal to receipts on additional output $(R_c - K_2)(q_c - q_0)$ minus the loss of revenues from the lower price $(R_0 - R_c)q_0$.[9]] But once the existence of R_c is known

[7] Given that $\pi = [(R_0 - K_2)(\alpha - \beta R_0) - K_1]$ and that, for maximum π, $R_0 = [(\alpha + \beta K_2)/2\beta]$, then $\pi = [(\alpha - \beta K_2)^2/4\beta - K_1]$. It should be recognized that prices are set for all of those market periods in which demand and cost conditions remain the same, so that the present value of profits is

$$[(\alpha - \beta K_2)^2/4\beta - K_1] \int_0^\infty e^{-\lambda t}\, dt = [(\alpha - \beta K_2)^2/4\beta - K_1]/\lambda$$

[8] Demand D' is thus defined as $q = \alpha - \beta R_0 + \gamma(R_0 - R_c)$ with $\gamma > \beta$; cf. also the appendix to this chapter.

[9] There are more general proofs that profits are greater at R_c than R_0 while R_0 is being quoted by the loyal association members. It can be asserted here, however, that if R_0 maximized the size of an equal share of joint profits, and if demand D' is more elastic and to the right of D at any $R_c < R_0$, then maximum profit price R_c for D' must be lower than R_0, and maximum profits on D' must be greater than on D.

to the other producers, R_0 is not likely to be maintained. Profits of the cheater at some lower official price are decreased since demand at R_c is decreased[10]; these profits *after detection* are less than the profits which would have followed from loyalty to the agreement in that period.[11] But this reduction in profits need not be so great as to cancel the gains from cheating during the period before R_c is acknowledged, so that there can conceivably be a net gain.[12]

The cartel might follow the policy of matching the lower price of a disloyal firm. The policy could be announced as a warning: R_0 is reduced to the nonadherent's price R_c when there is evidence that R_c exists. This would decrease the profits of the independent price setter since it decreases demand at R_c from q_c to q_c' (in Figure 2.2). This cannot reduce the profits from cheating to less than those from being loyal to the agreed price, however. It forces the cheater to revise the extent of his discount on the official price: rather than setting R_c to maximize profits before cheating is detected, this company now sets price to maximize the present value of

[10] For example, quantity demanded at R_c declines from q_c to q_c' when R_0 is set equal to R_c because, with all roads charging the same price, it is expected that the cheater receives no more than the allotted share.

[11] With the losses on the price decline equal to $(R_0 - R_c)q_0$ and the gains on output equal to $(R_c - K_2)(q_c' - q_0)$ in Figure 2.2, it can be seen that there are net losses. The official price R_0 provided maximum profits given demand curve D so that R_c with demand D results in less than maximum profits.

[12] This is to assume that price reductions by the cartel provide "penalties" — in the form of profit reduction — against the cheater. Money penalties might be levied directly on the cheater upon discovery of his activities; the level of the penalty could be set higher than the net profits at R_c (before detection). An association rule to the effect that "100 per cent of net revenues on output greater than an allotted share is to be paid into the general treasury" would cancel profits from nonadherence. Such penalties are effective in preventing cheating — because they prevent profits from cheating — but they cannot be collected easily since they are not legally enforceable. Cf. the decision of Justice W. H. Taft in *United States v. Addyston Pipe and Steel Company*, 85 Fed. 271 (1898), where the court reviewed the common law with respect to contracts involving agreements to set prices and apportion market sales. It was said that "covenants in partial restraint of trade are generally upheld as valid when they are agreements (1) by the seller of property or business not to compete with the buyer . . . ; (2) by a retiring partner not to compete with the firm; (3) by a partner pending the partnership not to do anything to interfere . . . ; (4) by the buyer of property not to use the same in competition with the business retained by the seller; and (5) by an assistant, servant, or agent not to compete with his master . . . Before such restraints are upheld, the court must find that the restraints attempted are reasonably necessary to the buyer to the enjoyment of property and good will . . . It would certainly seem to follow from the tests laid down . . . that no conventional restraint of trade can be enforced unless the covenant is merely ancillary to the main purpose of a lawful contract . . . If the restraint exceeds the necessity presented by the main purpose of the contract it is void." [*Ibid.*, pp. 281–282.] Under these standards, money penalties for nonadherence to a price agreement are not legally collectible.

profits to be received over the entire period from the first day of cheating to the last day in which R_c is matched by the loyal cartel members.[13] But the cheater can always find some price that increases profits (over those from R_0) before detection by a sufficient amount to compensate for the decreased profits (again, as compared to R_0) when that price is quoted by all the other companies.[14] As a consequence, the price-matching policy cannot serve to remove profit incentives for cheating.[15]

The association could react to evidence of cheating by setting a new official price which in turn undercuts the independent price. This increases the loyal firms' profits (*given* the secret rate, the demand for any loyal firm when undercutting is more similar to D' than D).[16] It also has direct effects upon the profitability of cheating: the cheater's demand curve D' is shifted to the left given the new official price below R_c (to demand D'', given that $q_0{}^*$ is the allotted share at official price $R_0{}^*$ in Figure 2.3). The disloyal firm, in this period of "price warfare," receives less profits than at the original official rate (as can be seen from comparing the rectangle $[(R_0 - R_c)q_0]$, the revenue lost from decreasing price, with the rectangle

[13] Such a cheater's price policy is necessary because this company sets price for all firms for the entire period: R_c holds before detection and is matched by the others after detection.

[14] The proof that cheating is always profitable when the association matches R_c is based on the following assumptions: consider that D and D' indicate not quantity at some prices R_0 and R_c but receipts above unit costs for any quantity (or that the vertical axis of Figure 2.2 indicates not the price, but $(R - K_2)$, with K_2 equal to marginal and average variable costs); that maximum profit R_0 is equal to $\alpha/2\beta$ and profit share $\pi_s = \alpha^2/4\beta\lambda$; and that the profits from cheating equal $R_c[\alpha - \beta R_0 + \gamma(R_0 - R_c)]$ before detection, and equal $R_c(\alpha - \beta R_c)$ after the independent price is matched. The profits from cheating equal

$$\pi_c = \frac{1}{\lambda}\{R_c(1 - e^{-\lambda m})[\alpha - \beta R_0 + \gamma(R_0 - R_c)] + e^{-\lambda m}R_c(\alpha - \beta R_c)\}$$

for interest rate "λ" and for the "m" periods before cheating is detected. Maximum profit price $R_c{}^*$ is set so that $\partial\pi_c/\partial R_c = 0$. The conditions under which $R_c{}^*$ leads to greater profits than R_0, or under which $\pi_s < \pi_c$, are that $m > 0$ (that R_c is not instantaneously detected), that $\alpha > 0$, $\beta < 0$, $\beta < \gamma$ (or that demand is finite and that shippers substitute the services of the independent firms at the lower rate for those of the loyal members of the cartel). These conditions are obtained from the value of $R_c{}^*$ at $\partial\pi_c/\partial R_c = 0$, from inserting this value in π_c and then solving for m, α, β *et al.* in $\pi_s = \pi_c$—as shown in the appendix to this chapter.

[15] Table A.1 of the appendix indicates that substantial errors could be made in setting the price given assumed linear demand curves *et al.*, without entirely eliminating the profits from cheating. For the examined instances, the time period before detection could have been one-half as long as expected without eliminating those profits.

[16] The loyal firm's demand curve is defined in the appendix as

$$q = \alpha - \beta R_0{}^* - \delta(R_0{}^* - R_c)$$

with $R_0{}^*$ the new cartel price undercutting R_c and with $\delta < \gamma$.

$[(q_c'' - q_0)(R_c - K_2)]$, the revenues gained from increased sales); these profit reductions can conceivably be larger than the profit increases from R_c in the period before detection. The sum total of profits from cheating can be less at cheater's price R_c, first at tonnage q_c, and then at tonnage q_c'', than the sum total of profits at the official price R_0 during the entire time period. There can be "reasons of profit" for remaining loyal to the agreement.[17]

The association's expected reaction to cheating, in other words, may be a deterrent in some cases. If the cheater's increase in demand is not "large,"

[17] Greater specification of conditions for losses from cheating — profits from loyalty — can be provided from more extensive consideration of the case of linear demand and cost curves. Assume that the association sets the official price $R_0 = [(\alpha + \beta K_2)/2\beta]$ with total demand $[Q = (n + 1)\alpha - (n + 1)\beta R_0]$, and marginal costs K_2 for each company in order to realize maximum group profits

$$(n + 1)\pi_s = \frac{(n + 1)(\alpha - \beta K_2)^2}{4\beta} - (n + 1)K_1$$

The potential cheater has demand equal to $[q = \alpha - \beta R_0 + \gamma(R_0 - R_c)]$ and sets R_c to obtain maximum net revenue

$$\pi_c = \frac{[\alpha - \beta R_0 + \gamma(R_0 - R_c)]^2}{\gamma} - K_1$$

for each of the m periods before cheating is recognized. When the association acts on definite evidence of cheating so as to maximize the group's profits (in accord with demand $[Q = n\alpha - n\beta R_0 - \delta(R_0 - R_c)]$) the rate R_0 is abandoned in favor of R_0^* for profits

$$n\pi^* = \frac{n[\alpha - \beta R_0^* - \delta(R_0^* - R_c)]^2}{\beta + \delta} - nK_1$$

during each of the periods after cheating is detected. The cheater's reaction to the lower price is to reset R_c at the lower level R_c^* for an indefinite period beginning at the end of the m periods. This company's profits in each period after detection are lower than in the earlier periods of undetected cheating and also are lower than if the original rate R_0 were being maintained. If profits are so much less than at R_0 that the gains before the detection of cheating are canceled, then cheating "doesn't pay." For two periods of equal length, one before detection and one after,

$$\pi_c + \pi_c^* = \frac{[\alpha - \beta R_0 + \gamma(R_0 - R_c)]^2}{\gamma} + \frac{[\alpha - \beta R_0^* + \gamma(R_0^* - R_c^*)]^2}{\gamma} - 2K_1$$

has to be greater than

$$\pi_s = \frac{(\alpha - \beta K_2)^2}{2\beta} - 2K_1$$

to have cheating for profit.

There are values of α, β, γ, and δ for which $\pi_c + \pi_c^* < \pi_s$. The potential cheater forming expectations not in terms of α, β *et al.*, but in terms of profits, realizes that the shift in tonnage (the value of γ) is not sufficient, the level of price high enough (the value of $R_c - R_0$), or the time lag before detection of such length that there is an inducement to cheat. Cf. the discussion in the appendix to this chapter.

then the losses during the expected "price war" are greater than the immediate gains. If the period before cheating is detected is "short" or if the price set by the cartel in the initial agreement is "low," then there are no gains from cheating. The association announcing a policy of undercutting an independent price in these instances need not experience cheating for profit — or need not experience any cheating at all (other than for reasons "of error").

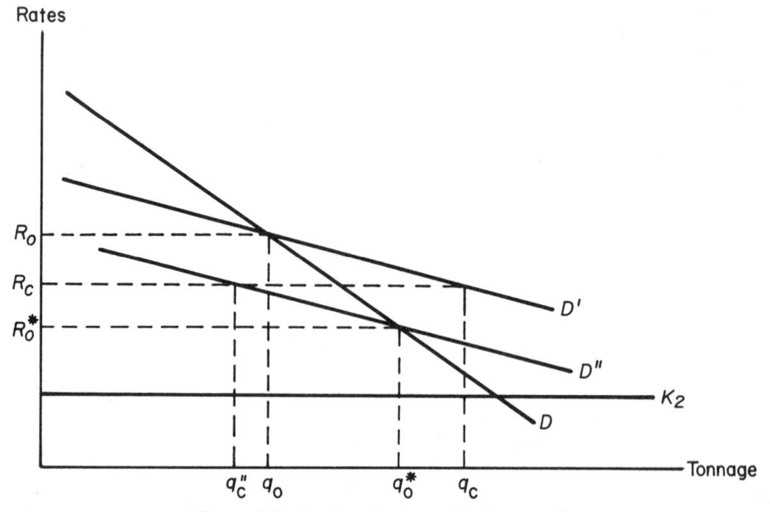

Figure 2.3. Profits during a "price war."

The association might operate a harmonious and prosperous agreement by merely announcing its intention to react to cheating in this particular manner. But an announcement is not always sufficient. If a "large gain" in sales before detection is expected (from large γ compared to β, or from a large number of periods m of undetected cheating), the potential cheater is rewarded with higher profits rather than deterred by losses. Cartel success depends upon "the conditions of the market" making for large relative changes in sales, for long periods before detection, and for high official prices. The cartel can seek to change such market conditions — to decrease the length of the period of undetected cheating, to decrease the cartel official price, or to decrease the shift in patronage at secret prices. If this is not successful, then market conditions determine failure.[18]

[18] Table A.2 of the appendix suggests that a shift of from 3 per cent to 10 per cent of total sales to the cheater for an 100-day period is sufficient for cheating to be profitable — at least when there are no more than four producers, constant marginal costs, linear demands, etc.

The Results of a Successful Cartel

The bureaucracy of a price-setting association can measure its success in terms of effects on the prices, market shares, and profits of individual members. The enthusiasm for any particular result may vary from one organization to another, but conventional standards of success include an increase in price, the allocation of market shares in accord with agreed market shares, and an increase in the total of the profits.

Prices are higher when all members of a cartel association conform to the agreement than when a firm deliberately nonconforms. The joint profit maximizing price R_0 is higher than the cheater's initial price R_c. The association's price undercutting the cheater's price, and the cheater's final price in reaction to being undercut, are in turn lower than R_c.[19] Consequently the collusive price R_0 is much greater than those at the low point of the "price war." [For example, with gains in sales from cheating roughly equal to twice the gain from a market-wide official price decrease, the cheater is expected to cut price so that his final price is one-third the original collusive price.[20]]

Market shares are more stable when cartel members are loyal to the agreement than when there is cheating. Cheating for profit can only take place when the price cut is more than compensated for by an increase in the cheater's percentage of sales. The association maximizes profits in the presence of cheating by undercutting the cheater's price so as to obtain a good part of the cheater's market share. As a consequence from preventing cheating from ever taking place, a successful cartel prevents such deliberate changes in market shares.

The successful cartel guarantees a greater total amount of profits, as well. The cartel price R_0 is the joint profit maximizing price; a reduction in R_0 by the cheater should have the effect of increasing the cheater's profits, but not by as much as the decrease in profits for the cartel members; the cartel's matching or undercutting of the cheater's price R_c increases profits of the loyal members relative to those when R_c is not acknowledged, but this reaction price cannot result in profits equal to those at R_0.[21] The

[19] More formal argument that R_0 is greater than R_c and greater than succeeding prices during the "price war" is provided in the appendix to this chapter. The argument continues to rely on linear demand curves, constant (or zero) marginal costs, and some shift of patronage to the price cutter from the loyal members of the cartel.

[20] Assuming that $\gamma = 2\beta$, $\delta = 2\beta$, and that there are five members of the cartel association, the cheater's final price R_c^* is approximately equal to $R_0/3$. This is indicated from solving the equations for the final cheater and cartel prices (in the appendix to this chapter) for R_0/R_c^* in terms of β.

[21] It has been indicated that R_0 is greater than price-matching R_c or the undercut R_0^* of the cheater's price. If R_0 is the joint profit maximizing price, then the subsequent lower prices result in lower joint profits.

cartel able to prevent price warfare can report to its members the highest consolidated income statement.

Success to the cartel is failure to purchasers of the cartel members' output. The maintenance of R_0 rather than $R_0{}^*$ has adverse effects upon all of those individuals and companies buying from the members of the cartel: unless there are compensating price decreases elsewhere in the economy, real incomes of purchasers are decreased. To be sure, there may be advantages for some buyers at the expense of others. Preventing cheating prevents the offer of R_c to some purchasers — to their disadvantage, but not to the disadvantage of other consumers not so offered. When the consumers are producers of goods competing in retail markets and purchasers of an input factor from the cartel, an offer of R_c may result in a firm having lower production costs than those firms purchasing at R_0 and a competitive advantage in the final market. No offers of R_c are to the advantage of buyers that would be excluded from purchasing at R_c, in other words. On the whole, however, sharp reductions of price, extensive fluctuations in market shares of producers, and reduction of producers' total profits are most agreeable measures of cartel failure from the buyer's point of view.

Measuring Reasons and Results

In the instances of cartels operated by the trunk-line railroads, success or failure was proclaimed at frequent intervals by the bureaucracy or by the managers of the individual railroads. The proclamations, in the form of correspondence and speeches, pointed to progress in gaining adherence to the agreement or otherwise took the form of bitter recriminations against cheaters responsible for breakdowns in the official rates. Shippers of agricultural commodities in particular were informed through the good offices of the *Railroad Gazette* and the New York or Chicago newspapers of the extent of success of any agreement on rates to the eastern seaboard.

The world was not informed of the reasons for cartel success, nor was there documentation of the effects. When there was a dissolution of a cartel organization, there were usually accusations against particular railroads of cheating so as to disrupt the entire organization. No explanation for the cheating was provided, however; the accused could have set lower rates by error, could have sought to deliberately destroy the agreement at his own expense, or could have cheated for profit. The effects from cartel success were mentioned in passing: railroad rates were said to have been increased, and railroad profits were said to have been at a more favorable level. Otherwise, there was no accounting for success and no statistical measure of its benefits.

Determining the actual reasons for success of the railroad cartels does not seem to be possible at this late date. If a breakdown in an agreement did take place, however, plausible reasons for failure might be offered. It would seem plausible that cheating had taken place for the reason that profits were expected; if it is estimated that increased profits were realized by the alleged cheater from cutting the official cartel rate, then this can be offered as the reason for the breakdown. An estimate of the profits from cheating equals $(R_c - K_2)Q_c$, the cheater's rate net of transport costs per ton multiplied by the tonnage of agricultural commodities actually delivered. The cheater's rate should decrease over time (from R_c to $R_c{}^*$, in reaction to having been undercut by the cartel's lower rate $R_0{}^*$).[22] If the profits under this sliding scale of rates are higher than $(R_0 - K_2)Q_0$, net receipts at official rates times the allocated share of the tonnage, then cheating would appear to have been profitable — and the cartel would appear to have broken down "for reasons of individual profit." The absence of profits in some one instance may suggest a plausible reason for cartel success after this instance — that the cheater demonstrated the existence of losses from such activity for all to see and to avoid.

Another plausible reason for profitless cheating might be that of deliberate destruction of the agreement as the first step in reforming the allocation of tonnage shares or the level of official rates. This reason can be substantiated by information on the reformation of the agreement; for example, if the reform seems to have been in favor of the alleged cheater, then it might be plausible to conclude the destruction of the agreement was for its own sake.

The effects of cartel success can be inferred in part from accounts of prices, profits, and shares of agricultural tonnage to the eastern seaboard. Levels of prices in periods of cartel operation can be compared with those during cartel breakdown. The official rate of the cartel R_0, and an estimate of the actual rate R_T can be contrasted in periods in which R_T was alleged to have been the cheater's rate R_c; if $R_0 > R_T$, and R_0 decreases, then cartel failure may imply a lower rate. Shares of the total tonnage of agricultural commodities transported from Chicago or received at some eastern seaport can be compared with allotted shares under the cartel agreement or with the long-term trend of the share of any railroad. If tonnage shares were "more stable," then it is assumed that the predicted effects from successful cartel activity were realized. Profits can be estimated from accounting statistics on annual income or from changes in common stock prices of the individual railroads (given that the common stock price is assumed to be the present discounted value of future profits for the stockholder).

[22] That is $(R_c - K_2)Q_c$ is a summary version of $(R_c - K_2)q_c + (R_c - K_2)q_c''$ in Figure 2.1 and in the most extended form of $(R_c - K_2)q_c + (R_c - K_2)q_c'' + (R_c{}^* - K_2)q_c{}^*$ in the discussion following Figure 2.1.

Estimated profits under cartel operation can be compared with those during periods of cartel inactivity, to ascertain whether breakdowns decreased total returns.

The effects of the cartel on consumers are the most difficult to measure. There is no testimony on the damage to shippers inflicted by the successful cartel. All that is attempted, in what follows, is a comparison of commodity shippers' gross margins with the rates for transportation so as to indicate the incursions made by higher rates. The inference is that the successful cartel was an imposition on shippers if the average gross margin increased with higher rates.

3

The Beginning of Cartel Control of Trunk-Line Railroad Rates, 1871 to 1879

In the early 1870's, it became the established custom for the New York Central and the Pennsylvania railroads to post the same rates for transport service from Chicago to the eastern seaboard. The other "trunk-line" railroads quoted these rates on transshipments from these two lines, or on freight shipped part of the way by Great Lakes steamer.[1] The formal means for setting the same rates were agreements at intermittent meetings of the freight agents of the companies; these invariably led to uniform posted schedules on the basis of cents per hundred pounds for grain from Chicago to New York or for fourth-class merchandise New York to Chicago.[2] For example, in August of 1872, "at a recent meeting of the Western General Freight Agents held in Toledo, divisions were agreed upon for east-bound traffic . . . to New York, Boston, Philadelphia and Baltimore based on

[1] The reliance of the Erie on the two major railroads is noted in a letter of November 25, 1874, from John King, vice-president, to the president of the Baltimore and Ohio Railroad: "President: . . . Much, if not most of the [Erie] freight has been sent [over] the Michigan Central road [of the New York Central]. You may remember that the effort on the part of the Erie to connect with the Suspension Bridge and the Canadian system was the cause of a fight with the New York Central road, some years ago . . . apart from this, the Fort Wayne [Pennsylvania] road has always been open to the Erie at Mansfield; and Jewett told me last summer that one of the difficulties he had to contend with was that Scott was urging him to send much of his business that way, and he did not care to commit himself in that direction." . . . From the *Archives of the Baltimore and Ohio Railroad* (1870–1879) of the Maryland Historical Society.

[2] The Interstate Commerce Commission's compilation of rates published in the 1870's, as reported in *Railways in the United States in 1902* (Washington: Government Printing Office, 1903): Part II "A Forty Year Review of Changes in Freight Tariffs," was "obtained from the older established companies whose records were found satisfactory for this purpose: such companies are the standard lines [the New York Central, Pennsylvania, and Baltimore and Ohio]." (Cf. p. 44 of Part III.) The rates shown for 1871, for example, for grain from Chicago to New York are as in the second column:

(*Footnote continued on next page.*)

the rate of 50 cents per 100 pounds grain from Chicago to New York,"[3] and later, "at the convention of general freight agents in Indianapolis on (April 2, 1874), rates were adopted on the basis of 40 cents from Chicago to New York . . . about 5 cents per hundred higher than the old rate." [4]

From all appearances the posted schedules were adhered to by the interested parties. Instances of cheating were few; the most serious was described as: "breaks in rates, when regular rates are not adhered to and in fact do not exist, are extremely infrequent in east-bound business . . . the tariffs for west-bound rates have remained about the same as 1869, but fell some 10 to 20 per cent in 1872 . . . with concessions made below the established tariff to load the larger number of empty cars which must be hauled back. . . ." [5] This was alleged to have continued in 1873: "the rates westward are in a terribly demoralized state, owing to a reckless competition . . . there is a constant inclination to do this from the fact that most cars go back empty and the addition of a carload in that direction is almost clear gain. . . ." [6] There is no indication that these shipments

---

	Interstate Commerce Commission's Published Rates (Cents per 100 Pounds)	*Lake Shore Published Rates (Cents per 100 Pounds)*	*New York Produce Exchange Rates*
March 4	50	50	50
April 7	45	45	45
June 26	40	40	40 (June 13)
July 10	45	45	45
August 11	50	50	50
September 21	55	55	55
October 2	60	60	60
October 25	65	65	65

The Lake Shore and Michigan Southern grain rates for this year are shown in the third column (as compiled by the *Railroad Gazette* from the Lake Shore published tariffs in July, 1884). Also, the same rates are shown as "All Rail Freight Eastward from Chicago in 1871" in the *Annual Report* of the New York Produce Exchange (New York, 1871), p. 401. The coincidence of dates for rate changes in the three sources suggests that the trunk lines were setting common tariffs.

[3] The *Railroad Gazette*, "New Freight Rates" (August 31, 1872), 386.

[4] The *Railroad Gazette*, "Eastbound Freight Rates" (April 11, 1874 and May 16, 1874), 133 and 187. The general freight agent was the manager of freight services, costs, and revenues for each railroad.

[5] The *Railroad Gazette*, "The Tendency of Freight Rates" (February 8, 1873), 56. Similar reports here, and in the following, can be found in the railroad notes of *The New York Times*, and *New York Herald*, or sporadically in the *Chicago Daily Commercial Bulletin* or the *New York Journal of Commerce*. Attention is centered on the *Railroad Gazette* because of the extended editorial comment on such occurrences; if the details of the accounts vary, however, other sources are cited as well.

[6] The *Railroad Gazette*, "Freight Rates Westward" (August 30, 1873), 352.

at below-tariff rates were of sufficient extent or duration to merit an investigation.

There is evidence of general adherence to official rates on large-volume shipments of agricultural products to the eastern seaboard. Posted rates on grain from Chicago during each week of the years 1871 to 1874 can be compared to the rates that members of the Chicago Board of Trade said that they paid for transporting grain to New York. The latter rates, compiled and published as a conscientious estimate of actual rates by the Board of Trade in the *Chicago Daily Commercial Bulletin*,[7] would have been lower if the cheating was taking place and the informal association of freight agents had not yet adjusted to it. In fact, they do not appear to have been lower. During the 21 weeks of the summer season of 1871, the official rates R_0 averaged 48.10 cents per hundred pounds, while the Board of Trade rates R_T averaged 47.62 cents per hundred pounds.[8] During the winter the average of the official rates increased to 63.93 cents, while the Board of Trade average rate increased to 63.75 cents. In 1872 the differences were small once again: the average official rate in the summer declined to 54.33 cents, while the Board of Trade average rate was .33 cent lower; the average official and Board of Trade rates were both 64.55 cents in the winter. The Board of Trade rate did not decline as much in the summer of 1873 as did the official rate (the former averaged 50.00 cents, the latter averaged 48.55 cents); during the winter of 1873–1874 and the summer of 1874, the Board of Trade and official rates were 46.05 cents per hundred pounds.

Loyalty to agreed rates cannot be inferred only from small differences between average official and Board of Trade rates; there should also be

[7] The rates were quoted by the *Bulletin* anonymously to protect transporter and shipper from charges of discrimination. It might be argued that the grain shippers should have quoted fictitious rates below the official rate, so as to bring about a rate decrease in reaction to (mythical) cheating. Such a policy on the part of the traders would have involved creating erroneous market information which would have destroyed the value of the reporting service. The Board of Trade seems to have been able to prevent this by reporting rates different from the official rates only when there were numerous testimonies, and by noting the particular importance to be attached to reports received from the railroads themselves (as to their own, and their competitors', activities).

[8] Since the standard deviations of R_0 and R_T were, respectively, 5.585 and 6.045 cents per 100 pounds, a test for differences of means disproves the hypothesis of a significant difference (using as the criterion for a *significant* difference a calculated difference more than twice the size of the standard deviation of a sample of such differences). The standard deviations for R_0 and for R_T are both greater than 2.5 cents in each of the seasons during 1871 to 1874 (except for the summer of 1874, when $R_0 = R_T$), so that the differences are "not significant." Given that reporting errors on very few shipments could have raised or lowered the average Board of Trade rate, it is assumed that the given differences are a reflection of R_T having been collected from large numbers of shippers and R_0 having been announced by the two railroads — that there was small, but observable, "measurement error."

evidence on the timing of changes in these rates. Posted and actual rates could have differed by only 1 or 2 cents while the official rate was repeatedly being broken, if rate changes each time began with a cut in the Board of Trade rate.[9]

Complete loyalty requires that changes in Board of Trade rates should have taken place only as a result of changes in official rates or that, with weekly rates R_0 and R_T, and $R_T = \alpha + \beta R_0$ with $\alpha = 0$ and $\beta = 1.00$, the percentage of explained variation in R_T as shown by the coefficient of determination $\hat{\rho}^2 = 1.00$. In contrast, in the presence of cheating, R_T should have decreased before R_0; in $R_T = \alpha + \beta R_0$, $\beta < 1.00$, and the coefficient of determination $\hat{\rho}^2 < 1.0$, since some proportion of the variation in R_T followed from cheating rather than from changes in R_0.

The lines $R_T = \hat{\alpha} + \hat{\beta} R_0 + u$, as estimated separately for the summer and winter seasons for each year from 1871 to 1874,[10] seem to indicate little or no cheating (as shown in Table 3.1). The equation for the summer season of 1871 is typical: $R_T = -2.863 + 1.049 R_0$; $\hat{\rho}^2 = .940$. The computed value $\hat{\beta}$ is close to the hypothesized $\beta = 1.00$[11]; and 94 per cent of the variation in R_T is explained by changes in R_0. For succeeding seasons, the lowest value of $\hat{\beta}$ is 0.746 (for the summer of 1873) and the highest value is 1.035 (for the summer of 1872), neither of which differs significantly from 1.00.[12] In the equation for the summer of 1873, only 75.7 per cent of variation in R_T is explained by changes in R_0; in the equation for the winter of 1871–1872, the proportion of explained variation is 88.6 per cent. In all other equations for seasons prior to 1875, the explained variation in R_T is greater than 96 per cent of total variation. The rates posted by the railroads, and those supposedly quoted to the members of the Chicago Board of Trade, seem to have changed concurrently in the years from 1871 to 1874.

[9] The cheater might have lowered his rate a cent or two under the official rate and experienced an increase in shipments followed by rate-matching by the other transporters. The rate difference would not have been more than 2 cents, but it might have been sufficient to render the agreement unstable — by providing higher profits from cheating.

[10] The term "u" is calculated so that mean $\bar{\mu} = 0$; S_u = standard error of estimate. It accounts for variation in R_T because of errors in measurement at the Board of Trade. These regression lines were computed to minimize the sum of the square of the deviations of the observations from the computed line.

[11] The computed $S_{\hat{\beta}} = 0.061$ so that the value of $t = [(1.049 - 1.000)/0.061]$ is "small" (or less than 2.0, the value leading to rejection of the hypothesis $\beta = 1$). The test is not accurate; successive values of $\{R_T - (\hat{\alpha} + \hat{\beta} R_0)\}$ are highly correlated, so that the computed $S_{\hat{\beta}}$ may well be a serious underestimate of $S_{\hat{\beta}}$ and, consequently, t is unduly large. But the bias in calculated "t" does not dispute that $\beta = 1.0$.

[12] Cf. the preceding footnote, however. All that can be said generally is that the seven compared values of $\hat{\beta}$ in the period 1871 to 1874 have a mean of 1.006 and a standard deviation of only 0.105.

TABLE 3.1

Official and Board of Trade Rates on Grain Transport from Chicago to New York, 1871 to 1874

Calculations of $R_T = \alpha + \beta R_0 + \nu$

R_T, the rate on transport of grain products by railroad from Chicago to New York as shown in the Board of Trade *Daily Commercial Bulletin;* R_0, the rail rate from compilations of archives of tariffs and from announcements at the meetings of the general freight agents of the railroads affiliated with the New York Central, Pennsylvania, Baltimore and Ohio, and Erie systems.

The standard error of the computed $\hat{\beta}$ coefficient is shown in parentheses below the coefficient; the calculated value of the coefficient of determination $\rho^2 = \hat{\rho}^2$ are also calculations of average values of $R_0 = \overline{R}_0$, $R_T = \overline{R}_T$. All statistics are in cents per 100 pounds of grain transported.

SUMMER SEASON, 1871:

(21 weeks); $R_T = -2.863 + 1.049R_0$; $\hat{\rho}^2 = .940$; $\overline{R}_T = 47.619$; $\overline{R}_0 = 48.095$
(0.061)

WINTER SEASON, 1871–1872:

(28 weeks); $R_T = -10.833 + 1.167R_0$; $\hat{\rho}^2 = .886$; $\overline{R}_T = 63.750$; $\overline{R}_0 = 63.929$
(0.082)

SUMMER SEASON, 1872:

(30 weeks); $R_T = -2.237 + 1.035R_0$; $\hat{\rho}^2 = .967$; $\overline{R}_T = 54.000$; $\overline{R}_0 = 54.333$
(0.036)

WINTER SEASON, 1872–1873

(22 weeks); $R_T = 0.000 + 1.000R = 64.545$

SUMMER SEASON, 1873:

(31 weeks); $R_T = 11.363 + 0.769R_0$; $\hat{\rho}^2 = .757$; $\overline{R}_T = 50.000$; $\overline{R}_0 = 48.548$
(0.132)

WINTER SEASON, 1873–1874:

(19 weeks); $R_T = 0.000 + 1.000R_0 = 56.053$

SUMMER SEASON, 1874:

(33 weeks); $R_T = 0.000 + 1.000R_0 = 44.545$

Sources: The official rate is that rate agreed upon by the freight agents of the relevant railroads as reported in three sources mentioned in footnote 2, as well as in the following three sources: (1) "All rail rates from New York to Chicago in cents per hundred pounds," the testimony of Mr. G. R. Blanchard, The New York State Legislature, Special Committee on Railroads, *Railroad Investigation 1879* (the Hepburn Committee, p. 2995); (2) "Wholesale Prices, Wages and Transportation," Report by Mr. Aldrich from the United States Congress, Senate Committee on Finance, March 3, 1893 (Vol. I, "Changes in competitive rates," pp. 429–609); (3) *Annual Reports*, the Chicago Board of Trade, 1870–1900 (the table entitled "All Rail Freights Eastward from Chicago"). The Board of Trade rail rate is that rate recorded in the "freights" column of the *Daily Commercial Bulletin* (Howard, Bartels & Company: Chicago, Illinois, daily except Sundays and holidays, 1870–1900); the column appears from inspection to have been devoted to recording the rates paid by anonymous members of the Board for grain shipments to New York, Philadelphia, and Boston as well as to points South and West, or else to have been the quotations of the railroads to shippers on the trading floor in Chicago.

The Effects of Collusive Rates, 1871 to 1874

These rates set by the general freight agents are difficult to characterize. They would be "cartel rates" for the two large and two small railroads if $(R_0 - K_2)Q$ was a maximum, with R_0 the official rate, K_2 the marginal transport costs, and Q the total tonnage; if the rates R_0 equaled marginal transport costs K_2, then the meetings were a mere formality in competitive

TABLE 3.2

Long- and Short-Distance Rates on the Erie Railroad, 1865 to 1878

Service	*Distance (Miles)*	*Fourth-Class Rate (Cents per 100 Pounds)*
New York City to Susquehanna, Pa.	235	24
New York City to Binghamton, N.Y.	265	20
New York City to Bath, N.Y.	350	36
New York City to Elmira, N.Y.	310	20
New York City to Addison, N.Y.	330	35
New York City to Howellsville, N.Y.	365	38
New York City to Genesee, N.Y.	415	41
New York City to Cuba, N.Y.	420	43
New York City to Olean, N.Y.	435	37

Sources: New York State Legislative Assembly, Special Committee on Railroads, "Railroad Investigation, 1879" (New York: Evening Post Steam Presses, 1879): Testimony, Volume 3, p. 2836 and "Exhibits, New York Central and New York, Lake Erie and Western Railroads," p. 222.

rate-setting. But direct examination of $(R_0 - K_2)$ is not possible since marginal transport costs K_2 cannot be estimated with accuracy. Consequently, rates R_0 can be considered successful "cartel rates" only in retrospect — in comparison with those that followed in the breakdowns of the agreements of the middle 1870's.

Some characteristics of through rates can be noted, however. The Erie Railroad as a participant in the meetings set rates in 1872 that averaged 58 cents per hundred pounds of grain from Chicago to New York and 29 cents from Buffalo to New York (50 per cent of the Chicago rate). This railroad published tariffs on shorter distance traffic not far different from this average; the Erie's schedule for transport from New York City to various upstate cities indicates rates up to 43 cents for 420 miles of transport. The higher short-distance rates could have followed from higher unit costs from transporting smaller volumes. But these local rates were not entirely in accord with distance: the rates to Susquehanna, Addison, and Bath, three points served exclusively by the Erie, were greater than those

TABLE 3.3

Long- and Short-Distance Rates on the Pennsylvania Railroad, 1876

	Distance (Miles)	*Fourth-Class Rate (Cents per 100 Pounds)*
Philadelphia to Altoona, Pa.	230	20
Philadelphia to Pittsburgh, Pa.	325	20
Philadelphia to Williamsport, Pa.	210	23
Philadelphia to Erie, Pa.	415	23

Source: U. S. Interstate Commerce Commission, *Railways in the United States in 1902:* "Part II: A Forty Year Review of Changes in Freight Tariffs" (Washington, 1903), p. 168.

from New York to Binghamton and to Elmira, points which were rail centers for other transporters but which were farther from New York (as shown in Table 3.2). Similarly, the rate to Olean was lower than that to Howellsville; Olean was 70 miles farther from New York City, but was served by two railroads. The presence of more than one railroad seems to have made for lower quotations by the Erie (rather than the presence of cost differences alone, since marginal transport costs could not have been negative for the additional miles from Cuba to Olean).[13]

Similar geographical rate patterns can be detected in some of the Pennsylvania Railroad freight tariffs. The Pennsylvania's rate to Altoona, Pittsburgh *et al.* from Philadelphia were proportionately less than the Chicago–New York rate (as shown in Table 3.3, in comparison with 1,000-mile rates of 58 cents in 1872). But the Altoona rate was the same as Pittsburgh, ninety-five miles farther. This railroad's rates from Philadelphia to Williamsport, and from Philadelphia to Erie, were both 23 cents per hundred pounds, while the city of Erie was almost twice as far from the point of origin. The lower rates to farther points are partly explainable in terms of the number of railroads: Pittsburgh was a major interior rail center, as well as a center for barge traffic from two rivers; Erie was a major port on the Great Lakes serviced by a number of roads; the cities of Altoona and Williamsport were major sources of traffic for the Pennsylvania road alone.

The level of the rates to Olean, or to Pittsburgh, seems comparable to

[13] It might be expected that transport costs were less for shorter distance and for shipment to the locations with the greatest total volume of receipts (since there could have been economies of large-scale unloading and storage). If the second factor compensated for the first — that is, if the farther location receiving a larger volume had lower marginal transport costs — it would have been more profitable to have shipped all of the short-distance traffic to the farther location. This would have resulted in no receipts at the closer location — as in the present instance of the depots in small towns surrounding New York City.

the Chicago–New York through rate on bulk commodities. But these rates were lower than the short-distance rates from New York or Philadelphia to "one-railroad towns." Some part of the difference may have been due to an inability of groups of railroads to set as high rate quotations as those of "monopoly" railroads. If this was the case, then at the beginning of collusive rate setting the joint profit maximizing rate was not set.[14]

The Effect of Collusive Rates Upon Grain Markets, 1871 to 1874

The consequences of rate control for the costs of actual shipments can be seen from rate-induced changes in the size of the differential between spot prices for grain or produce in Chicago and spot prices for the same goods in New York. The arbitrage of speculators in the two grain markets should have resulted in a differential between Chicago and New York prices for any grade of traded grain that was equal to the cost of transportation. The "spreaders" in the Chicago Board of Trade in particular should have purchased Chicago wheat for transport and sale to New York when the New York spot price was greater than the Chicago spot price plus the cost of transport between the cities.[15] Such behavior by a number of spreaders — or greatly expanded purchases in Chicago and sales in New York by one spreader alone — should have had the effect of increasing the Chicago price (because of increased demand) and decreasing the New York price (because of increased supply). Expanded purchases and sales can be expected to have continued until an equilibrium price difference was established — a difference equal to the cost of transport.[16] Then an

[14] The theory of Chapter 2 did not state that the "perfect" cartel would vary the level of rates with the number of members at any point of shipment. It is observed here that rates were higher where there was only one road than where there were two or more roads, so that the theory of the perfect agreement would not seem accurate in this instance. Rather, some explanation linking cartel stability with the number of members as to the level of rates would seem more helpful. Cf. B. Fog, "How are Cartel Prices Set?" *Journal of Industrial Economics* (November, 1956).

[15] They "should have" carried out such activities because a profit could have been made: $P_g' - P_g - R > 0$, where P_g' is the New York spot price per hundred pounds, P_g is the Chicago spot price for the same grade of grain, and R is the rail rate for 100 pounds of grain from Chicago to New York.

[16] If the trunk-line railroads had set an official rate of 50 cents per hundred pounds, and the spot price of No. 2 Chicago red wheat was $1.00 per hundred pounds in Chicago and $1.60 in New York, spreaders should have purchased in Chicago until the Chicago price increased (to some amount such as $1.07) and resold in New York until the price had fallen (to, for example, $1.57) so that the difference equaled the official rate. Aside from movements of price differences toward equilibrium, such as the decline of this example price difference from $.60 to $.50, changes in price differences should have followed from changes in the actual railroad rates.

informal cartel association setting rates should have determined costs for spreaders, and should have determined the equilibrium grain price difference.

With the agreed rates maintained during 1871 to 1874, the official rate or the Board of Trade rate should have determined the Chicago–New York differences in grain prices. The grain price differences should have responded to changes in either rate: during the winter, $P_g^* = \alpha + \beta R + \nu$, where P_g^* is the weekly average grain price difference, R is either one of the rates, and ν indicates the effect of short-term speculation moving price away from or toward equilibrium (mean $\bar{\nu} = 0$, standard deviation $\sigma\nu > 0$). When the Great Lakes were open for steamer shipping to Buffalo, then $P_g^* = \alpha + \beta R + \gamma R_L + \nu$ with R_L equal to the lake-rail rate from Chicago to New York (from Buffalo via railroad). The single important requirement is that spreaders held stocks in both cities and had instantaneous communication between the two locations. Then $P_g^* = \alpha + R$ for the weeks during the winter, where α is the total cost associated with loading and unloading, spoilage in transit *et al.*, and β equals one.[17] If the summer lake rate decreased as a consequence of the decrease in the rail rate, then with $P_g^* = \alpha + \beta R + \gamma R_L$ and $R_L = R - \Delta$, it follows that $P_g^* = (\alpha - \Delta\gamma) + (\beta + \gamma)R$,[18] or $P_g^* = (\alpha - \Delta\gamma) + R$. That is, the average grain price difference was greater than the rail rate, but changes in this difference equaled changes in the rate.

This single requirement may not have been realized. There would seem to have been stocks in both cities, but not always sufficient for affecting instantaneous price adjustments[19]; rather than immediate communication on price differences between Chicago and New York, there was a published report in both cities of the previous day's markets in the other city.[20] The absence of stocks, particularly during the late winter months, made actual shipments necessary in order to decrease New York grain prices, so that there might have been a lag of four days or more between the decrease in

[17] If $P_g^* = \alpha + \beta R + \nu$, then $\partial P_g^*/\partial R = \beta$ and instantaneous adjustment assumes $\partial P_g^*/\partial R = 1$. The average experience, then, is $P_g^* = \alpha + R$.

[18] The discount Δ of the lake-rail rate under the all rail rate equals the *additional* cost of spoilage by steamer, the additional costs of unloading the steamer and loading freight cars at Buffalo, and the additional costs of holding inventory in transit as a result of the longer shipping period by steamer.

[19] If New York stocks had been drawn down to the point where consumption demand accounted for all the holdings at the present price, then further receipts would be required to decrease price there. In 1871, for instance, stocks in New York fell to 270,000 bushels of corn per week in January, making actual shipments from Chicago necessary to lower New York prices, rather than increased sales from stocks. Receipts were half as large as stocks for a number of weeks.

[20] An unknown number of brokers had direct telegraph service between the Chicago and New York exchanges; the newspapers reported the telegraphed information the following day.

rates and the consequent Chicago–New York grain price adjustment. The lack of immediate communication might have increased the adjustment lag to one week or more. For the average $P_{g,t}$ in week t and $P_{g,t-1}$ in the previous week, the relationship $P_{g,t} = P_{g,t-1} + \epsilon(P_g^* - P_{g,t-1}) + \nu$ should have followed (P_g^* having been the equilibrium grain price difference, as defined above). With a one-week lag of grain prices behind a change in the rail rate, $\epsilon = .50$ and $P_{g,t} = P_{g,t-1} + .50(P_g^* - P_{g,t-1}) + \nu$ in contrast with instantaneous adjustment $\epsilon = 1$ and $P_{g,t} = P_{g,t-1} + (P_g^* - P_{g,t-1}) = P_g^*$.[21]

The estimate of these equations for P_g^* and $P_{g,t}$[22] for the summer and winter seasons of 1871 to 1874 indicate that the weekly averages of daily Chicago–New York price differences for spot wheat, corn, and oats correspond with the lake-rail and rail rates during the summer of 1871. The equations for instantaneous adjustment shown in Table 3.4, with first R_0 and then R_T as independent variables, seem sufficiently similar to establish the inference that grain did not move at (Board of Trade) rates undercutting the official rates. The rates averaged slightly less than 48 cents per hundred pounds, while grain price differences averaged 44.5 cents. Changes in the rail rate and the lake-rail rate[23] seem to have been followed by similar, or slightly larger, changes in the grain price differences with no more than a week's delay.[24] Changes in the Chicago–New York price differences were "explained" in good part by rate changes.[25]

The computed equations for the winter season grain price differences indicate the presence of considerable imperfection in arbitrage. The average price difference was approximately 60 cents, while the average rate was 64 cents. The equations for instantaneous adjustment indicate that, for every 10-cent increase in either the official rate or the Board of Trade rate, the grain price difference increased by approximately 6 cents per 100 pounds. Also, variance in grain price differences was equally "explained" by variance in the official rate as by variance in the Board of Trade rate:

[21] A proof that $\epsilon = .50$ results in a one-week average lag is found in J. V. Yance, "A Model of Price Flexibility," *American Economic Review*, 50 (June, 1960), 415–417.

[22] The method of least squares is used to fit the function to the observed values of $P_{g,t}$ for given R, R_L, $P_{g,t-1}$, etc., so as to minimize the sum of the square of the deviations of observed $P_{g,t}$ from the function.

[23] There is indication that the rates changed together: the simple correlation between R_0 and R_L in the computation of the lag equation is .920.

[24] However, the estimate of $\epsilon = .449$ in the lagged equation is not statistically different from 1.0 (the value in the presence of instantaneous adjustment). The test for significance is $t = \hat{\epsilon} - 1.000/S_{\hat{\epsilon}}$ where $\hat{\epsilon}$ and $S_{\hat{\epsilon}}$ are the least-squares estimates and $t > 1.96$ is considered a value disproving the hypothesis that $\epsilon = 1$.

[25] The computed $\hat{\rho}^2$, indicating the percentage of variance in the dependent variable that is associated with the independent variables, ranged from .324 to .768. The lag equation was computed in the form $P_{g,t} - P_{g,t-1} = \Delta P_g = -\epsilon P_{g,t-1} + \epsilon(\alpha + \beta R + \gamma R_L)$ so as to minimize additions to $\hat{\rho}^2$ due to lag; values as great as .500 are not highly likely. Cf. J. V. Yance, *op. cit.*

TABLE 3.4

Official Rail Rates, Board of Trade Rail Rates, and Differences Between Chicago and New York Grain Prices, 1870 to 1874

$$P_g^* = \alpha + \beta R_0 + \gamma R_L + \nu; \qquad P_g^* = \alpha + \beta R_T + \gamma R_L + \nu \tag{1}$$

P_g^*, the weekly average of the daily differences between Chicago and New York spot prices, in cents per hundred pounds, for the No. 2 "Chicago" grade of wheat, corn, and oats; the official rate R_0 or the Board of Trade rate R_T; the Board of Trade lake-rail rate R_L from Chicago to New York via steamship to Buffalo and rail from Buffalo to New York.

$$P_{g,t} = P_{g,t-1} + \epsilon(P_g^* - P_{g,t-1}) \tag{2}$$

$P_{g,t}$, the average of the daily Chicago–New York grain price differences during week t and $P_{g,t-1}$, the average for the previous week; $P_g^* = \alpha + \beta R + \gamma R_L$ for the summer season and $P_g^* = \alpha + \beta R$ for the winter season from an instantaneous adjustment to changes in transport rates. Calculations have been made of the form $P_{g,t} - P_{g,t-1} = -\epsilon P_{g,t-1} + \epsilon P_g^*$.

The standard error of each coefficient is shown in parentheses below the coefficient.

SUMMER SEASON, 1871:

$\overline{R}_0 = 48.095$; $\overline{R}_T = 47.619$; $\overline{R}_L = 37.823$; $\overline{P}_g = 44.450$

(17 weeks); $P_g = -3.351 - \underset{(0.339)}{0.394} R_0 + \underset{(0.479)}{1.763} R_L$; $\hat{\rho}^2 = .768$

(17 weeks); $P_g = -5.148 - \underset{(0.327)}{0.305} R_T + \underset{(0.434)}{1.694} R_L$; $\hat{\rho}^2 = .767$

(15 weeks); $P_{g,t} = 0.551 P_{g,t-1} + \underset{(0.314)}{0.449}(-17.239 - 0.605 R_0 + 2.444 R_L)$; $\hat{\rho}^2 = .324$

(15 weeks); $P_{g,t} = 0.541 P_{g,t-1} + \underset{(0.321)}{0.459}(-19.744 - 0.365 R_T + 2.199 R_L)$; $\hat{\rho}^2 = .324$

WINTER SEASON, 1871–1872:

$\overline{R}_0 = 63.929$; $\overline{R}_T = 63.750$; $\overline{P}_g = 59.935$

(25 weeks); $P_g = 15.595 + \underset{(0.304)}{0.698} R_0$; $\hat{\rho}^2 = .184$

(25 weeks); $P_g = 24.169 + \underset{(0.236)}{0.561} R_T$; $\hat{\rho}^2 = .197$

(22 weeks); $P_{g,t} = 0.602 P_{g,t-1} + \underset{(0.173)}{0.398}(-1.434 + 0.959 R_0)$; $\hat{\rho}^2 = .242$

(22 weeks); $P_{g,t} = 0.597 P_{g,t-1} + \underset{(0.168)}{0.403}(5.077 + 0.860 R_T)$; $\hat{\rho}^2 = .274$

SUMMER SEASON, 1872:

$\overline{R}_0 = 54.333$; $\overline{R}_T = 54.000$; $\overline{R}_L = 47.700$

(30 weeks); $P_g = 11.251 - \underset{(0.387)}{0.269} R_0 + \underset{(0.240)}{1.139} R_L$; $\hat{\rho}^2 = .819$

(30 weeks); $P_g - 9.606 - \underset{(0.368)}{0.222} R_T + \underset{(0.241)}{1.119} R_L$; $\hat{\rho}^2 = .819$

$P_g = 50.961$

SUMMER SEASON, 1872: (Continued)

(29 weeks); $P_{g,t} = 0.400P_{g,t-1} + 0.600(12.059 - 0.321R_0 + 1.199R_L)$; $\hat{\rho}^2 = .253$
(0.209)

(29 weeks); $P_{g,t} = 0.424P_{g,t-1} + 0.576(15.949 - 0.501R_T + 1.321R_L)$; $\hat{\rho}^2 = .264$
(0.208)

WINTER SEASON, 1872–1873:

$\overline{R}_0 = 64.545$; $\overline{R}_T = 64.545$

(22 weeks); $P_g = 28.552 + 0.646R_0$; $\hat{\rho}^2 = .069$
(0.531)

(22 weeks); $P_g = 28.552 + 0.646R_T$; $\hat{\rho}^2 = .069$
(0.531)

$\overline{P}_g = 70.255$

(20 weeks); $P_{g,t} = 0.782P_{g,t-1} + 0.218(-175.781 + 3.813R_0)$; $\hat{\rho}^2 = .311$
(0.157)

(20 weeks); $P_{g,t} = 0.782P_{g,t-1} + 0.218(-175.781 + 3.813R_T)$; $\hat{\rho}^2 = .311$
(0.157)

SUMMER SEASON, 1873:

$\overline{R}_0 = 48.548$; $\overline{R}_T = 50.000$; $\overline{R}_L = 40.407$

(17 weeks); $P_g = 8.335 - 0.052R_0 + 1.089R_L$; $\hat{\rho}^2 = .581$
(0.458) (0.427)

(17 weeks); $P_g = -2.124 + 0.434R_T + 0.743R_L$; $\hat{\rho}^2 = .598$
(0.554) (0.455)

$\overline{P}_g = 51.223$

Sample limited to 10 weeks, so that $P_{g,t} - P_{g,t-1} = \epsilon P_{g,t-1} + \epsilon(\alpha + \beta R + \gamma R_L) + \nu$ has not been estimated.

WINTER SEASON, 1873–1874:

$\overline{R}_0 = 56.053$; $\overline{R}_T = 56.053$

(19 weeks); $P_g = 28.144 + 0.504R_0$; $\hat{\rho}^2 = .316$
(0.186)

(19 weeks); $P_g = 28.144 + 0.504R_T$; $\hat{\rho}^2 = .316$
(0.186)

$\overline{P}_g = 56.388$

(15 weeks); $P_{g,t} = 0.436P_{g,t-1} + 0.564(30.774 + 0.452R_0)$; $\hat{\rho}^2 = .332$
(0.231)

(15 weeks); $P_{g,t} = 0.436P_{g,t-1} + 0.564(30.774 + 0.452R_0)$; $\hat{\rho}^2 = .332$
(0.231)

SUMMER SEASON, 1874:

$\overline{R}_0 = 44.545$; $\overline{R}_T = 44.545$; $\overline{R}_L = 28.636$

(31 weeks); $P_g = 18.284 - 0.073R_0 + 0.895R_L$; $\hat{\rho}^2 = .352$
(0.985) (0.378)

(31 weeks); $P_g = 18.284 - 0.073R_0 + 0.895R_L$; $\hat{\rho}^2 = .352$
(0.985) (0.378)

$\overline{P}_g = 40.808$

TABLE 3.4 (Continued)

SUMMER SEASON, 1874: (Continued)

(29 weeks); $P_{g,t} = 0.512P_{g,t-1} + 0.488(31.061 - \underset{(0.154)}{0.392}R_0 + 0.950R_L)$; $\hat{\rho}^2 = .291$

(29 weeks); $P_{g,t} = 0.512P_{g,t-1} + 0.488(31.061 - \underset{(0.154)}{0.392}R_T + 0.950R_L)$; $\hat{\rho}^2 = .291$

Sources: The official and Board of Trade rates are as described in Table 3.1. The Chicago and New York daily grain prices are those recorded by the *Daily Commercial Bulletin* and the *Annual Report* of the New York Produce Exchange (1871 to 1879) for No. 2 old Chicago spring wheat, No. 2 mixed western corn, and No. 2 mixed oats. Both the daily high and low prices were recorded, and the averages of the daily highs and lows for the week in Chicago and in New York were calculated to obtain the Chicago-New York "grain price difference."

the coefficients of determination for official rates and Board of Trade rates respectively were .184 and .197.[26] But these coefficients are so small that it would appear that most of the immediate changes in grain price differences were independent of known rate changes. This is not surprising: the Great Chicago Fire in the first week of October destroyed the storage facilities of the railroads and private warehousemen in the city itself, and also destroyed the Board of Trade facilities. The market listed no prices for the period October 9 to October 23. Shipments did not take place at all for three weeks in October, and were curtailed to some extent until the following April. The *Daily Commercial Bulletin* repeatedly asserted that "there was limited Chicago trading, mostly for local speculative accounts," during January and February of 1872. It would seem that decreased supply of storage increased the Chicago spot price for purchases of grain for local consumption, and also routed grain shipments around Chicago and into the other western exchanges. The Chicago price consequently increased relative to the New York price, and fluctuated independently of the New York price.[27]

[26] The two equations assuming some lag in adjustment indicate a relatively slow movement of prices in accord with rate changes. The estimates of ϵ are close to .40, indicating an average lag of $1\frac{1}{2}$ weeks, rather than the 1-week average of the previous summer. The importance of the previous week's price difference, for explaining changes in this week's difference, is indicated by multiple correlation coefficients close to .25 rather than from .184 to .197 (for the equations assuming immediate adjustment). It should be noted once again that the estimates of the equations are for the form

$$P_{g,t} - P_{g,t-1} = -\epsilon P_{g,t-1} + \epsilon(\alpha + \beta R + \gamma R_L + \nu),$$

so as to *minimize* the additions to $\hat{\rho}^2$ from the correlation of $P_{g,t}$ with $P_{g,t-1}$.

[27] Wheat shipments from Chicago fell to less than 10,000 bushels per week. Concurrently, New York receipts were more than 190,000 per week in December and more than 30,000 bushels per week in January and February of 1872. With Chicago prices governed by local consumption needs, the Chicago–New York grain price difference averaged little more than 59 cents per 100 pounds, while the official and Board of Trade rates were somewhat greater than 63 cents per 100 pounds.

The statistical pattern of price differences in accord with 50 to 65 cent rates seems to have been repeated during 1872 and 1873. Grain price differences were close to 50 cents; changes in the difference followed changes in lake-rail and all-rail rates the same week in the summer seasons,[28] with the sum of the computed coefficients of the two rates equal to from .85 to 1.17. More than one-half the variance in weekly grain price differences was explainable in terms of variance in rail rates and a short (one-half week) adjustment lag in the summer of 1873.

In the winter season of 1872–1873, however, grain price differences were more than 70 cents on average, while rates averaged 64 cents. The differences increased by only 65 per cent of any increase in either the official rate or the Board of Trade rate (assuming instant adjustment to rate changes and with computed coefficients of R_0 and $R_T = .646$); only 7 per cent of the variance in grain prices could be explained by variance in rail rates. In the computed lag equations, the estimated length of the adjustment process is 4 weeks and the change in grain price differences associated with a change in rate is much too large (since $\hat{\beta} = 3.8$). Once again there were special circumstances: it was alleged that "it is not possible to obtain transport to the Seaboard" in the Chicago *Daily Commercial Bulletin* during the period from January 25 to February 12, 1873. The *Gazette* also noted that "there is a continued lack of rail transportation" from February 15 to February 28, 1873. Without either of these statements being literally true, there must have been delays and additional storage required because of long periods in which severe storms blocked the lines. The costs of transport to the shippers were higher, and fluctuated more often, than the quoted rates throughout the winter given the cost of additional storage and delay in transit.[29]

For the remainder of the period through the summer of 1874, the official rates and Board of Trade rates did equally well in explaining changes in grain price differences. When the lakes were open, concurrent changes in lake-rail and all rail rates determined changes in the grain price difference. When the lakes were closed and transport was limited to the trunk-line railroads, the official rate and the Board of Trade rate were exactly the same; variance in either explained slightly more than 30 per cent of the variance in the grain price differences. Assuming a lag in the adjustment of grain prices to rates, an estimated (one-week) lag and the rail rates

[28] That the rail rates and lake-rail rates changed together is indicated by simple correlation coefficients of R_0 and R_L of .931 in the instantaneous adjustment equation, and of R_T and R_L of .930 in the lagged equations.

[29] This would explain the large computed values for β in the lagged equation: increased rates, at the time of blockage of the lines, were accompanied by increased costs of delay and storage (and increased grain price differences) three times as large. Also, as shown in Table 3.4, the average grain price difference was greater than the quoted rates, in contrast to the experience in the winter of 1871–1872.

explain more than half the variation in grain price differences. The rates decreased to 5 to 8 cents less than those of the previous year, and grain price differences had decreased by more than 10 cents on average over the two seasons.

In general, there would seem to have been a tendency for Chicago–New York grain price differences before 1875 to follow changes in official rates (and in Board of Trade rates, since these were approximately the same as official rates). There were two winters in which this was not the observed phenomenon, but there are at least plausible "special explanations" for the grain price differences in these periods. For each summer and for the winter of 1873–1874, the average grain price differences were close to the rates: the summer rail rates were from 44 to 55 cents per hundred pounds, the lake-rail rates were from 29 to 48 cents, and the summer average grain price differences were from 40 to 50 cents; the average winter rates were from 56 to 64 cents per hundred pounds, while the grain price differences were from 56 to 70 cents. Changes in rates were followed, seldom more than a week later, by similar changes in the grain price differences. That is, shipper's margins were 50 cents per hundred pounds because of official rates on transport.

The Entry of the Baltimore and Ohio into Chicago and the Collapse of the Agreement, 1874 to 1876

To maintain this admirable performance in the control of rates, the officers of the trunk lines agreed to form a permanent organization at Saratoga Springs, New York, in July, 1874. The Western Railroad Bureau was set up in New York in October, 1874, "[as] a bureau consisting of seven commissioners, whose duty it shall be to regulate and establish through rates for the transportation of passengers and freight over the several lines between competitive points East and West with [a cooperating] eastern bureau of commissioners representing the [eastern connecting roads of the trunk lines]." [30] The commissioners were not only to have the duty but also "The commissioners have the power to establish through rates from time to time based on mileage and distances now fixed by Graham's tables." [31] The power was granted without accompanying means for rapid detection of rates below those set by the bureau, and without a system of penalties against those setting the lower rates. All that was specified was that "In the event of any freight organization, company officer, or employee evading such regulations in any way, the commission-

[30] The agreement of the Western Railroad Bureau, Section (1) as reproduced in the *Railroad Gazette* (October 13, 1874), 403.

[31] Section (2), *ibid.*, 403.

ers shall prescribe penalties in addition to dismissal from service and shall prescribe the modes of enforcing the same as may be found necessary to make such regulations effective." [32]

Concurrent with the emergence of a formal cartel was the emergence of another full-length trunk line. The Baltimore and Ohio Railroad constructed an extension from its main line west of Pittsburgh to southern Lake Michigan and Chicago in 1874. Early in October it was noted that "the construction of the extension of the Baltimore and Ohio to Chicago is well advanced, and judging from present appearances the company may be able to run trains into Chicago over the new lines within three months." [33] This source of transport, by increasing the number of roads able to carry large volumes directly from each of the major western cities, provided "new problems" for the new rate-setting association.

The Baltimore and Ohio was conspicuous in its uncooperative attitude toward this organization. A well-publicized visit to Baltimore by the presidents of the New York Central, the Erie, and the Pennsylvania Railroads in November of 1874 led to a statement by the Baltimore company of "refusal . . . to enter into the Saratoga agreement." [34] There was, only, from the view of the other companies, "a pledge [of the Baltimore and Ohio] to maintain the rates made from time to time to and from Boston and New York and that those from Philadelphia and Baltimore should also be maintained, the proper geographical differences being preserved to those cities." [35] The pledge, if made, was abandoned as early as November of 1874. On November 24, 1874, soon after the official rate had been set by the commissioners of the Western Railroad Bureau at 45 cents,[36] the Baltimore and Ohio began operating: "two through passenger trains each way daily between Baltimore and Chicago . . . Baltimore papers report considerable shipments of freight for Chicago coming by steamer from Boston and freight is also coming from Chicago consigned to Norfolk, Charleston, and other Southern ports for which Baltimore is a very favorable distributing point . . . ;" [37] This company set a lower

[32] *Ibid.*

[33] "Some Notes on the Chicago End of the Baltimore, Pittsburgh and Chicago," the *Railroad Gazette* (October 10, 1874), 392.

[34] *The Great Railway Conflict: Remarks of John W. Garrett, President, Made on April 14, 1875 at the Regular Monthly Meeting of the Board of Directors of the Baltimore and Ohio Railroad Co.* (Baltimore: The Sun Book and Job Printing Co., 1875), p. 4.

[35] Letter of T. A. Scott, president of the Pennsylvania Railroad, to J. W. Garrett, February 15, 1875, as shown in an unpublished manuscript entitled "The Pennsylvania Railroad Company" in the Archives of the Smithsonian Institution.

[36] Cf. "Freight Rates, 1871–1902," *Railways in the United States in 1902*, *op. cit.*, p. 78.

[37] "The Baltimore and Ohio," the *Railroad Gazette* (December 5, 1874), 477. Also, "the Baltimore and Ohio direct line to Chicago was open late in November (so that all) reported receipts and shipments embrace the business of a few days (in November and the month of December). Cf. *The Annual Report of the Board of Trade of Chicago 1874*, *op. cit.*, p. 144.

rate when it began operations: "on the opening of its road . . . the company fixed 35 cents per hundred pounds as a proper and judicious rate for grain and other fourth class freights between Baltimore and Chicago." [38] The Railroad Bureau reduced official rates on December 11, 1874: "mainly on account of the competition of the B & O." [39]

There were accusations by the Pennsylvania Railroad that the Baltimore railroad continued to depart from the official rate: "Had the agreement been carried out by your company in good faith no one would have had cause of complaint, but it is a notorious fact well-known to everyone in railroad life that this has not been done, and abundant evidence of that fact can be given at any time." [40] In return, Mr. Garrett of the Baltimore and Ohio alleged that: "since [the entry of the Baltimore and Ohio into Chicago] disastrously low rates were from time to time made by your lines for the purpose of causing the B & O Company to abandon the low, but, according to its judgment, reasonable tariff it had adopted." [41] February's allegations were followed by the cessation of meetings of the Railroad Bureau in March and the demise of the organization by July.

The nonadherence to official rates in the winter of 1874–1875 had at least three consequences. First, there was a decrease in official rates given that the average rate was 40 cents rather than the 56 cents or more of the previous three winters. Second, average Board of Trade rates were lower than average official rates during the winter season of 1874–1875, the summer season of 1875, and the summer season of 1876. Board of Trade quotations on shipments to Baltimore were also lower than the official rates to that city in the winter of 1875–1876. Moreover, the lower Board of Trade rates seem to have led to decreased margins between Chicago and New York grain prices.

The averages of official and Board of Trade rates by season and the effect of these rates on Chicago–New York grain price differences are shown in Table 3.5. The average Board of Trade rate was approximately 3 cents per hundred pounds below the average official rate for the season beginning December 7, 1874, and ending May 3, 1875. Board of Trade rates changed frequently throughout the period independently of official rates (which remained at 40 cents throughout the period).[42] The grain price differences declined 16 cents throughout the season concurrent with the decline in the Board of Trade rates; these rates, assuming instantaneous adjustment of

[38] Letter of J. W. Garrett to T. A. Scott, February 15, 1875, the unpublished history, "The Pennsylvania Railroad Company," *op. cit.*, p. 22. The Western Railroad Bureau's rate to Baltimore was 40 cents, on the 45-cent basis to New York.

[39] Cf. the *Railroad Gazette* (December 19, 1874), 497.

[40] Letter of T. A. Scott to John W. Garrett, *op. cit.*, February 15, 1875.

[41] Letter of John W. Garrett to T. A. Scott, *op. cit.*, February 15, 1875.

[42] So that the relationship $R_T = \alpha + \beta R_0$ is such that $\hat{\beta} = 0.000$ in this period and the computed correlation coefficient $\hat{\rho}^2 = .000$ (indicating that none of the variance in R_T is explained by variance in R_0).

grain prices to rate changes, explained approximately 36 per cent of the variation in the grain price differences.[43] Assuming lagged price adjustments to rate changes, the combination of a computed lag of $1\frac{1}{2}$ weeks and changes in Board of Trade rates explained approximately 27 per cent of the variance in grain price differences [as shown by $\hat{\rho}^2$ in the calculations of $P_{g,t} = (1 - \epsilon)P_{g,t-1} + \epsilon(\alpha + \beta R_T)$ in Table 3.5]. The constant official rate explained none of the variance in grain price movements.

There were disparities to 6 cents between the official rates and those listed on the Board of Trade in the summer of 1875. The published rates of the Pennsylvania and New York Central, set by the freight agents of the two lines together after the demise of the Western Bureau, averaged 39.3 cents for one hundred pounds of grain shipped from Chicago to New York, while the Board of Trade listings averaged 32.8 cents per hundred pounds. Changes in the official rates explained only 2 per cent of the variance in the Board of Trade rates (given that the coefficient of determination $\hat{\rho}^2 = .022$ in the calculation of $R_T = \alpha + \beta R_0$).[44] There would seem to have been two railroad rates: the "announced" rate and a "secret" concession for some Board of Trade members.

These lower Board of Trade rates seem to have had some effect on volume shipments of grain. During this summer, as in the previous, there was a decided preference of shippers for lake-rail transport — for transport from Chicago to a lake port in the East and from there to New York City or Baltimore by railroad. It is evident that the lake-rail rate was predominant — more than 80 per cent of the summer tonnage from Chicago was transported by steamer[45] — but this rate seems to have followed Board of Trade rates (the correlation coefficient for R_T and R_L was .854).[46] The weekly differences between Chicago and New York grain prices averaged 34 cents, in accord with the Board of Trade rate of 33 cents and lake-rail rate of 28 cents, but not in accord with the official rate of 39 cents. Assuming simultaneous adjustment of P_g to rates, one cent changes in R_T and R_L led on average to a change of .84 cent in the average grain price difference (since the computed $\hat{\beta} + \hat{\gamma} = .838$ in Table 3.5); the independent change in R_0 brought forth a change of .41 cent in the grain price difference

[43] Testing the equations $P_g{}^* = \alpha + \beta R_0$ and $P_g{}^* = \alpha + \beta R_T$ with P_g the average weekly Chicago–New York grain price difference (assumed to adjust *this* week to *this* week's rate), Table 3.5 shows that $P_g = \alpha + \beta R_T$ "fits" the data better; the computed correlation coefficient $\hat{\rho}^2 = .363$ rather than $\hat{\rho}^2 = .000$ for $P_g = \alpha + \beta R_0$.

[44] Also, the calculated value of β in $R_T = \alpha + \beta R_0$ was .250 (as shown in Table 3.5) so that the change in Board of Trade rates was only 25 per cent of official rate changes.

[45] From comparison of total monthly shipments with shipments by steamer as shown in the 1875 *Annual Report* of the Chicago Board of Trade. Of the 2,790,000 bushels of wheat shipped in August, for example, 2,559,000 went by steamer.

[46] Official rates and lake-rail rates did not change at the same time, however; the simple correlation coefficient between R_0 and R_L was .091.

TABLE 3.5

Official Rail Rates, Board of Trade Rail Rates, and Differences Between Chicago and New York Grain Prices, 1875–1876

$R_T = \alpha + \beta R_0 + \nu$; R_T, the Board of Trade rate and R_0, the official rate; $P_g^* = \alpha + \beta R_0 + \gamma R_L + \nu$; $P_g^* = \alpha + \beta R_T + \gamma R_L + \nu$; P_g^*, the weekly average of the daily differences between Chicago and New York spot prices, R_L, the lake-rail rate

$$P_{g,t} = P_{g,t-1} + \epsilon(P_g^* - P_{g,t-1})$$

$P_{g,t}$, the average of the daily Chicago–New York grain price differences during week t and $P_{g,t-1}$, the average for the previous week; equilibrium $P_g^* = \alpha + \beta R + \gamma R_L$ for the summer season and $P_g^* = \alpha + \beta R$ for the winter season.

The standard error of each coefficient is shown in parentheses below the coefficient.

WINTER SEASON, 1874–1875:

(21 weeks); $R_T = 37.381 + \underset{(0.000)}{0.000} R_0$; $\hat{\rho}^2 = .000$; $\bar{R}_0 = 40.000$; $\bar{R}_T = 37.381$

(19 weeks); $P_g = 41.732 + \underset{(0.000)}{0.000} R_0$; $\hat{\rho}^2 = .000$; $\bar{P}_g = 41.372$

(19 weeks); $P_g = 28.832 + \underset{(0.110)}{0.343} R_T$; $\hat{\rho}^2 = .363$

(15 weeks); $P_{g,t} = 0.936 P_{g,t-1} + \underset{(0.249)}{0.064}(19.731 + 0.000 R_0)$; $\hat{\rho}^2 = .005$

(15 weeks); $P_{g,t} = 0.592 P_{g,t-1} + \underset{(0.272)}{0.408}(17.675 + 0.613 R_T)$; $\hat{\rho}^2 = .265$

SUMMER SEASON, 1875:

(30 weeks); $R_T = 22.500 + 0.250 R_0$; $\hat{\rho}^2 = .022$; $\bar{R}_0 = 39.333$; $\bar{R}_T = 32.800$; $\bar{R}_L = 28.160$

(25 weeks); $P_g = -5.515 + \underset{(0.106)}{0.413} R_0 + \underset{(0.281)}{0.824} R_L$; $\hat{\rho}^2 = .749$; $\bar{P}_g = 34.045$

(25 weeks); $P_g = +1.039 + 0.006 R_T + 0.832 R_L$; $\hat{\rho}^2 = .724$

(23 weeks); $P_{g,t} = 0.515 P_{g,t-1} + \underset{(0.159)}{0.485}(-26.672 + 0.839 R_0 + 1.004 R_L)$; $\hat{\rho}^2 = .452$

(23 weeks); $P_{g,t} = 0.524 P_{g,t-1} + \underset{(0.170)}{0.476}(2.436 + 0.238 R_T + 0.867 R_L)$; $\hat{\rho}^2 = .380$

WINTER SEASON, 1875–1876:

(22 weeks); $R_T = -6.530 + \underset{(0.063)}{1.146} R_0$; $\hat{\rho}^2 = .944$; $\bar{R}_0 = 41.681$; $\bar{R}_T = 41.227$

(21 weeks); $P_g = 27.134 + \underset{(0.129)}{0.391} R_0$; $\hat{\rho}^2 = .325$; $\bar{P}_g = 43.366$

(21 weeks); $P_g = 29.490 + \underset{(0.108)}{0.338} R_T$; $\hat{\rho}^2 = .339$

(18 weeks); $P_{g,t} = 0.577 P_{g,t-1} + \underset{(0.198)}{0.423}(21.502 + 0.500 R_0)$; $\hat{\rho}^2 = .280$

(18 weeks); $P_{g,t} = 0.593 P_{g,t-1} + \underset{(0.178)}{0.407}(21.685 + 0.501 R_T)$; $\hat{\rho}^2 = .363$

TABLE 3.5 (Continued)

SUMMER SEASON, 1876:

(31 weeks); $R_T = 19.258 + \underset{(0.000)}{0.000}R_0$; $\hat{\rho}^2 = 0.000$; $\overline{R}_0 = 20.000$; $\overline{R}_T = 19.258$; $\overline{R}_L = 18.154$

(26 weeks); $P_g = 12.122 + \underset{(0.000)}{0.000}R_0 + \underset{(0.253)}{0.676}R_L$; $\hat{\rho}^2 = .229$; $\overline{P}_g = 24.394$

(26 weeks); $P_g = 3.780 + \underset{(0.708)}{0.458}R_T + \underset{(0.273)}{0.736}R_L$; $\hat{\rho}^2 = .243$

(24 weeks); $P_{g,t} = 0.764P_{g,t-1} + \underset{(0.112)}{0.235}(33.374 + 0.000R_0 - 0.6064R_L)$; $\hat{\rho}^2 = .259$

(24 weeks); $P_{g,t} = 0.760P_{g,t-1} + \underset{(0.119)}{0.240}(4.135 + 0.448R_T + 0.529R_L)$; $\hat{\rho}^2 = .261$

(since $\hat{\beta} = .413$ in calculations of $P_g = \alpha + \beta R_0 + \gamma R_L$ in Table 3.5).[47] If a lag in adjustment is assumed, the combined effect of a 1-cent change in R_T and R_L was to change the grain price difference by 1.10 cents, while a change of R_0 alone changed P_g by .839 cent (as shown by the computations for $P_{g,t} = (1 - \epsilon)P_{g,t-1} + \epsilon(\alpha + \beta R + \gamma R_L)$ in Table 3.5).

The winter of 1875–1876, from the view of the established railroads into Chicago, was not an improvement. Rates on the Board of Trade averaged only one-half cent less than the published rates, and more than 90 per cent of the variance in these rates was explainable by changes in R_0 (as shown by $\hat{\rho}^2 = .944$ for $R_T = \alpha + \beta R_0$). But the unexplained reductions on the Board of Trade included a 5-cent decrease in February and another in April; both were quickly followed by reductions in official rates.[48] Both may have merely followed more extensive rate cuts on shipments to Baltimore, however. The Board of Trade listed rates for transport to Baltimore that averaged 34.52 cents while the official rates averaged 35.81 cents, so that the difference between the two rates was greater for transport to Baltimore than to New York. Variations in official rates do not explain changes in Board of Trade rates to Baltimore to the same extent as for the Chicago–New York rates: the least-squares regression equation for Chicago–Baltimore rates for the winter of 1875–1876 is [$R_T = -15.405 +$

[47] The combination of changes in R_0 and R_L, and in R_T and R_L, explained approximately the same percentage of variation in P_g, as shown by the computed values of $\hat{\rho}^2$. Distinction should be made (as above), however, between the combined effect of R_T and R_L and the separate effect of R_0.

[48] These two decreases had little effect on grain price differences, according to the equations for instantaneous adjustment, but, on the basis of a (computed) one and one-half week lag, changes in Board of Trade rate explained more of the variance in grain price differences ($\hat{\rho}^2 = .363$ for $P_{g,t}$ as a function of the lag and R_T, while $\hat{\rho}^2 = .280$ for $P_{g,t}$ as a function of the lag and R_0).

$1.394R_0]$ and the coefficient of correlation $\hat{\rho}^2 = .838$ (whereas the coefficient $\hat{\rho}^2 = .944$ for Chicago–New York rates).

These Board of Trade rates to Baltimore seem to have had their effect. The price differences between spot grain in Chicago and in the Baltimore Corn and Flour Exchange averaged 33.54 cents per hundred pounds. They varied more in accord with the Baltimore Board of Trade rates than the official rates as well: the equation predicting weekly average grain price differences from the official rate was $[P_g = 32.840 + 0.325R_0]$ with $\hat{\rho}^2 = .213$, while the equation with the Board of Trade rate was $[P_g = 34.898 + 0.256R_T]$ and $\hat{\rho}^2 = .288$. Approximately 7 per cent of the variance in grain prices unexplained by the official rates can be explained by variance in the Board of Trade rates.[49] The Baltimore and Ohio seems to have celebrated its Chicago entry with rates that undercut the official rates of the New York Central and Pennsylvania. The accused acknowledged some independent rate-setting; this and the rate-matching of the two larger roads reduced price margins first on shipments of grain to Baltimore, then on shipments to New York.

The question to ask of the accused is whether there were profits from cheating.[50] The Baltimore road's profits depended upon whether the lower rate increased its share of the shipments by such a great amount that subsequent rate-matching was tolerable.[51] In fact, the Baltimore and Ohio did receive far more in the first month of operation than its average for a month in that season over the period 1871 to 1879.[52] This company's market share subsequently was less than average for a number of months. During the winter season of 1875–1876, market shares initially were greatly in favor of the Baltimore road — the company more than doubled its average percentage — and subsequently were somewhat below the average for this railroad.

Percentages of grain shipments for the three railroads from Chicago are summarized in Table 3.6 and shown in detail in the statistical appendix to

[49] It might be noted that these calculations of the instantaneous-adjustment equation $P_g^* = \alpha + \beta R + \nu$ for Chicago-Baltimore are from the spot prices of No. 2 corn and oats in Baltimore as reported in "markets by telegraph" in the daily *Chicago Tribune* 1875–1876.

[50] As is argued on the preceding pages, profit from disloyalty is assumed to render the agreement ineffective, given that the individual railroad is expected to cut rates systematically.

[51] As is argued in Chapter 3, the shipments from cheating q_c had to be so much greater than the allotted share q_0 that (properly discounted) $(R_0 - K_2)q_0 < (R_c - K_2)q_c + (R_0^* - K_2^*)q_c^*$ for official rate R_0, independent rate R_c, and the group's rate in reaction to cheating R_0^* when the lag in detection is one-half the planning period.

[52] "Average share" is here used to designate the calculated value

$$S = 8.942 - 0.022T - 3.451\overline{X}$$

for the relevant months "T," season "$\overline{X}$" as shown in Table 3.6.

TABLE 3.6

Grain Shipments from Chicago, 1874 to 1879

The share of each railroad $S = \alpha + \beta T + \gamma \overline{X} + \delta \overline{Y} + \nu$ has been estimated by the method of least squares, with S = percentage of total tonnage from Chicago of wheat, corn, oats, and flour; T = month of shipment, numbered from 1 to 108 for the period 1871 to 1879; $\overline{X} = 0$ for the lakes closed to traffic or $\overline{X} = 1$ for the lakes open to shipping; $\overline{Y}$ = the change in shares with the entry of the Baltimore and Ohio ($\overline{Y} = 1$ for all months after the Baltimore road's entry). The estimates of trend are:

New York Central Railroad,
$S = 60.271 + 0.536T - 49.244\overline{X} - 2.861\overline{Y}$
$\hat{\rho}^2 = .917$ and $Su = 10.646$
Pennsylvania Railroad,
$S = 15.108 + 0.095T - 15.521\overline{X} - 2.211\overline{Y}$
$\hat{\rho}^2 = .798$ and $Su = 5.917$
Baltimore and Ohio Railroad (1875 to 1879 only),
$S = 8.942 - 0.022T - 3.451\overline{X}$
$\hat{\rho}^2 = .574$ and $Su = 4.491$

The "trend" share for any month is obtained by inserting the relevant values of T, $\overline{X}$, and $\overline{Y}$ in the correct equation. Actual shares for each month of 1874 to 1879 are given in the Statistical Appendix to Chapter 3 along with the difference between the actual and trend share "S" $= \hat{\alpha} + \hat{\beta}T + \hat{\gamma}X + \hat{\delta}Y$. The following consists of the average actual share and the average residual from computed "S" for each season.

	New York Central		*Pennsylvania*		*Baltimore and Ohio*	
	Per Cent of Shipments	*Excess or Deficit from Trend*	*Per Cent of Shipments*	*Excess or Deficit from Trend*	*Per Cent of Shipments*	*Excess or Deficit from Trend*
Winter 1874–1875	67.4	+6.7	18.1	0.0	4.8	−2.9
Summer 1875	6.4	−4.8	4.9	+2.0	1.3	−2.9
Winter 1875–1876	39.0	−21.7*	18.6	−0.2	17.2	+9.6*
Summer 1876	27.8	+8.0	3.5	−0.6	5.4	+1.4
Winter 1876–1877	57.5	+5.0	13.7	−3.1	11.4	+4.8*
Summer 1877	10.8	−1.2	2.4	−2.6	1.3	−2.9
Winter 1877–1878	58.8	+6.5	16.9	−1.1	6.2	−2.2
Summer 1878	15.8	+2.7	4.1	+2.0	1.4	−2.0
Winter 1878–1879	57.2	−5.4	32.3	+10.0*	3.8	−3.0
Summer 1879	18.2	+4.4	6.7	−0.8	3.3	+0.9

* Denotes three successive deviations from the trend line greater than the value of the standard error of estimate.

Source: From shipment figures in the *Annual Reports* of the Chicago Board of Trade (Chicago, 1871 to 1879) or the *Daily Commercial Reports* of the Board of Trade as shown in the "Weekly Commercial Bulletins" (1871 to 1879).

this chapter. The percentages reflect the annual summer loss to the steamer trade after the lakes opened to traffic in May. They also indicate some trend in relative size for each road; the estimates of trend values for each month (from fitting the equation $S = \alpha + \beta T + \gamma \overline{X} + \delta \overline{Y}$ to the observed shares "S," the monthly time periods "T," season "X," and period before and after the entry of the B & O "Y") indicates the two periods of extraordinary shares for the B & O. The railroad realized 13.2 per cent of total grain shipments during December of 1874 when, according to trend, it should have shipped 7.6 per cent from Chicago. For the succeeding twelve months, it should have shipped greater percentages than it did if it was to receive its "regular" share. In January, 1876, this railroad made 19.6 per cent of total grain shipments when according to trend it should have made 7.7 per cent. During February and March it accounted for 25.8 per cent and 28.4 per cent of shipments — both more than twice the trend percentage.[53]

The Baltimore and Ohio seems to have "succeeded" in one instance, and "failed" in the other, as a result of these changes in shares. The records of the railroad are not complete enough to show increased or decreased profits from shipping more of the grain from Chicago to the eastern seaboard[54]; assuming, however, that these shipments were at cheater's rates R_c equal to the Board of Trade rate R_T, and that marginal costs K_2 were no greater than the average costs of transport on all shipments, some inferences on profits can be made.[55] The imputed profits $(R_c - K_2)q_c$ each

[53] Some of the shift of tonnage to Baltimore should have taken place because, as part of the agreement of that time period, rates to these cities were reduced until the official differential between Chicago–New York and Chicago–Baltimore was 6 cents per hundred pounds rather than 5 cents. Official rates were 45 cents per hundred pounds to New York, and 39 cents to Philadelphia or Baltimore after December 13, 1875. This could not have been the only reason, however, because the lower rate applied to Philadelphia as well but shipments to Philadelphia declined sharply. The Baltimore and Ohio railroad received a larger percentage than it was to have at any other time before 1880 — and all of the traffic from the new differential.

[54] Some information is available on tonnage of various products over each of the railroads by year (in the *Annual Reports* of the Illinois Railroad and Warehouse Commission, for example) from all locations together. There are collections of total monthly receipts and expenses (in the same source, for 1871 to 1879) for any one railroad passing through Chicago. But these result in only average revenue and cost figures per annum.

[55] Variable costs include expenses listed in the *Annual Reports* of the individual railroads and of the state of Ohio for: (1) locomotive repairs; (2) locomotive service; (3) train service; (4) fuel; (5) oil and waste. Simple correlation analyses indicate that these cost components vary with train mileage each period. The totals were multiplied by the per cent of freight train miles to total train miles to obtain freight costs. The freight cost figures were then divided by total freight ton miles to obtain average variable costs per ton per mile. Ton-mile average costs were then multiplied by the appropriate line length, and by 100/2,000, to obtain average costs per 100 pounds for transport from Chicago to New York.

The average cost statistics have an upward bias as an estimate of marginal costs for long-distance transport because they are based upon a much shorter average distance and it is possible that total costs increase at a decreasing rate with distance; also, grain transport costs are less than those for nonagricultural commodities. The effect of the upward bias is to underestimate the profits from cheating on the official rate.

month are shown in Table 3.7.[56] The amount in December of 1874 was greater than that expected from the trend value of shipments at the official rate. The much larger market share that month more than compensated

TABLE 3.7

Profits from Independent Rate-Setting, 1875–1876

Total profits from an independent rate are $(R_T - K_2)q$ with "R_T" the Board of Trade rate, "K_2" the average variable costs of transporting 100 pounds from Chicago to Baltimore, and "q" the hundreds of pounds of grain or flour shipped from Chicago by the railroad in question.

Cartel profits for one railroad are returns calculated from the "expected" share of tonnage at the official rate R_0. These are $(R_0 - K_2)\cdot S\cdot Q$ with "R_0" the official rate, "K_2" the average variable costs, "S" the expected percentage of shipments for each railroad from $S = \hat{\alpha} + \hat{\beta}T + \hat{\gamma}X + \hat{\delta}Y$ (as described in Table 3.6), and "Q" the total shipments of wheat, corn, oats, and flour from Chicago in hundreds of pounds.

Month	*Year*	*Railroad*	*Imputed Total Profits (in Dollars)*	*One Share of the Cartel Profits (in Dollars)*
December	1874	Baltimore and Ohio	12,189	8,865
January	1875		1,634	13,893
February	1875		1,955	16,621
		Total	15,778	39,379
January	1876	Baltimore and Ohio	31,427	13,062
February	1876		61,822	19,444
March	1876		60,876	17,034
April	1876		16,395	36,805
May	1876		9,414	5,399
June	1876		6,190	1,450
July	1876		1,432	818
		Total	187,556	94,012

Source: Rates are from Table 3.4; market shares and total shipments from the Statistical Appendix to Chapter 3; average variable costs for each road are the sum of fuel expenses, oil and waste expenses, locomotive repair and service expenditures, and train service expenses for that road divided by the sum of total passenger train miles and freight ton miles, reduced to cents per 100 pounds Chicago to New York or Baltimore (the figures as listed in the *Annual Reports* of the railroad or of the Railroad Commission of the State of Ohio).

for the reduced 35-cent rate. The railroad's profits in January and February of 1875 were lower than those from an official rate of 40 cents and a share

[56] This is, again, a minimum estimate of profits. That is, if there were particular cost savings in large volume transport under an independent rate, then cheating could have been profitable while computed total revenue and shipment figures do not indicate such profits. It is difficult to conceive of an instance in which the computed profits in Table 3.7 overestimate returns from cheating.

of grain shipments equal to its long-term trend of shares.[57] The initial increase was not sufficient to compensate for the later losses: the sum of "profits" for the three-month period was less than that expected from loyal membership in the Western Railroad Bureau.

This "failure" was in contrast with the "success" of the winter of 1875–1876. The initial increase in tonnage for the Baltimore and Ohio was quite large, and the difference between official rates and Board of Trade rates to Baltimore was little more than 1 cent per hundred pounds so that immediate profits were more than $100,000 greater than the share from loyal adherence to official rates. The independent rate was probably matched in April (there were listings of reductions of official rates to the level of actual rates on the Board of Trade in the second week of April). The losses following the detection of cheating were taken after the lakes were open, when the normal shipments by rail were small. The imputed profits were approximately $90,000 from not adhering to the official rate.

Such large returns cast doubt on the assertion that the first independent venture of the B & O was a "mistake." The managers of the railroad immediately announced its 1874–1875 independent rates, so that there was no lag before detection and matching by the two larger transporters. The 1875–1876 independent rates were not announced but were discovered by the Board of Trade — at the time of large shifts in tonnage to the Baltimore and Ohio. The differences are so great that, rather than losses following from "error" in the first instance, perhaps they were taken deliberately so as to disrupt the agreement between the Pennsylvania and New York Central. After the agreement was disrupted, a differential official rate was installed that benefited Philadelphia and Baltimore. The Baltimore railroad "succeeded" in obtaining most of the benefits as its own by profitable cheating in the winter of 1875–1876. This did not escape the attention of the railroads and the trading interests in New York. As they saw it, "[Railroad] discriminations in favor of Baltimore and Philadelphia, commenced turning the trade from their competitors, the New York railways and the New York Canals . . . differences in the local charges further favored shipments to Baltimore . . . and resulted in a continuous freight warfare between the trunk railway lines during all the remaining portion of 1876 extending into 1877." [58]

[57] The railroad's "share of cartel profits" was calculated on the basis of a 40-cent rate for the entire three months. This is to assume that the *reason* for the reduction of the official rate in January was to meet the independent rate of 35 cents — not because of any change in group demand or costs.

[58] "The Division of Trade from New York," *Annual Report of the New York Produce Exchange* (New York, 1877), p. 220.

The New Trunk-Line Rate Agreements, 1876 to 1879

The relations between the trunk-line railroads after 1876 involved little more than intermittent agreement. The Baltimore and Ohio had demonstrated the profitability of independent action, and the lesson from the demonstration seems not to have been rendered obsolete by later collusive agreements. An attempt was made to restore through rates at a conference of the Baltimore and Ohio, the Pennsylvania, and the Erie railroads as early as October of 1876[59]; rates on grain from Chicago to the Atlantic seaboard "remained at the publicly acknowledged level of 20 cents per hundred pounds," because the "New York Central and Hudson River . . . regarded as a common enemy . . . was not present. . . ." [60] Somewhat later, in December of 1876, an agreement was made to increase rates to 30 cents per hundred pounds and to set all Chicago–Baltimore rates at 13 per cent less than Chicago–New York rates to favor New York in the traffic from Chicago and to set rates on tonnage from St. Louis at 14 per cent less to Baltimore than to New York ". . . giving Baltimore a greater advantage for freight on the southern part of the district in which all the trunklines competed." [61] Higher rates and market-sharing were part of the agreement: the "package" included also the prerogative for any railroad to pay rebates to international shippers through one city that found the total freight bill to Liverpool greater than through other cities.[62] There were requirements to limit rebates to this case, but there was a conspicuous lack of procedure by which the organization could ascertain whether rebates were being so limited.

Accusations of cheating were made from the date of birth of the agreement. Before it even went into effect, there was "a dispute as to whether when freight is received at a seaport at the local rate and afterward exported, the drawbacks shall consist of the difference between the railroad rate to Baltimore and that to New York, or of the difference in the rail rate plus the ocean rate from the respective ports." [63] Rates were advanced early in January of 1877 to the level of 40 cents per hundred pounds on grain to New York, even though "the Lake Shore and Michigan Southern maintains the old rates of 35 cents per hundred pounds for grain. . . ." [64] The other lines succeeded in this attempt for a few days because "under

[59] Cf. "Settling the Railroad War," the *Railroad Gazette* (October 6, 1876), 434.

[60] *Ibid.*

[61] "The End of the Railroad War," the *Railroad Gazette* (December 22, 1876), 556.

[62] Cf. "The Trunklines Agreement," the *Railroad Gazette* (December 29, 1876), 572.

[63] Cf. the *Railroad Gazette* (December 29, 1876), 571.

[64] "Differences in Rates between Transporters," the *Railroad Gazette* (January 19, 1877), 30.

existing circumstances when the snow blocks the roads so that they are unable to supply the demands upon them it may not make much difference with the traffic of the roads [that there were two rates to New York]." [65] But the grain rate was reduced once again to 35 cents per hundred pounds during the last week of this month.

The Baltimore and Ohio began complaining of nonadherence to the 35-cent rate in February. In a telegram to President Garrett in New York dated February 7, 1877, Vice-President King charges "it is positively stated in Chicago and believed by railroad men that the Lake Shore and New York Central made contracts with large shippers for 1,000 cars grain at 15 cents." [66] Later that month Mr. King informed the Baltimore office that "Ledyard says that cut rates are quoted on the exchange via the Erie and New York Central routes from here [Chicago] and that it is idle to talk about advancing rates when the Lake Shore will not adhere to the existing tariff. . . ." [67]

The complaints came to an end in April when a new rate contract was signed in which the differences between the Chicago–New York rates and the Chicago–Baltimore rates were canceled, and the payment of official drawbacks on shipments to Liverpool was done away with.[68]

There was established a "new spirit of harmony" at the same time on shipments westbound from the Atlantic seaboard. Negotiations had been started in April, and were continued through May, on an agreement to "pool" westbound tonnage that gave each line a percentage of the total and that transferred tonnage to any railroad not receiving its allotted share from another with a surplus over its allotted share. It was agreed on May 25, 1877, that the New York Central should receive 33 per cent, the

[65] *Ibid.*

[66] Atlantic and Pacific Telegram from John King, Jr., vice-president, to John W. Garrett, president. February 7, 1879, the Baltimore and Ohio Archives, the Maryland Historical Society, Baltimore, Maryland.

[67] Baltimore and Ohio Telegram from John King, Jr., vice-president in Chicago, to John W. Garrett, president, in Baltimore, February 22, 1879; the Baltimore and Ohio Company archives, the Maryland Historical Society, Baltimore, Maryland. The letter indicates that not only the shippers but also the transporters obtained information from the Board of Trade rates.

[68] By the terms of the new agreement, the rates on all traffic from Chicago to the seaboard were 3 cents per hundred pounds lower to Baltimore and 2 cents per hundred pounds lower to Philadelphia than the corresponding rates to New York. "These differences are to be maintained whatever may be the rates and . . . are to be subject to no rebate whatsoever." [The *Railroad Gazette* (April 13, 1877), 167.] Another change of some importance was that New York–Baltimore rate differences from all western points were made the same: "Baltimore no longer has an advantage on Cincinnati and St. Louis business greater than that on Chicago business." (*Ibid.*) The impression of the newspapers was that the agreement was to be easier to maintain because of the less complex structure of rates. [Cf. the *Railroad Gazette* (April 20, 1877), 180.]

Pennsylvania 25 per cent, the Erie Railroad 33 per cent, and the Baltimore and Ohio 9 per cent of the westbound freight tonnage.[69] To oversee the actual sharing of freight an administrative group was formed under Colonel Albert Fink, an acknowledged master of the procedures for operating a cartel agreement.[70] The contract was signed and a circular of serious intent was issued: "Notice is hereby given to all railroad companies and transportation lines connecting with the New York Central, the Erie, the Pennsylvania, and the Baltimore and Ohio Railroad that in order to secure uniformity in rates commencing with July 1, 1877, through bills of lading will be issued and through rates of transportation guaranteed on westbound freight from eastern seaboard cities only under the conditions that all connecting roads over which such freight is shipped will charge their full proportion of through rates, make no special rates with shippers or owners of goods, pay no rebates and hold out no inducements of any kind amounting practically to a reduction in the rates established by the above named roads. . . ." [71]

The agreement did not spread immediately to eastbound grain and provision traffic.[72] Rather, in the spring of 1877, there were frequent exchanges of accusations between the presidents of the trunk-line railroads as to the identity of the rate-cutter on the eastbound traffic. President Garrett of the Baltimore road accused employees of the New York Central of accepting freight from the Wabash Railroad (in Central Illinois) at reduced rates during the last week of June and the first week of July. President Vanderbilt replied, "Your dispatch this morning . . . is the first and only advice I have on the subject. . . . I cannot believe it possible [since] Mr. Rutter and my son W. K. are very active and constantly on the lookout, if anything turns up they are likely to know about it and will insist that no connection of ours be allowed to interfere with the arrangement made between the trunklines." [73] The accusation was repeated July 10, and

[69] Cf. "A Trunkline Pool," the *Railroad Gazette* (May 18, 1877), 224 and (May 25, 1877), 230; D. T. Gilchrist, "Albert Fink and the Pooling System," *Business History Review*, *XXXIV* (1960), 25–49.

[70] Colonel Fink's name had been long associated with the success of the Southern Railroad and Steamship Association, the combination of lines in the lower Mississippi and Ohio River Valleys. Cf. D. T. Gilchrist, *op. cit.*

[71] As reproduced in the *Railroad Gazette* (July 6, 1877), 306.

[72] It might be asked why it was easy to form an elaborate pooling arrangement on westbound traffic, but not on eastbound agricultural commodities. There is no convincing answer; the presence alone of the main trunk lines of two systems made traffic out of New York easier to administer, as compared to traffic over many lake steamer companies and five subsidiary railroads from Chicago, and the agreement practicable to operate may have been attempted first.

[73] Letter from J. W. Garrett to T. A. Scott, president of the Pennsylvania Railroad, dated July 13, 1877, and quoting in full a letter received from President Vanderbilt of the New York Central, dated July 9, 1877; the Baltimore and Ohio Railroad archives, the Maryland Historical Society, Baltimore, Maryland.

Mr. Rutter, after two further exchanges of correspondence, acknowledged that the president of the Wabash "said he had made the contract." [74] The New York Central was exhorted to charge the official rate as its share of the through rate on this Wabash traffic; the reply was that "Mr. Hopkins [the Wabash president] intimates to us that he has a right to do what he chooses with his own money . . . how do you suppose we can put on high charges . . . I doubt even your ability to do this." [75] This reply left it for the vice-president of the Baltimore road to state that "it seems clear to me that all that is needed is for the New York Central Company to assume and hold a firm position with its western connections . . . [to] effectively prevent the latter from destroying the trunk line agreement . . . when the 23 cent rate to which you refer was announced the Penn and B & O Companies promptly refused to accept it and if the New York Central had done the same the western line would not now have the excuse of that precedent." [76]

In contrast, there were no accusations at all of cheating on westbound rates. The Baltimore road feared that "Mr. Vanderbilt would attempt to reduce B and O quotas below 9 per cent" [77] on westbound traffic because the railroad had not been able to attain this assigned share; railroad officials complained that the failure was because "the attention to the transfer and its difficulties required so much time that there was little left in which to solicit freight . . ." and "that the 20 New York Central agents and the 29 agents of the western roads have all made extraordinary exertions . . . ," [78] but no reallocation took place, and rates were maintained. The agreement on westbound traffic "worked," so that the implication was obvious: pooling of tonnage should be extended to eastbound shipments as well. It seems to have taken another year for this to be achieved, however.

The initial step toward duplicating the acknowledged success on westbound traffic was to set up an administrative bureau to survey eastbound tonnage. In December of 1877, the general freight agent of the Baltimore and Ohio Railroad, Mr. John Guilford, was appointed "Commissioner" of general eastbound traffic "to hear and investigate complaints and report

[74] Dispatch from J. H. Rutter, the New York Central, to John King, Jr., of the Baltimore and Ohio, July 10, 1877 (7:43 P.M.); the archives of the Baltimore and Ohio, *op. cit.*

[75] Dispatch from J. H. Rutter to John King, Jr., dated July 11, 1877 (12:20 P.M.); the archives of the Baltimore and Ohio Railroad, *op. cit.*

[76] Letter from John King, Jr., to J. H. Rutter, dated July 12, 1877; the archives of the Baltimore and Ohio Railroad, *op. cit.*

[77] Letter to J. W. Garrett from A. C. Rose, New York Agent for the Baltimore and Ohio Railroad, dated September 17, 1877; the archives of the Baltimore and Ohio Railroad, *op. cit.*

[78] *Ibid.*

when the delinquent [railroad] should be punished . . . ,"[79] The second step was to punish cheaters for setting independent rates; it was not taken. There was a series of meetings "to secure the maintenance of the agreed rates on eastbound traffic which had been cut for two or three weeks past."[80] These meetings were "quite harmonious and all agreed that every effort should be made to maintain rates."[81] But they did not achieve the desired result: "A conference is being held [February 1, 1878] concerning eastbound rates which have been worse demoralized than before for the past week . . . freight is said to be taken freely at 25 cents per hundred pounds from Chicago and Milwaukee to New York and the time is not far distant when rates cannot be restored on grain due to the opening of navigation. . . ."[82] The Grand Trunk Railway of Canada, a line connecting with the New York Central at Detroit and with the Atlantic Ocean traffic at Portland, Maine, made bitter complaints against the results: "We have had to complain, and have supplied evidence, that the tariffs have been systematically disregarded by the lines presided over by Mr. Vanderbilt and working in connection with the New York Central road. Papers were sent [to the Commissioner] which proved that the rates at Milwaukee had been ruthlessly broken down and traffic which would have gone by Grand Trunk route was sent forward by the Lake Shore and by the Canada Southern Lines. The agents of this company were instructed to adhere to the tariff and we lost large quantities of traffic, and are now suffering from observance of engagements entered into with those who from the very commencement disregarded them."[83]

In February it was stated that "eastbound rates continue broken and are unsatisfactorily low . . . with no present prospect of their restoration."[84] An attempt was made to restore rates, but "the meeting was disappointing . . . rates were not restored and propositions to restore them to $\frac{1}{8}$ and $\frac{1}{4}$ lower were voted down . . . it was made to appear that there are a good many contracts outstanding to carry at greatly reduced rates for one, two, three, or more months to come . . . apparently the sole obstacle to the restoration of rates was refusal of a single company to commit their distribution of the business which had been secured ahead by time contracts . . . the company not agreeing to pool time contracts so as to be able to set a new rate was the Lake Shore and Michigan Southern."[85]

The third step, that of pooling eastbound traffic, was discussed in March.

[79] Cf. the *Railroad Gazette* (December 28, 1877), 576.

[80] "The East Bound Rate Meeting," the *Railroad Gazette* (January 18, 1878), 33.

[81] *Ibid.*

[82] Cf. the *Railroad Gazette* (February 1, 1878), 76.

[83] "A Letter from the Grand Trunk," the *Railroad Gazette* (February 1, 1878), 53

[84] "Rates and Combinations," the *Railroad Gazette* (February 15, 1878), 82.

[85] "The Meeting of the Western Railroads," the *Railroad Gazette* (February 15, 1878), 84.

Colonel Albert Fink, the Commissioner for westbound traffic, outlined the progress of the discussion in a letter to John W. Garrett of March 11: "At the last meeting of the presidents of the trunk lines held on February 11 . . . a resolution was passed directing me to devise a plan for pooling eastbound traffic, and to confer for that purpose with the representatives of the western roads. I take pleasure in reporting that the conference with and between the western roads has already resulted in an agreement to pool the eastbound business for a period of three months . . . [and] that the western roads have required me to assist in putting the plan into practical operation." [86] The shares of eastbound tonnage for each of the railroads were decided on March 26: ". . . after dividing into committees and examining figures on freight shipped over the lines during the last few years, . . . the percentage of apportionment of freight . . . from Chicago was [set] as follows: the Michigan Central [New York Central], 32 per cent; the Lake Shore and Michigan Southern [New York Central], 27 per cent; the Pittsburgh, Ft. Wayne and Chicago [Pennsylvania], 24 per cent; the Baltimore and Ohio, 10 per cent; the Pittsburgh, Cincinnati and St. Louis [Pennsylvania], 7 per cent." [87] The percentages were in effect on the eleventh of March and continued through the eleventh of June, 1878.[88] The first statement of actual percentages of tonnage transported was issued April 30, 1878; it showed that the subsidiaries of the New York Central had carried 63.5 per cent of the tonnage from Chicago, the Pennsylvania had transported 35 per cent, and the Baltimore and Ohio had transported only 5.8 per cent of the total tonnage[89] — or that the New York railroad "owed" the Baltimore and Ohio 4.2 per cent of the tonnage. Prior to this report eastbound rates had been broken: "There has been some controversy as to who began it, without any results, we believe . . . but the rates seem to have been reduced very generally from 25 cents to 20 cents per hundred pounds on grain and provision from Chicago to New York. . . ." [90] The New York Central was accused of taking the Baltimore

[86] Letter from Albert Fink, Commissioner, to John W. Garrett, dated March 11, 1878; the archives of the Baltimore and Ohio Railroad, *op. cit.*

[87] "The East Bound Freight Apportionment," the *Railroad Gazette* (March 28, 1878), 165. The combined share for the New York Central affiliates was 59 per cent; for the Pennsylvania affiliates, 31 per cent; the Baltimore and Ohio, 10 per cent. These same percentages are listed in *Proceedings Western Railroads, 1877 to 1880* (New York: Russell Bros., Printers, 1880), the Proceedings for March 26, 1878, "Report of the Chicago Committee."

[88] Cf. "A Review of the Operation of the East Bound Pool," the *Tenth Annual Report of the Massachusetts Railroad Commissioners* (Boston, 1879), p. 132. Each railroad was required to file periodic reports on the tonnage shipped, and these were used to compute tonnage percentages by Commissioner Fink's staff. (*Ibid.*)

[89] "Statement of Tons of Each Class of East Bound Freight from Chicago from March 11th to April 30th, 1878 . . . ," the *Proceedings Western Railroads, op. cit.*

[90] "East Bound Freight Rates," the *Railroad Gazette* (April 19, 1878), 201.

road's tonnage by cutting rates; the Central denied the charge and refused to transfer any tonnage. The rates to New York in May continued "to be demoralized, with no present prospect of their restoration. . . ." [91]

When the three-month trial period was completed June 11, 1878, there was little effort made to renew the agreement in that form. Pooling was declared to be unworkable: "the difficulty with it lay in the immense complication of the problems to be solved in the complete absence of any force to compel obedience to the decisions of the Commissioner. He found himself practically powerless." [92] The managers of the trunk lines settled once again for a simple agreement ". . . to maintain rates and report shipments precisely as if there were a pool, but not to make any distribution of traffic until some time in the future, when the results of . . . strictly maintained rates may serve as a basis for final division." [93]

The plan was put into operation late in November without a formal contract agreement. All rates were increased to 35 cents, but not for long: there were reports of nonadherence by the middle of December: "the rates have been fearfully cut during the last two weeks, especially on foreign freights . . . it had been denied up to a day or two that the rates to eastern points were being cut, but yesterday they were openly made by nearly all the lines at twenty-five cents per hundred pounds on grain from Chicago to New York . . . a cut of ten cents." [94] This continued: "most of the freight coming forward is probably carried on contracts at, we suppose, as much as ten cents per hundred pounds below the tariff of November which is still nominally enforced and under which doubtless some shipments are made." [95] Meetings were held in February at which complaints were heard against contract rates 10 cents below the official rates[96] arranged by the Pennsylvania lines: "the Vanderbilt managers contended that the Pennsylvania road's increase of business had been secured by cutting rates, while Mr. McCarr insisted that his roads had maintained rates as strictly as Vanderbilt's line maintained them. . . ." [97]

From the view of the managers of the trunk lines, there was no improvement during the spring of 1879. By late March traffic was heavy and "the demand for cars is brisk, [but] the Pennsylvania adhered to its determi-

[91] The *Railroad Gazette* (May 17, 1878), 245.

[92] *The Tenth Annual Report of the Massachusetts Railroad Commissioners*, *op. cit.*, p. 134.

[93] Cf. "Negotiations for an East Bound Pool," the *Railroad Gazette* (November 15, 1878), 550.

[94] "The Railroads: Important Meeting of Eastern and Western Managers," *Chicago Tribune* (December 19, 1878).

[95] "The East Bound Freight Agreement," the *Railroad Gazette* (February 5, 1879 and February 28, 1879).

[96] Cf. the *Railroad Gazette* (February 5, 1879 and February 28, 1879).

[97] *Ibid.*

nation not to enter into another agreement . . . the present rates are fifteen cents on grain and twenty cents on fourth-class freight to New York with no pretense of an agreement to continue them at these figures. . . ." [98] The complaints in May were stronger: "east bound rates were probably never before so unsatisfactory . . . it is said that a contract has been made to carry one million bushels of grain from Chicago to New York at twelve and one-half cents per one hundred pounds . . . if this is true then most of the grain is going at that rate. . . ." [99] If this was the lowest grain rate of the decade, then the precedent was not established for long: "east bound rates . . . have been quoted at ten and twelve and one-half cents per hundred pounds from Chicago to New York, with one shipment reported at seven and one-half cents." [100]

The Baltimore and Ohio Railroad expressed extreme dissatisfaction with this set of rates from December of 1878 through June of 1879. In a letter to Mr. J. H. Rutter, the General Traffic Manager of the New York Central, Mr. John King, Jr. vice-president of the Baltimore and Ohio, reviewed events beginning with a meeting of the general freight agents on December 19, 1878, when it was agreed "that tariff rates should be restored on that date." [101] There was an agreement to allow no rebates or drawbacks on the restored official rates, but soon thereafter reports reached Mr. King "going to show that the agreement was not being carried out by competing lines and that great irregularity in rates then existed." [102] No penalty or even proof of cut rates was provided when the northern roads denied the reports. "What followed is well-known: freight continued to be moved from Chicago at contract rates. Rates from all other points were consequently demoralized and went from bad to worse. Apparently no one had the power or desire to restore order." [103] At the meeting of March 19, (according to Mr. King) this condition of affairs was sharply criticized "and a determination expressed to restore rates at least so far as the trunk lines were concerned." [104] The resolutions, and the public notices of resto-

[98] The *Railroad Gazette* (March 24, 1879); the official rate had been reduced from 35 cents to 20 cents a week earlier, and continued to be listed at 20 cents per hundred pounds.

[99] Cf. the *Railroad Gazette* (May 28, 1879); the difficulty was not confined to grain shipments: ". . . the rates on livestock have gone to pieces . . . it has been divided through eveners in the past . . . but in spite of eveners the rates have now been cut." *Ibid.*

[100] Cf. the *Railroad Gazette* (June 6, 1879); the official rate was 20 cents per hundred pounds.

[101] Letters from John King, Jr. to J. H. Rutter, Esquire, dated July 3, 1879; the Archives of the Baltimore and Ohio Railroad, *op. cit.*

[102] *Ibid.*, p. 12.

[103] *Ibid.*

[104] *Ibid.*, p. 14. The plan was to charge the western connections cutting rates the full tariff for shipments east from the point of transfer.

ration of official rates had no effect: "if any consideration whatever was given [them] it has not been brought to my attention." [105] Meetings in the latter part of April and early May had no effect; each began with a statement that official rates would soon be put into effect and that only a small amount of freight under contract at cut rates remained to be forwarded, "yet the freight continued to come forward at the reduced rates under alleged old contracts." [106] There seems to have been widespread nonadherence to official rates in May: "this state of affairs (of monthly assurances that outstanding contracts were almost completed), taken in connection with the notice of the Pennsylvania Railroad (of withdrawal from the meetings of the general freight agents) resulted in a disgraceful contest during which property was transported from the far west to seaboard at rates insufficient to pay the cost of terminal expenses incurred at the delivery station. In fact a large amount of traffic was absolutely moved for nothing." [107]

Following this "disgraceful contest," it became evident to Mr. King, and to other freight agents, that nothing could be lost from drastically changing the contents of the rate agreement. Administrative changes were made as early as April 18, 1879, that simplified the procedures for setting rates and shares of the traffic: the trunk-line presidents agreed to delegate authority over all rates to the "Joint Executive Committee," of the managing freight agents of only the four eastern terminal lines of the trunk-line railroads and, rather than dividing traffic between the subsidiaries of the trunk lines in each western city, to divide the eastbound traffic in New York, Boston, Philadelphia, and Baltimore between the four roads.[108] Also, a second committee, the "Board of Arbitration," was organized to settle all questions of guilt of nonadherence to agreed rates and divisions.[109] The two committees provided "legislation" on rates and tonnage shares, and "judicial review" of departures from legislation.

By the first of August, 1879, the Executive Committee had agreed upon shares under this new method of apportionment. The Pennsylvania Company was to receive 35 per cent of the tonnage out of Chicago, while the New York Central was to receive 57 per cent, and the Baltimore and Ohio's own line into Chicago was assigned 8 per cent of the eastbound

[105] *Ibid.*, p. 15.

[106] *Ibid.*, p. 19.

[107] *Ibid.*, p. 20.

[108] Cf. "The Trunk Line Presidents Meeting," the *Railroad Gazette* (April 18, 1879), 213. Before the fact, it would seem that control of shipments into the east coast cities was sufficient: a good part of the tonnage was for shipment abroad so that the group controlled the "relevant (service) market"; if the western connections cut rates on their portion of the mileage, then the eastern terminal lines could raise their rates, so that control of through rates followed from control in the seaboard cities.

[109] "The Trunk Line Presidents Meeting," *ibid.*, 213.

tonnage.[110] There were no complaints of nonadherence to this agreement in the remaining months of 1879. When the Lakes closed in November, both the official grain rate and the Board of Trade rate were 40 cents per hundred pounds (in contrast to the acknowledged rate of 15 cents in May and June). This was the first time since 1876 that the "spirit of harmony" had carried over from the summer to after the Lakes traffic had ceased.[111]

From the spring of 1877 to the winter of 1879–1880, then, one railroad or the other was said to have quoted "secret rates" below the official rates. The New York Central refused to join the other roads in the official rates of February, 1877, and was accused of nonadherence to rates in the summer of that year as well. Once again, the New York railroad was accused of setting a rate lower than the official rate on eastbound traffic throughout the winter and spring of 1878. The Baltimore and Ohio accused that railroad of cheating in the winter of 1878–1879, while the other railroads accused the Pennsylvania of instigating the breakdown of rates in that period. There was hardly a month in which either the New York Central or the Pennsylvania was not being accused of disloyalty until the reorganized agreement was announced in August of 1879.

Many of the occurrences of rate-cutting were for unknown reasons and had little effect. The cause for the breakdowns in the winters of 1877–1878 and 1878–1879 seems to have been that profits were to be made from disloyalty, however. Occurrences of gains from cheating point to the existence of a profit motive in those seasons. The effects of the breakdowns in the two winters were substantial, as well.

The road first accused of cheating — the New York Central — received an increased share of the grain traffic during the winter of 1877–1878. According to the agreement of March, 1878, this road should have had 59 per cent of the total grain tonnage out of Chicago. In the months of December through March, the Central accounted for approximately 68 per cent of this tonnage (as shown in the statistical appendix to this chapter). This railroad continued to "do better" than 59 per cent of rail tonnage during the spring and early summer months (as can be seen from calculating the New York Central's percentage of total rail shipments from

[110] Cf. "Awards, 1879–1880," *Proceedings of the Joint Executive Committee, 1880* (New York: Russell Brothers, Printers, 1880), pp. 241–242. It is presumed that the trunk lines agreed to *accept* only these proportions from their western connections, or in their Chicago stations. The traffic out of St. Louis was divided evenly among five railroads: this guaranteed the western terminals of the Pennsylvania, New York Central, and Baltimore and Ohio 20 per cent apiece, while the independent lines (the Chicago and Alton, and the Wabash) divided the traffic among the three trunk lines in an undisclosed manner. (*Ibid.*)

[111] That is, "carried over" from an agreement on grain traffic made during the summer months (when lake steamers carried most of the tonnage) to the winter (when the railroads carried all of the grain).

the percentages of total lake and rail shipments shown in the statistical appendix). The company not only shipped more than allowed under the agreement, it also had more than its usual market share: given the Central's trend of increasing market share in 1877, 1878, and 1879, this company was more than 11 per cent over the trend in two of the spring months and averaged 6.8 per cent above trend (as shown in Table 3.6).

The New York Central, from all appearances, made a small amount of money from this increased share at the rates quoted below official rates of that season. Surplus shares at Board of Trade rates, rather than the assigned 59 per cent share at official rates, provided a $34,000 increase in profit in December, and an additional $70,000 in January (as shown in Table 3.8). Board of Trade rates declined sharply in February, March, and

TABLE 3.8

Profits from Independent Rate-Setting, Winter 1877–1878

Month	*Year*	*Railroad*	*Computed Profits (in Dollars)*	*59 Per Cent Share of the Profits at Official Rates (in Dollars)*	*"Trend" Share of the Profits at Official Rates (in Dollars)*
December	1877	New York	155,600	121,197	127,355
January	1878	Central	367,466	296,946	311,850
February	1878		338,377	373,528	393,091
March	1878		292,741	312,961	332,991
April	1878		92,836	135,126	166,460
			1,247,020	1,239,758	1,331,747

Sources: "Computed Profits" equal $(R_T - K_2)q$, with "R_T" the Board of Trade rate in Table 3.4, K_2 the average variable costs of shipping 100 pounds to New York City from the sources noted in Table 3.7, and "q" the hundreds of pounds shipped from Chicago by this railroad, as shown in the *Annual Reports* of the Chicago Board of Trade.

"59 Per Cent Share of Profits" approximates the allotted share for the New York Central in the agreement of that winter; profits are computed from $(R_0 - K_2)(.59Q)$, with R_0 the official rate and Q the total shipments from Chicago.

The "Trend Share of Profits" is an estimate of the net revenues from having conformed to the official rate while receiving the per cent share shown from the computed $\{S = \hat{\alpha} + \hat{\beta}T + \hat{\gamma}\overline{X} + \hat{\delta}\overline{Y}\}$ in Table 3.6. The trend "share" is from $(R_0 - K_2)(S \cdot Q)$.

April (from 28 to 30 cents to 20 cents per hundred pounds); the Central "lost" $35,000 in February, $20,000 in March, and $40,000 in April by transporting slightly more than its market share at these rates. The net gain for the five-month period, from the lower Board of Trade rate and the larger share of tonnage, would appear to have been approximately $8,000.

Perhaps, however, the Central assumed that as a matter of course it would receive its regular share of tonnage rather than the allotted 59 per cent share. This railroad had been receiving more than 60 per cent of the tonnage from Chicago, and its share was growing at the average rate of 0.2 per cent per month (as shown in the statistical appendix to this chapter). This might have been assumed to continue in spite of the agreement; in that case the profits from not adhering to the agreed rate would have been smaller in December and January, and losses in February, March, and April would have been greater. Computed profits from not adhering to the agreement were approximately $1,247,000, while computed profits from obtaining the usual share at official rates were $1,332,000 (as shown in Table 3.8).

This state of affairs, if actually encountered, must have been ironic. If the Central was expecting that the agreement would actually redistribute tonnage shares, it would have done well to have been disloyal. In contrast, if the Central expected no results from the new apportionment, it would have profited from remaining loyal. That is, the agreement was unstable[112] if it was expected to have been effective.[113]

The evidence of the breakdown in the winter of 1878–1879 suggests that grain was probably transported for less than full tariff rates and that the Pennsylvania Railroad profited from the cut rates. The summer of 1878 and the summer of 1879 were both seasons in which there was great similarity between official rates and those quoted privately to members of the Chicago Board of Trade. The averages of official rates and Board of Trade rates were within 1 cent of each other during both summers. Throughout the winter of 1878–1879, the average official rate was more than 4 cents higher than the average Board of Trade rate. The months in which the rate differences were greatest were those in which the Pennsylvania road experienced large increases in its share of the grain tonnage from Chicago. This transporter received 26 per cent of the tonnage of wheat, corn, and oats from Chicago in December of 1878, 5 per cent above its market share trend.[114] During January, February, and March, when Board of Trade rates were 5 cents below the official rates, the Pennsylvania received approximately 15 per cent more than expected (given the trend of its shares, as shown in the statistical appendix). After the opening of the lakes, official rates were 2 cents more than Board of Trade rates, and the Pennsylvania received from 4 per cent to 8 per cent more tonnage than expected.

[112] The term "unstable" is used according to the assumptions in Chapter 2: a profit incentive for not adhering to the agreed rate causes such systematic and persistent cheating that the cartel agreement is abandoned.

[113] The irony only follows from the New York Central having accepted in the new agreement a cut in its regular tonnage share.

[114] Since there were no changes in the agreed shares in 1878, the Pennsylvania was assigned and failed to receive 31 per cent of rail shipments. This was the case each month; it must have been apparent that the assigned share was not to be realized at official rates, so that profits $(R_0 - K_2)(.31Q)$ were not a relevant alternative.

This railroad would seem to have profited from the early increase in its share transported at Board of Trade rates. As shown in Table 3.9, profit

TABLE 3.9

Profits from Independent Rate-Setting, Winter 1878–1879

Month	*Year*	*Railroad*	*Computed Profits (in Dollars)*	*"Trend" Share of the Profits at Official Rates (in Dollars)*
December	1878	Pennsylvania	80,410	66,397
January	1879		104,898	64,541
February	1879		86,777	82,867
March	1879		61,840	66,395
April	1879		68,358	62,138
			402,283	342,338

Sources: "Computed Profits" equal $(R_T - K_2)q$, with "R_T" the Board of Trade rate in Table 3.4, "K_2" the average variable costs of shipping 100 pounds to New York City from the sources noted in Table 3.7, and "q" the hundreds of pounds shipped from Chicago by this railroad as shown in the *Annual Reports* of the Chicago Board of Trade.
The "Trend Share of Profits" is an estimate of the net revenues from having conformed to the official rate while receiving the per cent share shown from the computed $\{S = \alpha + \hat{\beta}T + \hat{\gamma}\bar{Y} + \hat{\delta}\bar{X}\}$ of Table 3.6. The trend "share" is from $(R_0 - K_2)(\hat{S} \cdot Q)$.

is estimated to have been $402,000 at Board of Trade rates, while net receipts at official rates on its trend share of total shipments were approximately $342,000. The gain from cheating was approximately $60,000 for the period from December to April of 1879.

Collusive control of grain rates seems to have been possible only for short and infrequent periods after the entry of the Baltimore and Ohio Railroad into Chicago. The operators of the trunk-line association, and the managers of the railroads, reported deviations from agreed rates by some railroad or other during most seasons of the year from 1875 to 1879. The deviations were sufficiently large that Board of Trade rates averaged 1 to 7 cents less than official rates during the winters (and part of the summers) of 1874–1875, 1875–1876,[115] 1877–1878, and 1878–1879. What was worse, in most cases the railroads accused of having disrupted the agreements appear to have benefited from such disruption: profits from cutting the official grain rates seem to have been greater than those from loyalty to such official rates. Consequently the agreements were undermined: departures were not only frequent, but (given a profit motive) deliberate and systematic.

[115] During this season, the Board of Trade rate averaged 1.3 cents less than the official rate on shipments to Baltimore, the terminal of the railroad accused of cheating.

The Consequences for the Shippers of Cartel Instability, 1875 to 1879

The most immediate effect of the breakdowns in the agreements was that there were large and abrupt decreases in grain rates. Board of Trade rates were cut to 31 cents per hundred pounds in the winter of 1877–1878, while official rates were approximately 38 cents; there were frequent changes in the Board of Trade rate independent of tariff rates that season as well ($R_T = \hat{\alpha} + \hat{\beta}R_0$ was such that $\hat{\rho}^2 = .340$, as shown in Table 3.10).

TABLE 3.10

Official and Board of Trade Rates on Grain Transport from Chicago to New York, 1876 to 1879

Calculations of $\{R_T = \alpha + \beta R_0 + \nu\}$ with: R_T the rate on transport of grain by rail from Chicago to New York as shown in the *Board of Trade Daily Commercial Bulletin;* R_0, the official rate of the associations or meetings of trunk-line officials on all-rail grain transport from Chicago to New York, as shown in the sources mentioned in Table 3.1 and the text. The standard error of the computed $\hat{\beta}$ is shown in parentheses below the coefficient; the calculated value of the coefficient of determination $\rho = \hat{\rho}^2$. Calculations of average $R_T = \overline{R}_T$, $R_0 = \overline{R}_0$. All statistics are in cents per 100 pounds of grain transported.

WINTER SEASON, 1876–1877:

(21 weeks); $R_T = 2.809 + 0.905R_0$; $\hat{\rho}^2 = .743$; $\overline{R}_T = 31.666$; $\overline{R}_0 = 31.905$
(0.122)

SUMMER SEASON, 1877:

(31 weeks); $R_T = 0.000 + 1.000R_0$; $\overline{R}_T = \overline{R}_0 = 33.226$

WINTER SEASON, 1877–1878:

(16 weeks); $R_T = -.21538 + 1.385R_0$; $\hat{\rho}^2 = .340$; $\overline{R}_T = 31.250$; $\overline{R}_0 = 38.125$

SUMMER SEASON, 1878:

(35 weeks); $R_T = -3.986 + 1.136R_0$; $\hat{\rho}^2 = .909$; $\overline{R}_T = 25.056$; $\overline{R}_0 = 25.572$
(0.063)

WINTER SEASON, 1878–1879:

(21 weeks); $R_T = 6.417 + 0.679R_0$; $\hat{\rho}^2 = .355$; $\overline{R}_T = 26.555$; $\overline{R}_0 = 30.833$
(0.210)

SUMMER SEASON, 1879:

(31 weeks); $R_T = -5.867 + 1.168R_0$; $\hat{\rho}^2 = .948$; $\overline{R}_T = 25.290$; $\overline{R}_0 = 26.677$
(0.051)

WINTER SEASON, 1879–1880:

(19 weeks); $R_T = 0.000 + 1.000R_0$; $\overline{R}_T = \overline{R}_0 = 37.895$

This was only the second time in the decade that winter Board of Trade rates were not higher than during the previous summer. Board of Trade rates in the winter of 1878–1879 were 4 cents less than the official rate, and were down to 26.5 cents on average; only 36 per cent of the variance in the Board rate was explained by variance in the published tariff (while changes in official rates the previous summer explained more than 90 per cent of the variance in Board rates that season).[116] The level of rates during the seasons of the worse disagreements varied from 25 to 33 cents; it was restored in the new agreement of 1879–1880 to an average of 37.90 cents.

The loss of collective control of long-distance rates had an effect upon both the level of rates and the differences between particular commodity rates. Official rates for transporting grain did not vary greatly from 65 cents per hundred pounds during the winters of 1871–1872 and 1872–1873 (as shown by the summary collection of average official and Board of Trade rates in Table 3.11), and the lower limit in the winter of 1873–1874 was

TABLE 3.11

Official and Board of Trade Rates in the 1870's

Season and Year	*Average Official Rate*	*Average Board of Trade Rate*
Summer 1871	48.10	47.62
Winter 1871–1872	63.93	63.75
Summer 1872	54.33	54.00
Winter 1872–1873	64.55	64.55
Summer 1873	48.55	50.00
Winter 1873–1874	56.06	56.05
Summer 1874	44.55	44.55
Winter 1874–1875	40.00	37.38
Summer 1875	39.33	32.80
Winter 1875–1876	41.68	41.23
Summer 1876	20.00	19.26
Winter 1876–1877	31.91	31.67
Summer 1877	33.23	33.23
Winter 1877–1878	38.12	31.25
Summer 1878	25.57	25.06
Winter 1878–1879	30.83	26.55
Summer 1879	26.67	25.59
Winter 1879–1880	37.90	37.90

Source: Table 3.1, 3.5, and 3.10.

close to 55 cents per hundred pounds. The summer rates during these years were approximately 45 to 55 cents per hundred pounds. The Board

[116] The coefficients of determination from $R_T = \hat{\alpha} + \hat{\beta} R_0$ were $\hat{\rho}^2 = .909$ in the summer of 1878, $\hat{\rho}^2 = .355$ the winter of 1878–1879, $\hat{\rho}^2 = .948$ the summer of 1879, as shown in Table 3.10.

of Trade rates followed these official rates quite closely; any season's average was within 1 or 2 cents of the average of official rates (and was as often higher as lower than the official rates). Rates decreased abruptly upon the entry of the Baltimore railroad, and there were further abrupt decreases in 1876 and 1877 consequent from reactions to cheater's rates. During the period 1874 to 1879, the winter average of official rates was from 30 cents to 42 cents per hundred pounds. Average winter rates quoted on the Board of Trade undercut official rates from 0.2 to 6.8 cents per hundred pounds. Rate reductions were notable, if not as extensive, during the summer season as well; average official rates were from 20 to 40 cents, and Board of Trade rates were slightly less. In general, the breakdown of group control had the effect of reducing rates on grain transport by 20 to 25 cents per hundred pounds.[117]

[117] An alternative hypothesis is that these rate changes followed (competitive) changes in transport costs. The reduction in rates followed from the general reduction of costs of short-term transport during the 1870's. Variable transport expenses (for locomotive repairs and service, train service, fuel, oil, and waste) decreased approximately 10 to 15 cents on each hundred pounds carried from Chicago to New York; at least, the recorded expenses for the Lake Shore subsidiary of the New York Central and the Pittsburgh, Fort Wayne and Chicago of the Pennsylvania Railroad were as follows:

	Lake Shore and Michigan Southern (New York Central) (Cents per 100 Pounds Chicago to New York)	*Pittsburgh, Fort Wayne and Chicago (Pennsylvania)*
1871	26.5	22.8
1872	28.2	22.1
1873	25.4	22.7
1874	24.6	20.8
1875	19.5	18.7
1876	16.4	18.9
1877	14.9	15.9
1878	13.4	13.5
1879	10.6	13.5

(Fiscal years, from July 1 of the year indicated to June 30 of the following year)

The decrease in average variable costs is not equal, for either railroad, to the decrease in official rates. Also, the timing of the rate changes was not concurrent with cost decreases; at least, the 8-cent decrease in Lake Shore costs took place during 1876–1877, approximately a year after the 15-cent decline in official rates, and a year before the further 5-cent decline from the breakdown of pooling. The Fort Wayne cost decreases were continuous while the rate changes took place in quite a discontinuous fashion. But the decrease in operating costs should have led to some decrease of official rates with complete cartel control of rates (in order to have increased profits for the members of the cartel). "Some portion" of the decline in official rates can be attributed to cost changes, even if the timing and extent of decline would seem to have followed from increasingly ineffective control by the various rate-setting associations.

The rate reductions were followed by decreases in the differences between Chicago and New York prices for transportable grades of grain. During the winter seasons of 1877–1878 and 1878–1879, grain appears to have been transported at below tariff rates, since grain price differences declined with rate cuts on the Board of Trade while official rates remained constant. There is no evidence of effects from nonadherence to cartel rates during any of the summer seasons from 1877 to 1879, or during the winter of 1876–1877; at the least, cutting of official rates did not seem to have affected the spread between Chicago and New York grain prices in these seasons.

This first period of shipments at lower than tariff rates was the winter of 1877–1878. There had been a full year of rates on the Chicago Board of Trade which were never more than 1 cent different from the average official rates. This winter the Board of Trade rate was 8 cents lower than the official rate on average, and changes were made independent of the tariff. The average grain price difference was below the average official rate, and the Board of Trade rates explained a far greater portion of the changes in the difference than did the less frequent changes in official rates. Chicago–New York grain price differences following immediately from official rates were calculated as $\{P_g^* = 6.054 + 0.786R_0\}$ with a coefficient of determination $\hat{\rho}^2 = .361$; grain price differences from Board of Trade rates were $P_g^* = 19.556 + 0.526R_T$ with a coefficient of determination $\hat{\rho}^2 = .914$ (as shown in Table 3.12). Grain price differences following rates with some lag in adjustment are shown in Table 3.12; it would appear that the lag was much shorter, and the variance explained by rates much greater, with Board of Trade rates rather than official rates.

The summer of 1878, and the summer of 1879, were seasons in which there was great similarity between official rates, Board of Trade rates, and week-to-week grain price differences. Average rates were within 1 cent of each other during both summers, lake rates were 4 to 7 cents lower, and grain price differences were between. In the winter of 1878–1879, the average official rate was more than 4 cents greater than the average Board of Trade rate and more than 3 cents greater than the average price difference. Also, the difference in the grain prices declined when Board of Trade rates declined, but before official rates declined. During the winter of 1878–1879, 52 per cent of the variance in Chicago–New York grain price differences was explained by immediate adjustment to Board of Trade rates, while only 32 per cent was explained by immediate adjustment to official rates.[118] The estimated lag in the adjustment to rate changes was shorter for the

[118] The relevant coefficients of determination were $\hat{\rho}^2 = .519$ and $\hat{\rho}^2 = .322$, in Table 3.12. For every 1-cent decline in Board of Trade rates, grain price differences declined by 0.25 cents per hundred pounds; at the same time, a 1-cent decline in official rates decreased the grain price difference by 0.21 cents per hundred pounds.

TABLE 3.12

Official Rail Rates, Board of Trade Rail Rates, and Differences Between Chicago and New York Grain Prices, 1876 to 1880

$$P_g^* = \alpha + \beta R_0 + \gamma R_L + \nu; \qquad P_g^* = \alpha + \beta R_T + \gamma R_L + \nu \tag{1}$$

P_g^*, the weekly average of the daily differences between Chicago and New York spot prices, in cents per hundred pounds, for the number 2 "western" grade of corn and oats; the official rate R_0 or the Board of Trade rate R_T; the Board of Trade lake-rail rate R_L from Chicago to New York via steamship to Buffalo and rail from Buffalo to New York.

$$P_{g,t} = P_{g,t-1} + \epsilon(P_g^* - P_{g,t-1}) \tag{2}$$

$P_{g,t}$, the average of the daily Chicago–New York grain price differences during week t and $P_{g,t-1}$, the average for the previous week; $P_g^* = \alpha + \beta R + \gamma R_L$ for the summer season and $P_g^* = \alpha + \beta R$ for the winter season from an instantaneous adjustment to changes in transport rates.

Calculations have been made of the form $P_{g,t} - P_{g,t-1} = -\epsilon P_{g,t-1} + \epsilon P_g^*$

The standard error of each coefficient is shown in parentheses below the coefficient.

WINTER SEASON, 1876–1877:

$\bar{R}_0 = 31.905$; $\bar{R}_T = 31.666$; $\bar{P}_g = 29.246$

(14 weeks); $P_g = 9.295 + \underset{(0.175)}{0.635} R_0$; $\hat{\rho}^2 = .522$

(14 weeks); $P_g = 9.288 + \underset{(0.173)}{0.628} R_T$; $\hat{\rho}^2 = .523$

(12 weeks); $P_{g,t} = 0.539 P_{g,t-1} + \underset{(0.227)}{0.461}(14.753 + 0.430 R_0)$; $\hat{\rho}^2 = .447$

(12 weeks); $P_{g,t} = 0.539 P_{g,t-1} + \underset{(0.227)}{0.461}(14.753 + 0.430 R_T)$; $\hat{\rho}^2 = .447$

SUMMER SEASON, 1877:

$\bar{R}_0 = 33.226$; $\bar{R}_T = 33.226$; $\bar{R}_L = 27.389$; $\bar{P}_g = 34.171$

(18 weeks); $P_g = 2.672 + \underset{(0.234)}{0.317} R_0 + \underset{(0.360)}{0.739} R_L$; $\hat{\rho}^2 = .830$

(18 weeks); $P_g = 2.672 + \underset{(0.234)}{0.317} R_T + \underset{(0.360)}{0.739} R_L$; $\hat{\rho}^2 = .830$

(15 weeks); $P_{g,t} = 0.052 P_{g,t-1} + \underset{(0.405)}{0.948}(3.593 + 0.327 R_0 + 0.708 R_L)$; $\hat{\rho}^2 = .341$

(15 weeks); $P_{g,t} = 0.052 P_{g,t-1} + \underset{(0.405)}{0.948}(3.593 + 0.327 R_T + 0.708 R_L)$; $\hat{\rho}^2 = .341$

WINTER SEASON, 1877–1878:

$\bar{R}_0 = 38.125$; $\bar{R}_T = 31.250$; $\bar{P}_g = 36.005$

(16 weeks); $P_g = 6.054 + \underset{(0.279)}{0.786} R_0$; $\hat{\rho}^2 = .361$

(16 weeks); $P_g = 19.556 + \underset{(0.043)}{0.526} R_T$; $\hat{\rho}^2 = .914$

(14 weeks); $P_{g,t} = 0.980 P_{g,t-1} + \underset{(0.129)}{0.020}(199.045 - 7.317 R_0)$; $\hat{\rho}^2 = .108$

(14 weeks); $P_{g,t} = 0.444 P_{g,t-1} + \underset{(0.198)}{0.556}(19.279 + 0.519 R_T)$; $\hat{\rho}^2 = .419$

TABLE 3.12 (Continued)

SUMMER SEASON, 1878:

$\overline{R}_0 = 25.572$; $\overline{R}_T = 25.056$; $\overline{R}_L = 17.731$; $\overline{P}_g = 22.378$

(26 weeks); $P_g = 0.729 + \underset{(0.230)}{0.400}R_0 + \underset{(0.255)}{0.684}R_L$; $\hat{\rho}^2 = .688$

(26 weeks); $P_g = 2.453 + \underset{(0.163)}{0.355}R_T + \underset{(0.228)}{0.660}R_L$; $\hat{\rho}^2 = .707$

(23 weeks); $P_{g,t} = 0.539P_{g,t-1} + \underset{(0.134)}{0.461}(-2.639 + 0.621R_0 + 0.563R_L)$; $\hat{\rho}^2 = .443$

(23 weeks); $P_{g,t} = 0.504P_{g,t-1} + \underset{(0.134)}{0.496}(0.734 + 0.437R_T + 0.639R_L)$; $\hat{\rho}^2 = .453$

WINTER SEASON, 1878–1879:

$\overline{R}_0 = 30.833$; $\overline{R}_T = 26.555$; $\overline{P}_g = 27.589$

(18 weeks); $P_g = 20.939 + \underset{(0.078)}{0.216}R_0$; $\hat{\rho}^2 = .322$

(18 weeks); $P_g = 20.787 + \underset{(0.062)}{0.256}R_T$; $\hat{\rho}^2 = .519$

(14 weeks); $P_{g,t} = 0.711P_{g,t-1} + \underset{(0.236)}{0.289}(19.300 + 0.201R_0)$; $\hat{\rho}^2 = .121$

(14 weeks); $P_{g,t} = 0.648P_{g,t-1} + \underset{(0.301)}{0.352}(20.444 + 0.220R_T)$; $\hat{\rho}^2 = .114$

SUMMER SEASON, 1879:

$\overline{R}_0 = 26.677$; $\overline{R}_T = 25.290$; $\overline{R}_L = 21.033$; $\overline{P}_g = 25.141$

(30 weeks); $P_g = 2.859 + \underset{(0.125)}{0.903}R_0 - \underset{(0.154)}{0.067}R_L$; $\hat{\rho}^2 = .853$

(30 weeks); $P_g = 6.779 + \underset{(0.084)}{0.677}R_T + \underset{(0.127)}{0.076}R_L$; $\hat{\rho}^2 = .871$

(29 weeks); $P_{g,t} = 0.834P_{g,t-1} + \underset{(0.114)}{0.166}(4.451 + 1.879R_0 - 1.234R_L)$; $\hat{\rho}^2 = .257$

(29 weeks); $P_{g,t} = 0.795P_{g,t-1} + \underset{(0.120)}{0.205}(11.450 + 1.183R_T - 0.637R_L)$; $\hat{\rho}^2 = .274$

WINTER SEASON, 1879–1880:

$\overline{R}_0 = 37.895$; $\overline{R}_T = 37.895$; $\overline{P}_g = 41.947$

(18 weeks); $P_g = 8.272 + \underset{(0.359)}{0.875}R_0$; $\hat{\rho}^2 = .270$

(18 weeks); $P_g = 8.272 + \underset{(0.359)}{0.875}R_T$; $\hat{\rho}^2 = .270$

(12 weeks); $P_{g,t} = 0.644P_{g,t-1} + \underset{(0.319)}{0.356}(6.001 + 1.012R_0)$; $\hat{\rho}^2 = .372$

(12 weeks); $P_{g,t} = 0.644P_{g,t-1} + \underset{(0.319)}{0.356}(6.001 + 1.012R_T)$; $\hat{\rho}^2 = .372$

Board of Trade rate than for the official rate. The effect of the unofficial rate on the grain prices, assuming the lag, was also greater than that of the official rate (as shown by the last two equations for the winter of 1878–1879 in Table 3.12).

The narrowing of the differences between Chicago and New York prices in these seasons of rate warfare is striking in comparison with the differences in the earlier winters of the decade. The average grain price margin had not fallen below 40 cents per hundred pounds before 1876. In the three following winters, it was successively 29, 36, and 27 cents.[119] With the new agreement in force in the winter of 1879–1880, the margin was more than 41 cents per hundred pounds once again.

The breakdowns of collective control of rates lowered grain price margins but did little for the costs of other transport services. There were generally lower rates for transport to the eastern seaboard,[120] but there was little change in the local rates for short-distance transport between points on any one line. Short-distance rates on the Erie Railroad in upstate New York held during the 1870's at the level set in the late 1860's. Consequent from the decline in official rates, the Erie was quoting lower rates for transport in 1876 from New York to Chicago than from New York City to northern New York State: the official rate on fourth-class traffic to Chicago was 20 cents per hundred pounds and to Genesee or Olean, New York, was more than 40 cents per hundred pounds.[121] The Pennsylvania Railroad rates in the summer of 1876 were as high for shipping from Philadelphia to Pittsburgh as from Chicago to New York through these cities (as can be seen from comparing fourth-class rates in Table 3.3 and Table 3.12). Charges for shipping grain along the Pennsylvania's line from Cessna, Pennsylvania, to Cumberland, Maryland, were 12 cents per hundred pounds from April, 1876, to February, 1877, and 8 cents per hundred pounds from April, 1877, to April, 1879 (as shown in Table 3.13). Through rates for a comparable distance fluctuated greatly in this period: the disparity between local rates and a comparable through

[119] The rates during the winter of 1876–1877 were characterized by rigidly maintained official rates at the quite low level of 31–32 cents per hundred pounds. Official tariffs at such low levels, as well as cut rates at these levels, had the effect of reducing grain price margins by 10 cents or more.

[120] The established difference between first- and second-class had been 35 cents from October, 1871, to March, 1874, and 40 cents from April, 1874, to February, 1878. This was reduced to 30 cents in April of 1878 and to 20 cents for two months in the fall of 1879. Similar declines in the difference between second- and third-class rates took place at the same time. Cf. *Railways in the United States in 1902*, *op. cit.*, p. 78, Table 63, "Freight Rates Charged for the Transportation of Classified Traffic from Chicago to New York from October 22, 1871 to April 1, 1902."

[121] Cf. *Railways in the United States in 1902*, *ibid.*, and Table 3.2. As noted in Chapter 1, the Erie received freight in New York for transfer to the New York Central, Pennsylvania, or Baltimore and Ohio to Chicago.

TABLE 3.13

Local Grain Rates on the Pennsylvania Railroad from Cessna, Pennsylvania, to Cumberland, Maryland, in the 1870's

Date of Shipment: Month and Year	*Grain Shipped (Hundreds of Pounds)*	*Recorded Rate (Cents per 100 Pounds)*	*The Rate for Comparable Distance, as a Percentage of the Chicago–New York Official Grain Rates*
April, 1876	546	12.0	0.78
June, 1876	220	12.0	0.78
January, 1877	793	12.0	1.24
February, 1877	260	12.0	1.24
April, 1877	440	8.0	1.29
May, 1877	740	8.0	1.29
June, 1877	220	8.0	1.29
March, 1878	480	8.0	1.48
December, 1878	251	8.0	1.20
January, 1879	760	8.0	1.20
February, 1879	1,780	8.0	1.20
March, 1879	220	8.0	1.20
April, 1879	220	8.0	1.03

Source: The *Manifest Book* of the Pennsylvania Railroad for Cessna, in which were recorded shipment tonnages and revenues for the station by the freight agent for each day of the year. This one station book survives in the Pennsylvania Company archives, Philadelphia, Pennsylvania.

rate was greatest in 1876, but there were decreases in the comparable through rate in the spring of 1879 that widened the disparity at the end of the period.[122] All of these local rates became "relatively higher" since they

[122] The absolute size of the disparity at any one point of time may have followed from the ability of the Pennsylvania to control all transport service between these two small towns, and from differences in cost of transport associated with the total volume of shipments between these two towns as compared to Chicago–New York. Monopoly control of transport should have increased the Cessna-Cumberland rate over the Chicago–New York rate for comparable distance (assuming ineffective cartel control of through rates). The smaller volume transported from Cessna to Cumberland may have prevented the adoption of cost-saving techniques for loading, unloading, and storage of large-volume shipments; because of higher variable costs for any individual shipment from Cessna, either competitive or monopoly rates should have been higher, per hundred pounds per mile from Cessna than from Chicago. Since both circumstances — differences in ability to control rates, and differences in cost associated with total volume transported — imply higher rates for shipments from Cessna, it is not possible to conclude that the breakdown of control of through rates resulted in the Cessna rate being 7 cents higher than a through rate for comparable distance. It would seem possible, however, that *increased disparity* in 1876, and again in the winter of 1878 and 1879, followed from the thoroughly publicized breakdown of group control of through rates.

did not change while through rates declined in the railroad wars of 1875–1876, 1877–1878, and 1878–1879.

The over-all impression is that destruction of the trunk-line railroad agreements from the winter of 1875–1876 henceforth was productive of lower rates on long-distance transport which were part of an increasingly discriminatory pattern of rates. The shippers received lower rates on through transport of bulk commodities and, from most indications, no lower rate on short-distance transport than during the early 1870's.[123] Consequently there was increased disparity between short- and long-distance rates for comparable service, and, in some instances, higher rates for short-distance transport than on shipments from Chicago to New York.

The Consequences for the Railroads of Cartel Instability, 1875 to 1879

Departures from the agreed rate were the rule not only for grain traffic but for all through traffic. It was not possible to isolate grain rates from the rest so that grain agreements could have been broken while agreements on provision and merchandise rates were not. Grain rates were the basis for setting rates on higher class traffic; there was "shading" on both the basis and on the premium for first or second class. There were intermittent notices, in Board of Trade reports on particular commodity markets, of discounts on official rates on provisions, cattle and sheep, and westbound dry goods during the winters of the breakdowns in the grain rates.

[123] It should be mentioned that not all local traffic was monopoly transport for one trunk line or other; to the contrary, much tonnage was "shared" between two or more railroads. When "sharing" took place, problems of control were encountered of the nature, if not the frequency, experienced by the trunk-line associations. For example, the Baltimore and Ohio encountered the gains and losses from not adhering to a two-firm agreement in the instance mentioned in the following letter.

J. W. Garrett, Prest.
7 South Street

Mr. Smith and I have just been conferring about rates for Carnegie on ores from Lynchburg to Bessemer via Strasburg. He offers a rate equivalent to six and five tenths mills, whereas the lowest rate you have authorized to be made is three-quarters of a cent per ton per mile. Should he not be able to ship at this rate? The ore will come to Baltimore by canal and schooner, and in that event we will only get our per centage under the pooling arrangement from Baltimore, and a little less rate per ton per mile, besides losing the carrying of the ore on the Virginia Midland road,— about 60 per cent of the tonnage. Will it not be well to authorize the acceptance of Mr. Carnegie's proposition at once? The amount of shipments will be 600 tons per week.

Robert Garrett
3rd VP

October 25, 1879

The Archives of the Baltimore and Ohio, *op. cit.*

The rate disruptions seem to have been accompanied by a redistribution of freight tonnage in favor of the rate-cutter. The Baltimore and Ohio railroad received approximately 5 per cent more of total eastbound tonnage from Chicago in 1876 than in previous and succeeding years (as shown in both Table 3.14 and Table 3.15). The New York Central railroad

TABLE 3.14

Shares of Eastward-Bound Freight from Chicago

	New York Central		*Pennsylvania*		*B & O**	*Total*
	Lake Shore†	*Michigan Central*	*Fort Wayne‡*	*P., C. & St. L.§*		
1873						
Tonnage	356,938	325,747	260,719	136,850	—	1,080,254
Percentage of Total	33.04	30.16 Total 63.20	24.13	12.67 Total 36.80		
1874						
Tonnage	331,825	322,869	256,253	119,879	—	1,030,826
Percentage of Total	32.19	31.32 Total 63.51	24.86	11.63 Total 36.49		
1875						
Tonnage	285,217	367,163	259,761	119,951	68,092	1,100,184
Percentage of Total	25.93	33.37 Total 59.30	23.61	10.90 Total 34.51	6.18	
1876						
Tonnage	445,855	518,911	388,812	154,779	185,631	1,693,988
Percentage of Total	26.32	30.63 Total 56.95	22.95	9.13 Total 32.08	10.96	
1877						
Tonnage	351,921	449,798	251,333	138,470	73,578	1,265,100
Percentage of Total	27.82	35.55 Total 63.37	19.87	10.95 Total 30.82	5.82	
1878						
Tonnage	563,354	599,998	438,379	203,546	108,189	1,913,466
Percentage of Total	29.44	31.36 Total 60.80	22.91	10.64 Total 33.55	5.65	

Source: Report of G. M. Gray, Assistant General Freight Agent of the Lake Shore and Michigan Southern Railway to the Office of the Joint Executive Committee, July 28, 1879; contained in *Argument Regarding the Division of East Bound Freight from Chicago* (New York: Russell Bros. Printers, 1879).

* Baltimore and Ohio Railroad.
† Lake Shore & Michigan Southern.
‡ Pittsburgh, Fort Wayne and Chicago.
§ Pittsburgh, Cincinnati and St. Louis.

TABLE 3.15

Statement Showing Tonnage of Eastbound Freight Forwarded from Chicago During Three Years Ending December 31, 1877

	1875		*1876*		*1877*		*Total*	
	Tons	*Per Cent of Total Tonnage*	*Tons*	*Per Cent of Total Tonnage*	*Tons*	*Per Cents*	*Tons*	*Per Cents*
Michigan Central	419,225	36.6	579,460	32.3	481,478	38.4	1,480,163	35.3
Lake Shore and Michigan Southern	296,972	25.8	456,265	25.4	338,380	27.0	1,091,617	26.0
Pittsburgh, Fort Wayne and Chicago	274,850	23.9	428,841	23.8	253,351	20.3	957,042	22.8
Pittsburgh, Cincinnati and St. Louis	73,985	6.5	114,526	6.3	89,619	7.2	278,130	6.6
Baltimore and Ohio	82,342	7.2	219,679	12.2	88,414	7.1	390,435	9.3
	1,147,374	100.0	1,798,771	100.0	1,251,242	100.0	4,197,387	100.0

Source: Reprinted in its entirety from the general freight agent's memorandum to the President of the Baltimore and Ohio, the Baltimore and Ohio Archives, *op. cit.*, January, 1879.

succeeded in increasing its share in 1877 by more than 7 per cent of the total tonnage. The Pennsylvania Railroad captured a portion of the Central's increased share in 1878, given that the Central's share declined by 3 per cent of the total while the Pennsylvania's share increased by a similar amount.[124]

Moreover, there is some indication of profit gains for the accused cheater, and of losses for others, during periods of rate demoralization. Short-term accounting profits, as shown in Table 3.16, indicate a shift toward the Baltimore and Ohio during the winter of 1875–1876, a large increase for the New York Central during the winter of 1877–1878, and a subsequent small increase for the Pennsylvania in the winter of 1878–1879.[125] The Baltimore road experienced losses in its first year in Chicago

[124] Again, as shown in Tables 3.14 and 3.15. It should be noted that the tonnages shown in the records of the Joint Executive Committee do not agree with those shown in the Baltimore and Ohio archives. In particular, the Baltimore and Ohio figures indicate a larger percentage share for that railroad. The changes in the percentages of total shipments from year to year for each railroad seem comparable, however. Since the shift of tonnage to the railroad accused of cutting rates is all that is of concern here, the two sets of figures are approximately of the same usefulness.

[125] These figures are grossly inaccurate, so that the changes of reported net income are no more than indications of the probable direction of change of profits from cartel breakdown. The figures account for revenues when received, while the transport services may have been rendered in the previous season. Short-term transport costs in each month are $\frac{1}{12}$ of annual transport costs which follow from separating freight costs from passenger transport costs on the basis of freight mileage as a percentage of total mileage. This estimating procedure must result in inaccurate figures in a number of instances. If freight transport costs rose quite sharply as tonnage increased because (for example)

TABLE 3.16

Accounting Profits from Freight Transportation, 1872 to 1879

Date	*New York Central Average per Month (in Dollars)*	*Pennsylvania*[b] *Average per Month (in Dollars)*	*Baltimore and Ohio Average per Month (in Dollars)*
Summer 1872	325,234	196,719	—
Winter 1872–1873	461,349	185,184	—
Summer 1873	341,525	167,259	—
Winter 1873–1874	460,870	235,270	—
Summer 1874	283,624	112,966	—
Winter 1874–1875	397,046	126,542	−7,895
Summer 1875	237,366	161,143	−6,724
Winter 1875–1876	397,378	199,143	10,899
Summer 1876	258,811	121,203	10,266[c]
Winter 1876–1877	285,519	—	—
Summer 1877	300,135	—	—
Winter 1877–1878	502,784	227,053	—
Summer 1878	451,153	168,058	—
Winter 1878–1879	353,179[a]	277,398	31,085
Summer 1879	340,128[a]	168,421	43,231

[a] Lake Shore figures only, because of the absence of revenue reports for the Michigan Central.
[b] Pittsburgh, Fort Wayne and Chicago only.
[c] April, May, and June only.
Source: Profits for each railroad were calculated by subtracting "freight operating expenses" from "gross freight revenues." Total expenses are the sum of expenditures for motive power, conducting transportation, maintenance of cars, and maintenance of rights of way (as shown in the *Annual Reports* of the railroad or the *Annual Reports* of the New York State engineers, the Illinois Railroad and Warehouse Commission, and the State Engineers of Ohio). The expenses for transporting freight are estimated by multiplying total expenses by the ratio of freight miles to total train miles. Gross freight revenue statistics are shown in the *Annual Reports* of the Illinois Commission.

that were followed by profits of almost $11,000 per month in the winter of 1875–1876.[126] The New York Central gained from $50,000 to $100,000

---

of a shortage of specific types of freight cars, while freight mileage remained roughly the same percentage of total mileage, then freight transport costs are underestimated. The effect of using annual costs per month is probably to overestimate both losses and gains of profits from tonnage changes. That is, when a large gain in tonnage is experienced, costs of transport most likely increase more than is shown by Table 3.16; when large losses of tonnage took place, immediate costs of transport most likely declined that month, while this is not shown by the table.

[126] The figures are for the Baltimore, Ohio and Chicago railroad only (that is, for the new extension into Chicago from the Baltimore and Ohio's terminus in the state of Ohio). Profits, consequently, were limited to receipts from through transport.

per month in the winter of 1877–1878.[127] And (except for the winter of 1871–1872) the Pennsylvania realized its largest profits of the 1870's during the "rate war" of the winter of 1878–1879.[128]

The profits realized from destruction of the agreement were ultimately paid to stockholders of railroad securities. If it is assumed that stockholders made rough estimates of the profitability of each railroad and paid no more for a share of stock than the present value of estimated future profits per share, then estimates of profits to be received from cheating affected a railroad's common stock price at the time of the cartel breakdown. During the latter half of the 1870's, those railroads that profited from disrupting cartel agreements should have been the lines with increased security prices before and during the "rate wars," while the other railroads in the process of losing money from remaining loyal should have experienced sharply declining prices during these periods. The Baltimore and Ohio common stock price increased appreciably in the period from the summer of 1874 to the winter of 1876–1877 (as shown in Table 3.17); the stockholders held high hopes for expanded profits from full use of the new facilities into Chicago and the upper Midwest. There was an extensive increase in the price of B & O common at the time of the railroad's entry into Chicago in the summer of 1874, to a level more than $8.00 above the 1870 to 1879 trend of all the weekly B & O prices. The price increased another $3.00 in the winter of 1874–1875, and another $8.00 in the summer of 1875,

[127] The increase might be measured by comparison with the winters of 1872–1873 and 1873–1874, when the Central averaged net receipts of approximately $460,000 per month on freight transport between Chicago and Buffalo, New York (for the Lake Shore and Michigan Southern, and Michigan Central). In this comparison the increase was approximately $40,000 per month. If the comparison is made between the winters of 1874–1875 and 1875–1876 (when there was alleged demoralization of rates on account of the expansion of the Baltimore and Ohio railroad), then the gain in profits was approximately $100,000 per month. Approximately one-half to one-quarter of the gain can be attributed to a general increase in shipments in this year, since profits for all roads would have been approximately $25,000 per month less if shipments were the same total volume in 1878 as in 1872.

[128] The profit figures are those for the Pittsburgh, Fort Wayne and Chicago, and the Pittsburgh, Cincinnati and St. Louis lines transporting between the first-named city and Chicago. The particular figure for 1878–1879 follows in part from increased tonnage (presumably, from increase in the demand for transport at lower rates) as well as from a lower rate on this line alone. In order to separate these two cases, assume that total shipments of all lines equal those for the respective seasons of 1871–1872 plus a percentage increase equal to the percentage decrease of the official rate (or that elasticity of total transport demand is equal to one). Then profits for the Pennsylvania would have been approximately $213,000 per month in the winter of 1878–1879, and (for example) approximately $189,000 the previous winter. The gain from the alleged rate-cutting may have been limited to $25,000 per month, with the remaining gain following from the increase in demand.

TABLE 3.17

Stockholders' Profits from Freight Transport, 1875 to 1879

(1) The trend of stock prices: $P = \alpha + \beta T + \gamma T^2$
for the Lake Shore and Michigan Southern (New York Central),

$$P = 100.607 - \underset{(0.012)}{0.065T} - \underset{(0.00002)}{0.0001T^2}; \quad \hat{\rho}^2 = .781$$

for the Pittsburgh, Fort Wayne and Chicago (Pennsylvania),

$$P = 92.347 + \underset{(0.0091)}{0.0034T} + \underset{(0.0001)}{0.0002T^2}; \quad \hat{\rho}^2 = .132$$

for the Baltimore and Ohio Railroad,

$$P = 124.056 + \underset{(0.026)}{0.353T} - \underset{(0.0005)}{0.008T^2}; \quad \hat{\rho}^2 = .524, \text{ with } T = \text{time, 0, 1, 2, 520}$$

for the 520 weeks in the period 1870 to 1879 in which the Wednesday closing prices for the relevant stocks were listed.

(2) Deviations from trend: average seasonal difference between actual price and price "P" expected from $P = \alpha + \beta T + \gamma T^2$

	Average Deviation from Trend: (Dollars per Share)		
	Lake Shore	*Pittsburgh, Fort Wayne and Chicago*	*Baltimore and Ohio*
Summer 1874	−1.823	−2.056	+8.313
Winter 1874–1875	+0.981	+1.035	+11.152
Summer 1875	−10.147	+2.439	+19.715
Winter 1875–1876	−4.722	−5.124	+17.327
Summer 1876	−8.072	−5.687	+6.664
Winter 1876–1877	−6.338	+2.557	+1.752
Summer 1877	+3.303	−6.854	−9.546
Winter 1877–1878	+10.613	−8.095	−24.276
Summer 1878	+18.295	−3.991	−30.947
Winter 1878–1879	−8.594	+3.017	−10.626
Summer 1879	−10.387	+2.893	+7.429

Source: Wednesday closing prices for common stock of these railroads in New York as listed in the *Financial Chronicle* (Chicago, Illinois).

however, when further trade expansion was not taking place. The last increase took place at exactly the time that the New York Central's (Lake Shore and Michigan Southern) common stock price was declining to $10.00 less than the trend of New York Central 1871 to 1879 prices. The Baltimore railroad's security price remained above trend during both seasons of 1875–1876, while the common stock prices of the New York Central and of the Pennsylvania Railroad were appreciably below their respective trends. These large, short-term price gains for Baltimore and

Ohio stock occurred when the road was twice accused of cutting rates and when there were profit gains from cutting the grain rate.[129]

The New York Central was accused particularly of cutting rates during 1877 and 1878. In fact, rate-cutting which took place on grain shipments during this period had an appreciable effect on prices in the Chicago and New York grain markets, and seems to have bestowed large increases in tonnage shares upon the Central's subsidiary lines. The stockholders in this company seem to have looked upon this particular period with great favor, as well: prices for common stock of the Lake Shore and Michigan Southern in the winter of 1877–1878 were more than $10.00 above the 9-year trend of such prices, and were an additional $8.00 higher than trend in the summer of 1878.

The common stock prices of the Pennsylvania Railroad were slightly greater than the long-term trend during the winter of 1878–1879, while both the New York Central prices and the Baltimore and Ohio prices were considerably less than their respective trends. Moreover, these prices had been rising: from −$8.0 of trend price in the winter of 1877–1878, to −$3.9 of trend price in the summer of 1878, to +$3.0 of trend price the winter of 1878–1879. The Pennsylvania was accused of cutting rates during this last winter. The railroad appears to have gained sufficient tonnage (of all commodities) in the process to convince stockholders of increased expected profits.

The pattern of both the accounting and the stock price estimates of profits suggests that rate-cutting after 1874 was profitable for a few disloyal members but quite unprofitable for those remaining loyal to the last day of an agreement. The railroad first to cut the official rate seems to have profited because of the lag in the reaction of the association to secret cutting and because the association did little more "in reaction" than match the rate of the cheater. The loyal members of the associations did not profit: in the words of an official having had the benefits of loyalty, "it will thus be seen that while rates were fairly remunerative the volume of the traffic was slight, and that while business was comparatively large

[129] It should be noted that there were much greater fluctuations in Baltimore and Ohio stock prices above and below the Baltimore and Ohio trend line of 1871 to 1879 stock prices than of other roads' prices around their respective trend lines. This might have followed from the close attention paid to revenues on that line from through traffic as compared to local traffic. Through transport accounted for more than two-thirds of the Baltimore road's receipts, while the other railroads rarely received more than 40 per cent from this source. Receipts on such transport were subject to the extent of effectiveness of association control, while receipts on local transport were generally for services within separate areas of each one of the trunk lines and were not subject to sudden change. It might be speculated, then, that larger changes in Baltimore and Ohio stock prices followed from changes in receipts on through traffic having a greater effect on over-all line profits.

in volume the rate per ton a mile was extremely low."[130] The experience of the New York Central and the Pennsylvania must have been that the sum of the profit declines was greater than the profit gain of the B & O in 1875–1876, (as shown in Tables 3.16 and 3.17). The experience of the loyal members in 1878 — the Pennsylvania and B & O — was that the losses were again greater than the New York cheater's gains. The Pennsylvania's modest gains in the war of 1878–1879 must have been expensive for the other trunk-line managers as well.

[130] A report of the second vice-president of the Baltimore and Ohio Railroad Company to the president of the railroad, January, 1878, on the revenue for the year from the Chicago division. The Archives of the Baltimore and Ohio Railroad, the Maryland Historical Society, *op. cit.*

4

Elaborate Trunk-Line Cartels and the Control of Rates, 1880 to 1886

The trunk-line railroads achieved a measure of harmony from the agreement of August, 1879. The first year, at least, was free of discounts on through rates and of large disparities between through rates and local rates. This was achieved in the face of the entry of the Grand Trunk Railway into Chicago and of accompanying disagreements on the tonnage shares of each of the railroads.

The measure was short. Official rates were cut during most of 1881 and throughout the years 1884, 1885, and 1886. These disruptions had all of the effects, for better or worse, on shippers and transporters noted in the 1875 to 1879 railroad wars. Shippers of grain and provisions from Chicago got sharp reductions in actual transport charges (with the effect of decreasing the margin between Chicago and New York commodity prices). Shippers in local transport markets competing with those in the Chicago–New York commodity markets got some rate reductions as well; other local shippers continued to pay the same rates as in the early 1870's. Thus one result of the railroads' difficulties with cartel agreements was to decrease long-distance transport charges relative to those on short-distance traffic. The railroads themselves experienced both losses and gains: in the periods of extensive rate warfare, the railroad initiating the rate reductions seems to have made some gain in profits, while those that remained loyal to official rates experienced large decreases in profits. The gains of the cheaters would not seem to have compensated for the losses of the loyal railroads, so that general profits declined.

These results provide impressive evidence that the trunk lines had not found the means for maintaining a cartel agreement: the extensive rate-cutting during the years 1884, 1885, and 1886 make it clear that the existing organizations could not set grain rates above 25 cents per 100 pounds (while the agreement of March, 1879, had resulted at first in the posting of rates from 30 to 40 cents per 100 pounds). There was some question as to whether it was possible to maintain 25-cent rates — whether the agreements were stable at all for rates more than the 20 cents received during the breakdown periods. Such extensive failure from 1881 to 1886 exhausted the obvious, and most of the elaborate, techniques for control.

Harmony in 1880 and Rate Wars in 1881

The agreement of 1879 provided the rules for rates in an extremely successful winter and summer season. The official rate for transporting grain from Chicago to New York was increased from 35 cents per 100 pounds to 40 cents during the first week of November of that year; it continued until the first week of March and then was followed by the 35-cent rate. When the lakes opened for steamer traffic, both the all-rail through rates and the lake-rail rate were established at 30 cents per 100 pounds, and the all-rail rate remained at this level while the lake-rail rate gravitated to approximately 25 cents per 100 pounds for the remainder of the summer. These rates were not only posted, but seem to have been quoted on almost all shipments. There were newspaper reports of "cut rates from Chicago" having been discussed at a meeting of the Trunkline Executive Committee in January[1]; there were reports also in September that "the competition of the Erie and the New York Central for Boston and New England business . . . [has led to] irregularities of eastbound rates . . . [which] have been greatly exaggerated in the newspapers." [2] But these reports were followed within a week by further statements that such discounts had been terminated and official rates were being maintained. During the remaining weeks of the year there were no further stories of deviations from the tariff.

The rates listed in the Chicago Board of Trade as "the quoted rates" for transport of grain to New York were the same each week as the official rates. With both Board of Trade and official rates higher than in the

[1] *The Proceedings and Circulars of the Trunkline Executive Committee, 1880–1881* (the Archives of the Association of American Railroads, the Bureau of Railway Economics, Washington, D.C.), meeting of January 17, 1880. The road forming the new extension of the Grand Trunk Railway into Chicago was the company accused of cutting rates. The accusation was described in detail in the *Railroad Gazette:* "A number of charges were made against the Grand Trunk Railway and the Central Vermont Railway for cutting rates on through transport. On the twenty-second of January, 1880, the representatives of these lines did give the promised assurance that rates should hereafter be maintained. The New York Central alleged that the Grand Trunk had violated these agreements by persistingly transporting property for several months past between the seaboard and the western points for less than the agreed rate. The investigating committee found these allegations fully substantiated." ["The Joint Executive Committee February Meeting," the *Railroad Gazette* (March 5, 1880), 128.]

[2] "The Irregularities in Eastbound Rates," the *Railroad Gazette* (September 10, 1880), 480.

previous season, it followed that the margin between Chicago and New York spot grain prices was higher. The average grain price difference for the winter season of 1879–1880 was 41.9 cents per 100 pounds while the average difference during the summer of 1880 was 31.7 cents per 100 pounds. The winter season's average was approximately 14 cents higher than during the previous winter, and the summer season's average 6 cents higher than in the previous summer (while the average rates, respectively, were 7 cents and 4 cents higher than in the previous seasons). Variations in week-to-week grain price differences were in accord with changes in official and/or Board of Trade rates. During the winter season, approximately 37 per cent of the variance in grain price differences was explained by variance in official rates (subject to a three-week lag in adjustment to rates).[3] During the summer season there was no variation in the official or Board of Trade rates, but variance in grain price differences was explained in good part by adjustment to changes in lake-rail rates; these lake-rail rates were, by agreement, an average of 5 cents less than the official all-rail rates (as shown in Table 4.1).

The trunk-line railroads were evidently delighted with this experience in rates. Each railroad transported (approximately) its share of grain tonnage from Chicago during the winter season at higher rates.[4] Each railroad received less than its usual share of the tonnage during the summer; the higher all-rail rates and the presence of "a larger fleet of vessels which can do nothing but carry grain" [5] cut into the railroads' shares of the traffic. This was not viewed with alarm, however, since the trunk lines made up for the loss of Chicago tonnage with increased tonnage from the lake ports to eastern seaboard cities (as shown in Table 4.3, the trunk-line railroads' shares of grain traffic into Montreal, Boston, New York, Philadelphia, and Baltimore were not less than the 1880 to 1886 trend of such shares). This resulted in the view that "even if we select a period when the rail shipments were larger, we do not find good reason for reducing rates to make them so. . . . Last year during the same two weeks the shipments were 10 per cent more but the rates were 20 cents rather than 30 cents . . . (so that) gross receipts were 13½ per cent more this year while 10 per cent less was carried." [6]

The trunk-line railroads' success was achieved in the face of entry of a new transporter. The agreement had been between the Baltimore and

[3] The price difference changed 1.01 cents for each 1-cent change in the official rate on average, as shown in the computed lagged adjustment equation in Table 3.12.

[4] Tonnage shares were somewhat greater than the long-term trend of shares from Chicago for most of the railroads, as shown in Table 4.2, as a result of decreased shipments to interior points to the South and West.

[5] "Eastbound Rates and Lake Shipments," the *Railroad Gazette* (August 20, 1880), 444.

[6] *Ibid.*

TABLE 4.1

Official Rail Rates, Board of Trade Rail Rates, and Differences Between Chicago and New York Grain Prices, 1880 to 1883

$$R_T = \alpha + \beta R_0 + \nu \tag{1}$$

R_T, the rail rate, in cents per hundred pounds Chicago to New York, from the Board of Trade *Daily Commercial Bulletin;* R_0, the rail rate from compilations of posted tariffs and from announcements of the various traffic associations.

$$P_g{}^* = \alpha + \beta R_0 + \gamma R_L + \nu; \qquad P_g{}^* = \alpha + \beta R_T + \gamma R_L + \nu \tag{2}$$

$P_g{}^*$, the weekly average of the daily differences between Chicago and New York spot prices, in cents per hundred pounds, for the number 2 "western" grade of corn and oats; the official rate R_0 or the Board of Trade rate R_T; the lake-rail rate R_L from Chicago to New York via steamship to Buffalo and rail from Buffalo to New York.

$$P_{g,t} = P_{g,t-1} + \epsilon(P_g{}^* - P_{g,t-1}) \tag{3}$$

$P_{g,t}$, the average of the daily Chicago–New York grain price differences during week t and $P_{g,t-1}$, the average for the previous week; $P_g{}^* = \alpha + \beta R + \gamma R_L$ for the summer season and $P_g{}^* = \alpha + \beta R$ for the winter season from an instantaneous adjustment to changes in transport rates.

$$\left.\begin{aligned} P_{g,t} - P_{g,t-1} &= -\epsilon P_{g,t-1} + \epsilon(\alpha + \beta R_0) + \nu \\ P_{g,t} - P_{g,t-1} &= -\epsilon P_{g,t-1} + \epsilon(\alpha + \beta R_T) + \nu \end{aligned}\right\} \text{Winter Season} \tag{4}$$

$$\left.\begin{aligned} P_{g,t} - P_{g,t-1} &= -\epsilon P_{g,t-1} + \epsilon(\alpha + \beta R_0 + \gamma R_L) + \nu \\ P_{g,t} - P_{g,t-1} &= -\epsilon P_{g,t-1} + \epsilon(\alpha + \beta R_T + \gamma R_L) + \nu \end{aligned}\right\} \text{Summer Season}$$

from (2) and (3).

The standard error of each coefficient is shown in parentheses below the coefficient.

SUMMER SEASON, 1880:

$R_0 = R_T = 30.00$ cents each week; $\quad \overline{R}_L = 26.814; \quad \overline{P}_g = 31.671$

(27 weeks); $P_g = 21.419 + 0.000R + 0.383R_L$; $\hat{\rho}^2 = .065$
(0.289)

(27 weeks); $P_{g,t} = .283P_{g,t-1} + 0.717(21.038 + 0.000R + 0.398R_L)$; $\hat{\rho}^2 = .394$
(0.181)

WINTER SEASON, 1880–1881:

$R_T = 0.918 + 0.967R_0$; $\hat{\rho}^2 = .950$; $\overline{R}_0 = 33.809$; $\overline{R}_T = 33.619$; $\overline{P}_g = 37.282$
(0.051)

(21 weeks); $P_g = 13.349 + 0.708R_0$; $\hat{\rho}^2 = .487$
(0.167)

(21 weeks); $P_g = 12.725 + 0.730R_T$; $\hat{\rho}^2 = .510$
(0.164)

(21 weeks); $P_{g,t} = 0.564P_{g,t-1} + 0.436(-3.439 + 1.197R_0)$; $\hat{\rho}^2 = .495$
(0.148)

(21 weeks); $P_{g,t} = 0.551P_{g,t-1} + 0.449(-3.526 + 1.207R_T)$; $\hat{\rho}^2 = .515$
(0.135)

TABLE 4.1 (Continued)

SUMMER SEASON, 1881:

$R_T = -5.335 + 1.176R_0$; $\hat{\rho}^2 = .951$; $\overline{R}_0 = 19.107$; $\overline{R}_T = 17.142$; $\overline{R}_L = 17.071$; $\overline{P}_g = 17.480$

(28 weeks); $P_g = -4.769 + \underset{(0.344)}{0.434}R_0 + \underset{(0.545)}{0.817}R_L$; $\hat{\rho}^2 = .489$

(28 weeks); $P_g = -2.368 + \underset{(0.309)}{0.359}R_T + \underset{(0.590)}{0.802}R_L$; $\hat{\rho}^2 = .484$

(28 weeks); $P_{g,t} = 0.560P_{g,t-1} + \underset{(0.161)}{0.440}(-1.002 + 0.720R_0 + 0.243R_L)$; $\hat{\rho}^2 = .261$

(28 weeks); $P_{g,t} = 0.572P_{g,t-1} + \underset{(0.159)}{0.428}(4.671 + 0.738R_T - 0.257R_L)$; $\hat{\rho}^2 = .220$

WINTER SEASON, 1881–1882:

$R_T = -4.204 + \underset{(0.099)}{1.192}R_0$; $\hat{\rho}^2 = .888$; $\overline{R}_0 = 19.631$; $\overline{R}_T = 19.158$; $\overline{P}_g = 19.952$

(19 weeks); $P_g = -0.355 + \underset{(0.216)}{1.034}R_0$; $\hat{\rho}^2 = .574$

(19 weeks); $P_g = 5.993 + \underset{(0.194)}{0.729}R_T$; $\hat{\rho}^2 = .454$

(19 weeks); $P_{g,t} = 0.746P_{g,t-1} + \underset{(0.218)}{0.255}(-14.012 + 1.842R_0)$; $\hat{\rho}^2 = .200$

(19 weeks); $P_{g,t} = 0.849P_{g,t-1} + \underset{(0.208)}{0.151}(-10.346 + 1.806R_T)$; $\hat{\rho}^2 = .131$

SUMMER SEASON, 1882:

$R_0 = R_T = 25.00$ each week; $\overline{R}_L = 18.343$; $\overline{P}_g = 23.22$

(33 weeks); $P_g = 16.600 + 0.000R_0 + \underset{(0.940)}{0.348}R_L$; $\hat{\rho}^2 = .007$

(33 weeks); $P_{g,t} = 0.598P_{g,t-1} + \underset{(0.212)}{0.402}(-21.512 + 1.831R_L)$; $\hat{\rho}^2 = .206$

WINTER SEASON, 1882–1883:

$R_0 = R_T = 30.00$ each week; $\overline{P}_g = 28.056$

SUMMER SEASON, 1883:

$\overline{R}_0 = \overline{R}_T = 25.303$; $\overline{R}_L = 19.636$; $\overline{P}_g = 20.719$

(34 weeks); $P_g = -24.371 + \underset{(0.211)}{0.819}R + \underset{(0.270)}{1.226}R_L$; $\hat{\rho}^2 = .539$

(33 weeks); $P_{g,t} = 0.702P_{g,t-1} + \underset{(0.131)}{0.298}(-9.966 + 0.367R + 1.070R_L)$; $\hat{\rho}^2 = .168$

TABLE 4.2

Grain Shipments from Chicago, 1880 to 1886

The trend of the percentage share of grain tonnage of each railroad has been defined as

$$S = \alpha + \beta T + \gamma X + \nu$$

with S = a railroad's percentage of total tonnage of Chicago shipments; T = month of shipments, numbered from 1 to 84 for the period 1880 to 1886; X = "1" for the winter months in which the lakes were closed to traffic, "0" for the summer months. The estimates of trend, as calculated by the method of least squares, are:

The New York Central Railroad	$S = 30.056 + 0.045T + 15.451X$;	$\hat{\rho}^2 = .283$
The Pennsylvania Railroad	$S = 12.358 + 0.061T + 2.979X$;	$\hat{\rho}^2 = .074$
The Baltimore and Ohio Railroad	$S = 3.628 + 0.023T + 1.662X$;	$\hat{\rho}^2 = .125$
The Erie Railroad (from June, 1883 to December, 1886)	$S = 5.437 + 0.122T + 1.394X$;	$\hat{\rho}^2 = .124$
The Grand Trunk Railway	$S = 5.831 + 0.036T + 3.768X$;	$\hat{\rho}^2 = .165$

The "trend" share for any month is calculated by inserting the relevant values of T and X in the computed equation. The average of actual shares for each season is shown in the table, along with the difference between average actual and average "trend" shares for that season. The actual shares and the difference between actual and "trend" shares for each month are shown in the Statistical Appendix to this chapter.

Season, Year	*The New York Central Railroad System*		*The Pennsylvania Railroad System*		*The Baltimore and Ohio Railroad*		*The Erie Railroad System*		*The Grand Trunk Railroad System*	
	Average Actual Share (Per Cent)	*Actual Share Minus "Trend" Share (Per Cent)*	*Average Actual Share (Per Cent)*	*Actual Share Minus "Trend" Share (Per Cent)*	*Average Actual Share (Per Cent)*	*Actual Share Minus "Trend" Share (Per Cent)*	*Average Actual Share (Per Cent)*	*Actual Share Minus "Trend" Share (Per Cent)*	*Average Actual Share (Per Cent)*	*Actual Share Minus "Trend" Share (Per Cent)*
Winter 1879–1880	50.8	+5.3	16.6	+1.1	8.5	+3.1			5.2	−4.5
Summer 1880	12.9	−17.6*	3.8	−9.1*	1.2	−2.6*			3.4	−2.7
Winter 1880–1881	63.0	+16.8*	15.6	+0.6	6.4	+0.8			10.4	+0.3
Summer 1881	21.1	−9.8	6.8	−6.7	2.1	−2.0			4.6	−1.9
Winter 1881–1882	51.1	+4.4	21.8	+4.8	4.5	−1.4			15.1	+4.6
Summer 1882	37.2	+5.6	27.9	+13.7*	7.4	+3.0*			11.5	+4.5
Winter 1882–1883†	52.5	+5.4	21.9	+4.4	8.2	+2.0			10.7	−2.3
Summer 1883‡	38.8	+6.7	20.3	+5.3	5.0	+0.4	13.4	+2.6	12.1	+4.7
Winter 1883–1884	44.7	−3.0	14.9	−3.5	4.5	−1.8	12.5	+2.3	13.3	+1.9
Summer 1884	41.8	+9.2	16.7	+0.9	6.9	+2.0	11.9	−0.3	8.7	+0.9
Winter 1884–1885	40.6	−7.7	22.0	+2.9	5.5	−1.2	10.8	+0.8	10.1	−1.7
Summer 1885	40.8	+7.6	17.3	+0.8	5.7	+0.6	11.7	−2.1	7.1	−1.2
Winter 1885–1886	36.1	−12.8*	15.0	−4.8	7.4	+0.5	11.0	−2.0	11.4	−0.9
Summer 1886	32.8	−0.9	13.2	−4.0	4.0	−1.3	18.5	+3.2	5.1	−3.6

* Greater than the Standard Error of Estimate during three successive months.
† Two months only.
‡ Six months only.
Source: Statistics on total monthly shipments of wheat, corn, oats, and flour out of Chicago, and by railroad, as shown in the *Annual Reports* of the Chicago Board of Trade 1879 to 1887. Monthly shares are shown in the Statistical Appendix to Chapter 4.

TABLE 4.3

Grain Shipments into the Eastern Seaboard, 1880 to 1886

The trend of shares $S = \alpha + \beta T + \gamma X + \delta Y + \nu$ has been estimated by the method of least squares with S = percentage of total shipments of wheat, corn, and oats into New York, Montreal, Boston, Baltimore, and Philadelphia; T = month of shipment, numbered from 1 to 84 for the period 1880 to 1886; X = "1" for the months when lakes were closed to traffic; Y = the change in trend with the entry of the Lackawanna Railroad ($Y = 1$ for all months after this railroad's entry). The estimates of trend are:

The New York Central Railroad	$S = 10.81 + 0.05T + 8.65X - 5.75Y$; $\hat{\rho}^2 = .533$; $S_u = 4.33$
The Pennsylvania Railroad	$S = 24.64 - 0.05T + 9.84X - 0.53Y$; $\hat{\rho}^2 = .487$; $S_u = 5.44$
The Baltimore and Ohio Railroad	$S = 6.93 + 0.00T + 6.91X + 1.50Y$; $\hat{\rho}^2 = .429$; $S_u = 4.15$
The Erie Railroad	$S = 9.37 + 0.00T + 3.45X - 3.19Y$; $\hat{\rho}^2 = .275$; $S_u = 3.68$
The Delaware, Lackawanna and Western Railroad	$S = 0.10 + 0.01T + 2.91X$; $\hat{\rho}^2 = .354$; $S_u = 2.02$

The "trend" share for each month is obtained by inserting the relevant values of T, S, and Y in the correct equation. The average of the actual monthly shares is shown for each season during 1880 to 1886 along with the average difference between monthly "trend" and actual shares. The monthly actual shares and the residuals from "trend" each month are shown in the Statistical Appendix to Chapter 4, Table A.2.

The source is: total receipts into the East Coast seaports each month for 1880 to 1886 are given in the *Annual Statistical Reports* of the New York Produce Exchange, 1880 to 1884 and the *Annual Reports* of the Baltimore Corn and Flour Exchange, 1880 to 1886. Receipts into New York each month for each railroad are given in the *Annual Statistical Reports* of the New York Produce Exchange, 1880 to 1884; these statistics were compiled for 1885 to 1886 from the daily reports of receipts, by shipper and by railroad, in the *New York Journal of Commerce* for these years. Receipts into Baltimore by railroad are given in the *Annual Reports* of the Baltimore Corn and Flour Exchange, 1880 to 1886; receipts into Philadelphia by line are not available in this or Philadelphia sources, so that the Pennsylvania Railroad is given credit for all receipts into this city. Receipts into Boston and Montreal are not given by railroad, and no railroad is given credit for these volumes. Percentages for each railroad equal $\Sigma\Sigma R_{ji}/\Sigma C_i$ where R is railroad "j's" receipts into city i, and C_i is total receipts of city i. (Thus for Boston and Montreal, $R_j = 0$, but $C_i > 0$.)

Season, Year	*The New York Central Railroad System*		*The Pennsylvania Railroad System*		*The Baltimore and Ohio Railroad*		*The Erie Railroad System*		*The Lackawanna Railroad*		*The New York, West Shore and Buffalo Railroad*
	Average Actual Share (Per Cent)	*Actual Share Minus "Trend" Share (Per Cent)*	*Average Actual Share (Per Cent)*	*Actual Share Minus "Trend" Share (Per Cent)*	*Average Actual Share (Per Cent)*	*Actual Share Minus "Trend" Share (Per Cent)*	*Average Actual Share (Per Cent)*	*Actual Share Minus "Trend" Share (Per Cent)*	*Average Actual Share (Per Cent)*	*Actual Share Minus "Trend" Share (Per Cent)*	*Average Actual Share (Per Cent)*
Winter 1879–1880	19.5	—	38.4	+4	13.6	—	11.7	−1			
Summer 1880	10.3	−1	23.9	—	9.2	+2	6.3	−3*			
Winter 1880–1881	17.4†	−2	34.1	—	16.5	+2	14.6	+1			
Summer 1881	14.4	+3*	25.7	+2	6.8	—	10.7	+2			
Winter 1881–1882	24.4	+4*	28.4	−5	6.1	−6*	14.6	+1			
Summer 1882	13.2	+1	22.5	−1	8.4	+1	8.7	−1			
Winter 1882–1883	18.9	+2	33.3	+1	13.5	−1	10.9	+1	4.4†	+1	
Summer 1883	6.6	−1	19.8	−2	8.2	—	8.2	+2	1.1	—	
Winter 1883–1884	19.2	+3	26.3	−5	16.2	+1	9.2	−3	2.8	−1	
Summer 1884	8.9	+1	20.8	—	9.3	+1	6.1	+1	1.0	—	
Winter 1884–1885	17.6	+1	33.8	+3	18.0	+3	6.8	−2	3.5	—	4.4
Summer 1885	9.4	+1	20.4	—	5.5	−3	7.0	+1	0.3	—	2.9
Winter 1885–1886	15.9	−2	30.0	—	19.1	+4	11.0	+1	5.0	+2	
Summer 1886	8.0	−1	29.1	+1	9.5	+1	7.0	+1	1.8	+1	

* Greater than the Standard Error of Estimate for three successive months.
† Four months only.

Ohio, the Pennsylvania, and the New York Central. During the month of February, another trunk-line railroad began operations out of Chicago without an allotted share of the traffic from that city: the Grand Trunk Railroad extended its main line at Port Huron, Michigan 330.5 miles to Chicago, with the construction of the "Chicago and Grand Trunk Railway," for the purpose of making its Chicago transport independent of transfers from the existing trunk-line railroads.[7]

The new railroad was convinced of the advantages of adherence to official rates.[8] After five months of loyalty to rates and of tonnage shares from 2 to 7 per cent of total Chicago grain tonnage, the Grand Trunk became an enrolled member of the Executive Committee and "settled" for 10 per cent of the traffic. The division of tonnage of August, 1879, was suspended as of June 1, 1880, and a new division allowed the Chicago and Grand Trunk 10 per cent, the New York Central system 49 per cent, the Pennsylvania Railroad system 33 per cent, and the Baltimore and Ohio 8 per cent.[9] Thus an expansion in the number of members and in the capacity for through transport was made without noticeable breakdown in the control of rates.

The Executive Committee dealt with other serious matters with some success. There was a controversy concerning realized and allotted shares of westbound freight from New York, Philadelphia, and Baltimore which was settled by transferring tonnage in the cities in which the deficits occurred.[10] There was dissatisfaction with shares of livestock shipments, which led to reports in September that contracts had been made "for carrying hogs at a rate 5 cents less than the regular rate." [11] These rates

[7] This was noted in a review of the first two years of operation by Sir Henry Tiller, president of the Grand Trunk Railway of Canada: "It is a matter of great interest to us as we go on developing our traffic over the Chicago and Grand Trunk to see what we are doing at the same time with the Michigan Central (of the New York Central system). Mr. Vanderbilt did all he could . . . by controlling the Michigan Central to prevent our obtaining independent connection with Chicago. . . . This move was to cut us off from through booking of passengers with the Michigan Central. . . . We had to through book over the Chicago and Grand Trunk and the result has been a large loss to them and a large gain to us." ["Sir Henry Tiller on the Railroad War," the *Railroad Gazette* (April 14, 1882), 221.]

[8] There were some reports of cut rates over the Grand Trunk, but in northern New England. There were none of the testimonials or the shifts in shipments to this line that constituted evidence of systematic deviations from official rates.

[9] Cf. "Re-apportionment of Freight," the *Railroad Gazette* (October 19, 1883), 692; the same figures are shown in "Percentages for the Division of Eastbound Dead Freight from Chicago," File 50114, June 20, 1891, the Archives of the Pennsylvania Railroad system, Philadelphia, Pennsylvania.

[10] *Proceedings and Circulars of the Trunk-line Committee*, *op. cit.* (1880–1881), June 17, 1880.

[11] "Irregularities in Eastbound Rates," the *Railroad Gazette* (September 10, 1880), 480.

seem to have been restored, as well, within one month with no radical departure from previous shares.

There was no success whatsoever, in contrast, in finding a solution to new difficulties between the New York Central system and the Erie Railroad on shares of traffic into New England and New York City. As early as September of 1880 there was "competition (between) the Erie and the New York Central for Boston and New England business." [12] The cause for competition was that "imperfection of the organization for division of eastbound traffic leaves the roads over which the traffic passes [interested] in diverting shipments to their lines." [13] Little more was said during the winter, but by early spring there were reports that the Erie was offering discounts and "the general feeling among railroad men was dismay and disgust when it was found that in the face of the formal agreement numerous contracts had been made to carry grain at 5 cents per hundred pounds less than the regular rate." [14] The discounts were matched by the Vanderbilt lines and within ten days there was a "meeting held at Mr. Vanderbilt's house from which resulted an order to restore the April 1 rate — 30 cents per hundred pounds from Chicago to New York — to take effect the following Monday (April 18, 1881)"; this rapid reaction seems to have had the effect of restoring rates, but it did not solve the problem: "The effective settlement will be found in dividing the business." [15]

The official rates continued in effect for perhaps five or six more weeks, at which time discounts on eastbound shipments reappeared on the Board of Trade. At the end of the first week in June, the discounts were thought to have been 5 cents per 100 pounds; by the second week they were more than 10 cents per 100 pounds. Official rates were cut from 30 cents to 17 cents per 100 pounds during this second week, but further discounts were offered which brought the Board of Trade rate to 15 cents per 100 pounds at the same time. The 17-cent official rate and the 15-cent Board of Trade rate continued through July and the first week of August. At that time the Board of Trade rate was cut to 13 cents per 100 pounds for three weeks, then restored to 15 cents per 100 pounds for three weeks, then cut to between 11 cents and 13 cents per 100 pounds each week until the first of November.[16]

[12] *Ibid.*

[13] *Ibid.*

[14] Rates of 23 to 27 cents per 100 pounds are shown in the "Railroad Freights" column of the *Chicago Daily Commercial Bulletin* during most of the period from April 4, 1881, to April 18, 1881; cf. "The Break in Eastbound Rates," the *Railroad Gazette* (April 15, 1881), 209.

[15] "The Restoration of Eastbound Rates," the *Railroad Gazette* (April 22, 1881), 220.

[16] The lake-rail rates announced on the Board of Trade were close to these Board of Trade all-rail rates each week until the 11 cent–13 cent levels were reached, at which point the lake rate was not decreased.

The disruption was accompanied by a greatly increased share for the Erie of the rail shipments from the Great Lakes to the eastern coast. This shift in tonnage, in fact, was given as a reason by the New York Central for the rate cuts: "The cause seems to be a diversion of traffic from the Vanderbilt lines and the New York Central generally to the advantage of the Erie and its lines . . . The New York Central has been losing business and this it does not propose to permit . . . [the New York Central's] reduction to 25 cents from Chicago last week probably brought . . . [its] rate down to the level of the cut at that time. . . ."[17] Tonnage transfers to the Erie gave this railroad more of the total grain tonnage into the seaboard cities during the period July to October, 1881, than during the same period in the preceding and following years (as shown in Table 4.3).[18]

The Erie did not profit from the rate reductions, however. The increases in tonnage shares, while from 20 to 50 per cent of the trend of Erie shares 1880 to 1886, were achieved in periods when rates declined so sharply and quickly that, in most months, there was not even an increase in gross receipts.[19] There might have been benefits for the Erie from indicating to the other railroads its willingness to accept large volumes of traffic at low rates ("as part of the argument for larger market shares") but there were not the immediate profits that would have rendered the cartel agreement unstable.

Whatever the Erie's reasons, there were effects upon shippers from this war. Grain speculators shipping quantities of wheat, corn, oats *et al.* to the eastern seaboard reacted strongly to the lower rates. Grain prices in Chicago increased and those in New York declined, so that the price margin between collection and receiving centers declined sharply; the average grain price difference in the summer of 1880 was more than 31.6 cents per 100 pounds, while the average difference in the summer of 1881 was 17.5 cents per 100 pounds. The grain price differences in the two

[17] "A Trunkline Railroad War," the *Railroad Gazette* (June 17, 1881), 334.

[18] The New York Central road received more than its regular share, as well, but not before the lowest level of rates had been reached. The Central received substantially less than its share of tonnage out of Chicago in the months of May, June, July, and August of 1881. This is as expected from the Erie having offered discounts of the lake-rail rate so as to have drawn traffic from the New York Central's all-rail transport. The later increase could have been concurrent with undercutting the Erie's discounts in rates.

[19] The increases in tonnage shares over the long-term trend of shares are as shown in Table 4.3; the rate decreases resulted in an average Board of Trade rate of 17.1 cents, and an average official rate of 19.1 cents, as compared to 30-cent rates the previous summer (as shown in Table 4.1). The percentage increase in tonnage would not seem to have been greater than the average percentage decline in rates. In particular months, such as September, there was a large percentage increase in tonnage but official rates declined from 30 cents (in the previous year) to from 14 to 17 cents per 100 pounds so that gross receipts decreased even in those months.

winter seasons of 1880–1881 and 1881–1882 were 37.3 cents per 100 pounds and 20 cents per 100 pounds (as shown in Table 4.1). Fluctuations in grain price differences were extensive in both the summer and winter seasons of 1881–1882, and were generally in accord with the sharp downward movements of all-rail and lake-rail rates. The weekly grain price difference adjusted within two or three weeks to any change in official or Board of Trade rates[20] (as shown by the lagged adjustment equation in Table 4.1); the adjustments to both rates were similar and involved a change of approximately .72 to .74 cent per 100 pounds for each 1-cent change in the rate.[21]

These conditions spread to related transport markets. It was reported that, by the middle of July, "there has been some irregularity in westbound rates . . . [as well as] cutting of passenger rates from New York and Boston to western cities, and irregularities on passenger rates in the other direction [with] rebates of more than $5 or $6 from Chicago." [22] There were "some four independent railroad wars continuing," [23] all of which were in rates for through transport services. There were no notices of disruptions in local rates on any one of the individual railroads, nor were there indications of any results from such. The short rate wars of 1881, then, led to large reductions in rates on some part of through transport and no noticeable reductions in any local rates.

The Pooling of Revenues and Joint Control of Rates During 1882 and 1883

The trunk-line railroads tolerated this "state of war" not only in the fall, but also during most of the winter of 1881–1882. A first attempt was made during January and February to repair the damage done to official rates; conferences were held to increase rates to their former levels and to find means for maintaining the increased rates. As a first step, the 15-cent official rate on grain was increased to 20 cents on January 23. The increase

[20] The quickness with which changes were made in official rates made for a small systematic difference between Board of Trade and official rates (as indicated by the computed values for $R_T = \alpha + \beta R_0$). That is, there were large discounts in rates resulting in a lower Board of Trade rate, but the discounts quickly reduced official rates. In this instance it might not be expected that the Board of Trade rates would have much more effect upon grain price differences than official rates.

[21] Board of Trade rates, and ultimately official rates, were in reaction to (suspected) discounts in lake-rail rates so that only indirect effects of the Erie's rate disruptions on grain price differences were recorded. These indirect effects seem to have included lower transport costs for shippers and lower grain price differences from week to week.

[22] "The Continued Railroad Wars," the *Railroad Gazette* (July 22, 1881), 402.

[23] *Ibid.*

was "a test of good faith" which apparently succeeded — Board of Trade rates increased by the same amount on the same day and were maintained at this level throughout the month of February.

The second step was a new pact on setting rates — "[an] agreement [which] has been called a truce and not a peace, and to a certain extent this is true . . ." [24] which was to serve as an experiment. There was to be a division of all through traffic over the trunk lines and money payments when tonnage receipts were less than the allotted shares. The traffic received at the eastern seaboard terminal points of the trunk lines was divided, rather than the tonnage out of Chicago; this was to give each of the trunk lines final responsibility so as, in particular, to reduce "the interest of the Erie [in] solicit[ing] freight in Chicago against the New York Central and Pennsylvania." [25] The basis of shares was tonnage percentages in the successful cartel year of 1880: "It is agreed that the division of eastbound freight including livestock shall be based generally on the actual distribution of traffic during the year 1880, when rates were well maintained throughout." [26]

The "money pool" was to be the means for assuring each line of its division of the tonnage. The members of the Trunkline Executive Committee made deposits with the Commissioner at the time the contract agreement was signed in March of 1882. There were to have been "monthly payments of money [to the deficit railroads] if the traffic is not divided in accordance with the agreed apportionment," [27] either voluntarily by the railroad with the tonnage surplus, or from the deposit of this railroad. Such payments were set at the level of the gross revenues from the surplus

[24] And that the eastbound rate was to be increased to 25 cents per 100 pounds after March 13, 1882. "The Trunkline Negotiations," the *Railroad Gazette* (February 17, 1882), 104.

[25] *Ibid.* This was alleged to have resulted from the Erie having had no part in the Chicago tonnage since it had no Chicago terminal. This road gained from offering such a low Buffalo–New York City rate that Chicago tonnage was directed to lake carriers for transfer to the Erie at Buffalo.

[26] *Ibid.* The westbound freight was to have been divided, as in previous years, as shares of the total freight tonnage from Boston, New York, Philadelphia, and Baltimore. "The division of the westbound freight usually gave the New York Central about 35 per cent of the earnings from the freight and the Erie 31 per cent. The Pennsylvania got 25 per cent of the traffic and 9 per cent went to the Baltimore and Ohio. The division is to be based on this or changes according to the 1880 pattern." (*Ibid.*) Changes in tonnage shares followed from an individual road presenting convincing information on "capacity" or "ability" to transport a larger share before the Advisory Commission of the Trunkline Freight Agents; such presentations were made successfully as is evidenced from the allowed increase to 11 per cent of the tonnage of the Grand Trunk Railway after March 13, 1882. [Cf. "Re-apportionment of Freight," the *Railroad Gazette* (October 19, 1883), 692.]

[27] "The Trunkline Agreement as to Shares of the Traffic," the *Railroad Gazette* (March 31, 1882), 198.

tonnage, so that there was an implicit penalty against the railroad with the surplus equal to the cost incurred in transporting this excess tonnage.

A 25-cent rate was announced, the new contract was signed, and the new arrangement on tonnage shares went into effect the first of April, 1882. Shares were allotted in accord with 1880 shares, and there were no reports of deviations during the summer or winter of 1882. All of the railroads received larger shares of the summer grain tonnage out of Chicago (to the detriment of the percentage of the lake steamers, presumably because of the 25-cent rate, rather than the 30- and 35-cent rates previous to the railroad war of 1881). There was an increased allotment to the New York Central system because of an appreciable expansion of its capacity with the opening of the "New York, Chicago and St. Louis Railway" (the "Nickel Plate" railroad between Buffalo and Chicago) during October.[28] Otherwise there was little change in assigned shares, and realized shares were quite close to those designated.[29]

Shortly after the beginning of 1883 the money pool began to encounter difficulties. The first of the problems followed from the extension of a number of the railroads beyond previous terminal points to the Great Lakes region or into Chicago itself. The Delaware, Lackawanna and Western Railroad, long considered an important "coal railroad" along the Pennsylvania–New York border, leased the New York, Lackawanna and Western Railway in October, 1882, so that its line extended from New York City to Buffalo. This "coal railroad" thereby became a transporter of grain and provisions from the Great Lakes to the eastern seaboard. At approximately the same time, the Erie Railroad entered Chicago by the construction and operation of the "Chicago and Atlantic Railway" from the affiliated Atlantic and Great Western to the center of the city; now there was an independent Erie line all the way from the midwest grain center to Jersey City, New Jersey. These two railroads had the "capacity" to transport appreciable portions of the assigned shares of the long-established lines; the problem was to obtain these new shares by peaceful agreement.

The second of the difficulties was presented to the Executive Committee by the New York Central Railroad. This railroad was clearly concerned

[28] The increase in the New York Central's allowance took place in the reapportionment of July 1, 1883, which delegated to this company 51.75 per cent of eastbound tonnage from Chicago, as compared to 45.5 per cent of the apportionment of March 13, 1882. This was at the expense of the Pennsylvania lines, since their share fell from 35.5 per cent to 27.25 per cent. [As shown in "Re-apportionment of Freight," the *Railroad Gazette* (October 19, 1883), 692.]

[29] As shown by the lack of three or more successive months of surpluses of deficits differing, by more than the standard deviation, from the trend of shares for individual roads, on grain from Chicago in Table 4.2 and also of grain tonnage into the eastern seaports in Table 4.3.

with not having received its share of the traffic: "The Vanderbilt roads have not carried as much as awarded them; they have carried 764,292 tons which is 41,412 tons short of their quota. The Pennsylvania roads are 1,497 tons ahead having carried 630,085 tons. But the Chicago and Grand Trunk has carried 40,827 tons more than its proportion, or 13½ per cent of the whole instead of the 11 per cent to which it is entitled." [30] The complaint required a shift of tonnage to the New York Central at precisely the time that changes had to be made in assigned proportions in favor of the Erie and the Lackawanna.

These difficulties were overcome in fact without the breakdown of the agreement. The Erie applied for a share of the tonnage in the Chicago pool and, in the apportionment of July 1, 1883, the new company received 11 per cent of total eastbound tonnage.[31] This share consisted of tonnage mostly from the New York Central, as was noted at the time: "The reduction on the two Vanderbilt roads made within a little more than three years has been something enormous; they have suffered more in proportion than the roads South of them because all the new roads carry to New York and New England chiefly." [32] The Central accepted the new conditions without breaking its promises, at least; and the problem of the railroads' tonnage deficit seems to have solved itself: "There has been some uneasiness because of the recent change in proportions of business going by the old lines (with the New York Central's) Nickel Plate taking much more than proportion. . . ." [33]

The official rates on grain transport were well maintained throughout. The 25-cent rate was in effect without exception throughout the summer season of 1882; Board of Trade rates did not differ from this official rate, and the announced lake-rail rate averaged slightly more than 18 cents per 100 pounds.[34] The winter official rates were based on a grain rate from Chicago to New York of 30 cents per 100 pounds, and this again was the announced Board of Trade rate for such transport each week of the season. The 1883 summer season rate was first 30 cents per 100 pounds, then was decreased to 25 cents early in May and maintained at this level until November. These official rates were matched each week by the quotations of the Chicago Board of Trade; announced lake-rail rates averaged 19.6

[30] "The Distribution of Chicago Shipments," the *Railroad Gazette* (February 23, 1883), 127.

[31] Cf. "Re-apportionment of Freight," the *Railroad Gazette* (October 19, 1883), 692.

[32] *Ibid.*

[33] "A New Chicago Apportionment," the *Railroad Gazette* (July 27, 1883), 499.

[34] Actually the expected difference between official and lake-rail rates of 5 cents per 100 pounds was too low by approximately 1.7 cents. Most of this difference occurred during July when lake-rail rates were 16 cents per 100 pounds. During the months of May, June, and most of September, October, and November, the difference was close to 5 cents per 100 pounds.

cents per 100 pounds and were within 2 cents of this average each week of the season. There were no variations from the Trunkline Association's announced rates so that the only disadvantage from this money pool was that the announced rates themselves were 10 to 15 cents per 100 pounds less than during "the successful cartel operations" of 1880.

Successful, if limited, pooling had an effect upon shippers' costs in Chicago and New York grain markets. The average difference between Chicago and New York spot grain prices increased by more than 6 cents during the summer of 1882 over that in the summer of 1881 (as shown in Table 4.1). The margin was 9 cents greater in the winter of 1882–1883 than in the winter of 1881–1882 and 3 cents higher again in the summer of 1883 (given that the average difference was 20.6 cents per 100 pounds, as compared with 17.5 cents during the summer of 1881). Week-to-week changes in grain price differences were small — the standard deviation of the weekly differences declined from approximately 5 cents during 1881 to less than 4 cents in 1882 and 2.5 cents in 1883 — and part of the small variance was explained by changes in the cartel's lake-rail rates during both of the summer seasons.[35] Altogether, higher rates for shippers between midwest and east coast grain markets were reflected in higher price differences between the exchange markets, and rigid adherence to official rates had the effect of limiting large changes in the price differences to those following announced rate changes.

Independent Railroads and a Breakdown in the Rate Agreement, 1884–1885

The closing of navigation on the Great Lakes in November of 1883 marked an end to the immediate success of the pool of the Trunkline Executive Committee. The organization had been able to deal with demands of established railroads for "equitable" shares and with reapportionment required for new shares for the Grand Trunk and the Erie railroads. It was not able to induce two other new railroads to accept memberships; dissatisfaction with the consequent state of affairs led to cheating on rates, and finally to a breakdown of the agreement.

The first new railroad not brought "into the fold" was the Delaware,

[35] Given the correspondence of Board of Trade lake-rail rates with "the ideal lake-rail rate" 5 cents less than the official all-rail rate, it might be inferred that changes of grain price differences with changes in lake-rail rates were in accord with the expected effects from a well-functioning cartel agreement. As shown by computed $\hat{\rho}^2$ in the lagged equations of Table 4.1, week-to-week changes in grain price differences were partially in adjustment to changes in lake-rail rates and partially a process of adjustment toward an equilibrium which took slightly more than two weeks to complete.

Lackawanna and Western — the new transporter of grain from Buffalo to New York City. The general freight agents of the trunk-line roads met on January 4 to act on evidence that the Lackawanna road was setting a rate on grain shipments below the official rate of the members. The committee decided not to react by lowering rates, but rather to maintain existing tariffs and undertake a boycott by "refusing to exchange through business with the Delaware, Lackawanna and Western or to be party to any through bills of lading of that line." [36] This did not force the Lackawanna to conform, and rates on other lines began to fall. The official grain rate of 30 cents per 100 pounds "was not being maintained, (so that) on January 4, 1884, a reduction was made to the basis of 20 cents per 100 pounds." [37] After some weeks of boycott, an attempt was made to restore rates to the November tariff but "this did not succeed . . . it was necessary on February 7 to hold another meeting at which it was agreed to enforce . . . the (existing) tariff," [38] and on March 21 to reduce the 20-cent rate to 15 cents per 100 pounds.[39] As a consequence of non-adherence to official rates (initiated, at least, by the Lackawanna road), the level of rates had decreased by 50 per cent between the first of January and March 21, 1884.

These abrupt changes in rates had the same effects upon the markets for the goods being transported as did the rate war of 1881. Rates quoted on the Board of Trade declined when official rates declined and were 5 cents less than official rates the last week of January and the first two weeks of March as well (so that changes in official rates "explained" 72 per cent of the variance in Board of Trade rates during the winter of 1883–1884, as shown by $\hat{\rho}^2$ in Table 4.4). The margin between Chicago and New York grain prices declined precipitously from an average of 28.1 cents per 100 pounds in the winter of 1882–1883 to 19.3 cents per 100 pounds in the winter of 1883–1884,[40] but cheating did not have as widespread an effect

[36] For the initial discussion of the charges against the Lackawanna, cf. *Proceedings and Circulars of the Trunkline Committees (freight department) 1884:* Trunkline Sub-Executive Committee, January 4, 1884 (*op. cit.*); for the decision to boycott the Lackawanna on shares of through traffic, cf. "The Trunklines and the Lackawanna," the *Railroad Gazette* (January 11, 1884), 37.

[37] "Proposed Remedy for the Defects of the Eastbound Pool," the *Railroad Gazette* (March 21, 1884), 226. The change in official rate is also shown, as in the other mentioned instances, in the Interstate Commerce Commission, *Railways in the United States in 1902*, *op. cit.*, Table XLIV — "Rates in Cents Per Hundred Pounds Upon Grain All Rail from Chicago to New York from March 28, 1864 to April 1, 1902," p. 79.

[38] "Proposed Remedy for the Defects of the Eastbound Pool," the *Railroad Gazette*, *loc. cit.*

[39] Cf. the Interstate Commerce Commission, *Railways in the United States in 1902*, *op. cit.*

[40] This extensive decline should have followed from a 6-cent decrease in Board of Trade rates, and is comparable to the experience during 1881. This decline followed from

TABLE 4.4

Official Rail Rates, Board of Trade Rail Rates, and Differences Between Chicago and New York Grain Prices, 1883 to 1886

$$R_T = \alpha + \beta R_0 + \nu \qquad (1)$$

R_T, the rail rate, in cents per hundred pounds Chicago to New York, from the Board of Trade *Daily Commercial Bulletin;* R_0, the rail rate from compilations of posted tariffs and from announcements of the various traffic associations.

$$P_g^* = \alpha + \beta R_0 + \gamma R_L + \nu; \qquad P_g^* = \alpha + \beta R_T + \gamma R_L + \nu \qquad (2)$$

P_g^*, the weekly average of the daily differences between Chicago and New York spot prices, in cents per hundred pounds, for the number 2 "western" grade of corn and oats; the official rate R_0 or the Board of Trade rate R_T; the lake-rail rate R_L from Chicago to New York via steamship to Buffalo and rail from Buffalo to New York.

$$P_{g,t} = P_{g,t-1} + \epsilon(P_g^* - P_{g,t-1}) \qquad (3)$$

$P_{g,t}$, the average of the daily Chicago–New York grain price differences during week t and $P_{g,t-1}$, the average for the previous week; $P_g^* = \alpha + \beta R + \gamma R_L$ for the summer season and $P_g^* = \alpha + \beta R$ for the winter season from an instantaneous adjustment to changes in transport rates.

The standard error of each coefficient is shown in parentheses below the coefficient.

WINTER SEASON, 1883–1884:

$R_T = 3.977 + \underset{(0.114)}{0.811} R_0$; $\hat{\rho}^2 = .728$; $\overline{R}_0 = 24.857$; $\overline{R}_T = 24.143$; $\overline{P}_g = 19.304$

(21 weeks); $P_g = 18.335 + \underset{(0.082)}{0.039}\overline{R}_0$; $\hat{\rho}^2 = .012$

(21 weeks); $P_g = 16.883 + \underset{(0.084)}{0.100} R_T$; $\hat{\rho}^2 = .069$

(21 weeks); $P_{g,t} = 0.846 P_{g,t-1} + \underset{(0.132)}{0.154}(9.018 + 0.396 R_0)$; $\hat{\rho}^2 = .154$

(21 weeks); $P_{g,t} = 0.825 P_{g,t-1} + \underset{(0.120)}{0.175}(17.439 + 0.066 R_T)$; $\hat{\rho}^2 = .077$

SUMMER SEASON, 1884:

$R_T = 0.955 + \underset{(0.079)}{0.927} R_T$; $\hat{\rho}^2 = .840$; $\overline{R}_0 = 21.893$; $\overline{R}_T = 21.250$;

$\overline{R}_L = 16.821$; $\overline{P}_g = 18.679$

(28 weeks); $P_g = 17.102 + \underset{(0.144)}{0.345} R_0 - \underset{(0.382)}{0.355} R_L$; $\hat{\rho}^2 = .309$

(28 weeks); $P_g = 20.323 + \underset{(0.132)}{0.325} R_T - \underset{(0.353)}{0.508} R_L$; $\hat{\rho}^2 = .317$

(28 weeks); $P_{g,t} = 0.456 P_{g,t-1} + \underset{(0.246)}{0.544}(32.524 + 0.262 R_0 - 1.150 R_L)$; $\hat{\rho}^2 = .332$

(28 weeks); $P_{g,t} = 0.442 P_{g,t-1} + \underset{(0.231)}{0.558}(32.792 + 0.289 R_T - 1.190 R_L)$; $\hat{\rho}^2 = .345$

WINTER SEASON, 1884–1885:

$R_T = -8.999 + \underset{(0.160)}{1.319} R_0$; $\hat{\rho}^2 = .808$; $\overline{R}_0 = 23.611$; $\overline{R}_T = 22.166$; $\overline{P}_g = 25.925$

TABLE 4.4 (Continued)

WINTER SEASON, 1884–1885: (Continued)

(18 weeks); $P_g = -11.076 + \underset{(0.242)}{1.567}R_0$; $\hat{\rho}^2 = .725$

(18 weeks); $P_g = 2.517 + \underset{(0.169)}{1.057}R_T$; $\hat{\rho}^2 = .709$

(18 weeks); $P_{g,t} = 0.359P_{g,t-1} + \underset{(0.151)}{0.641}(-20.108 + 1.934R_0)$; $\hat{\rho}^2 = .648$

(18 weeks); $P_{g,t} = 0.240P_{g,t-1} + \underset{(0.201)}{0.760}(0.399 + 1.143R_T)$; $\hat{\rho}^2 = .533$

SUMMER SEASON, 1885:

$R_T = -18.846 + \underset{(0.436)}{1.753}R_0$; $\hat{\rho}^2 = .374$; $\overline{R}_0 = 20.517$; $\overline{R}_T = 17.138$;
$\overline{R}_L = 15.241$; $\overline{P}_g = 15.512$

(29 weeks); $P_g = 0.332 + \underset{(0.363)}{0.184}R_0 + \underset{(0.187)}{0.748}R_L$; $\hat{\rho}^2 = .541$

(29 weeks); $P_g = 3.526 + \underset{(0.152)}{0.350}R_T + \underset{(0.225)}{0.393}R_L$; $\hat{\rho}^2 = .615$

(29 weeks); $P_{g,t} = 0.788P_{g,t-1} + \underset{(0.148)}{0.212}(14.496 - 0.708R_0 + 1.027R_L)$; $\hat{\rho}^2 = .105$

(29 weeks); $P_{g,t} = 0.715P_{g,t-1} + \underset{(0.132)}{0.283}(4.168 + 0.884R_T - 0.245R_L)$; $\hat{\rho}^2 = .258$

WINTER SEASON, 1885–1886:

$R_0 = 25.000$ each week; $\overline{R}_T = 24.235$; $\overline{P}_g = 22.808$

(17 weeks); $P_g = -3.882 + \underset{(0.422)}{1.109}R_T$; $\hat{\rho}^2 = .329$

(17 weeks); $P_{g,t} = 0.745P_{g,t-1} + \underset{(0.114)}{0.254}(-22.329 + 2.262R_T)$; $\hat{\rho}^2 = .378$

SUMMER SEASON, 1886:

$R_0 = 25.000$ each week; $R_L = 20.000$ each week; $\overline{R}_T = 24.818$;
$\overline{P}_g = 17.791$

(16 weeks); $P_g = 28.706 - \underset{(0.585)}{0.443}R_T$; $\hat{\rho}^2 = .039$

(12 weeks); $P_{g,t} = 0.742P_{g,t-1} + \underset{(0.142)}{0.259}(3.882 + 0.637R_T)$; $\hat{\rho}^2 = .350$

WINTER SEASON, 1886–1887:

$R_0 = 30.000$ each week from December 20 to March 14; $R_0 = 25.000$ thereafter;
$\overline{R}_T = 24.214$; $\overline{P}_g = 27.451$

(14 weeks); $P_g = 29.756 - \underset{(0.158)}{0.095}R_T$; $\hat{\rho}^2 = .030$

(14 weeks); $P_{g,t} = 0.927P_{g,t-1} + \underset{(0.194)}{0.073}(-29.452 + 2.359R_T)$; $\hat{\rho}^2 = .287$

on grain price differences as previously. Week-to-week changes in grain price differences were not as responsive to changes in either official or Board of Trade rates (the "explained" variance in price differences was between 6.9 and 15.4 per cent, for the instantaneous and lagged adjustment equations in Table 4.4; moreover, the percentage of "explained" variance was not greater in the Board of Trade equation than in the equation using official rates). This slow response must have resulted from the rate discounts having taken place first on the Buffalo–New York City traffic and second on the Chicago–New York traffic. By the time the cuts were posted on the Board of Trade, they were official or an acknowledgment of actual rates in effect for some weeks.

The results extended beyond immediate changes in grain price margins. The trunk-line roads hitherto engaged in boycotting the Lackawanna announced another invitation to this railroad to join the pooling agreements at the end of February; there was offered the inducement of a larger share of the westbound freight pool[41] but the company decided to remain outside of the agreement.[42] By doing so, the Lackawanna received approximately the same share of the total grain tonnage to the eastern seaboard as in the previous year (as in Table 4.3, the road received 3.2 per cent of total tonnage in the winter of 1882–1883, and 3.0 per cent in the winter of 1883–1884).[43]

The second new railroad followed the Lackawanna's example and declared its independence. The "New York, West Shore and Buffalo Railway" introduced service from Buffalo to New York City in the early months of 1884 along a new road parallel to the New York Central system the length of the state of New York. It managed to operate in harmony with the Central and the other roads during the winter months of 1884; during the summer season, however, the West Shore seemed to have become dissatisfied with the shares it was receiving of both eastbound and westbound tonnage when it quoted the official rates. In August "the West Shore is cutting rates in both directions . . . So far as westbound

---

earlier changes in Buffalo–New York City rates by the Lackawanna Railroad, and in Buffalo–New York grain price differences, so that it is only an indirect indicator of the effects of rate-cutting.

41 "The Lackawanna and the Trunklines," the *Railroad Gazette* (February 22, 1884), 154.

42 The absence of the Lackawanna road was noted at the Trunkline Executive Committee meeting in the third week of June, 1884 [cf. The *Railroad Gazette* (June 20, 1884), 465]; it was again mentioned at the Trunkline Executive Committee meeting of the second week of November [cf. the *Railroad Gazette* (November 14, 1884), 821].

43 It is not possible to determine whether this railroad gained profits from independence, because there is no information on the discount on the Buffalo–New York rate. Without some estimate of R_c ($\sim R_T$, for example), then $\pi_c = (R_c - K_2)Q_c$ cannot be estimated.

shipments are concerned the road has been permitted to secure what proportion of traffic it can and there is no doubt that it makes concessions from regular rates to secure it . . . Cuts in eastbound rates have been made by a freight line that works over the West Shore and to connecting roads west of Buffalo. . . ." [44] An invitation to membership was issued, but the new company demanded 10 per cent of the New York receipts and "the West Shore claims full rates if it does not carry its full share," [45] and the offer was withdrawn. The West Shore "began an aggressive movement intended to secure a position for that road." [46]

At that point the managers of the Grand Trunk Railway withdrew the Canadian road from the agreement. This railroad had managed to transport far more than assigned monthly shares of tonnage out of Chicago so that, by October of 1884, "according to the decision of the arbitrators, the Grand Trunk owed other roads for freight in excess of its share about $100,000." [47] The Grand Trunk refused to pay this amount[48] and, after repeated prodding by the other railroads, finally in February of 1885, "Notice has been given that the Chicago and Grand Trunk and the Grand Trunk Railways are withdrawing from the pools." [49] The company stated "that it has lost faith in the value of such an agreement." [50]

Three independent railroads were sufficient to bring about a breakdown of the agreement on eastbound rates.

There were disruptions of posted rates on provisions from Chicago to New York early in February, which resulted in Board of Trade rates having been (on average) 2 cents less than official rates. Official rates then decreased from 35 cents per 100 pounds to 28 cents per 100 pounds on seventh-class provisions; Board of Trade quotations decreased from 33 to 25 cents. The 25-cent rate was quoted from April through June. The attempt to increase Board of Trade and official rates to 30 cents in July, was successful, but an official 35-cent rate in August was not realized, given that the Board of Trade quotation had declined to 30 cents in Sep-

[44] "Rumors of Cuts in Trunkline Rates," the *Railroad Gazette* (August 15, 1884), 605.

[45] "The Trunkline War," the *Railroad Gazette* (October 3, 1884), 719.

[46] "The West Shore Line," the *Railroad Gazette* (October 24, 1884), 768.

[47] "The Trunkline War," *op. cit.*

[48] No arrangements were made for payment during 1884, and even with continued nonpayment an arbitrator awarded the Chicago and Grand Trunk 1½ per cent more of the livestock and dressed beef traffic (and the New York Central and Baltimore and Ohio systems 1½ per cent less). "New Apportionments of Chicago Shipments," the *Railroad Gazette* (January 6, 1885), 26.

[49] "The Trunkline Combination," the *Railroad Gazette* (February 6, 1885), 90. The *Railroad Gazette* also noted that "the two new railroads between New York and Buffalo which have so far remained outside of the eastbound pool have continued to complicate matters." (*Ibid.*)

[50] "The Grand Trunk's Position as to Pools," the *Railroad Gazette* (May 15, 1885), 314.

tember and October, and was matched by an official rate of 30 cents during November.

Some of the effects of these disruptions in provision rates can be indicated by changes which occurred in the New York and Chicago spot markets for "Chicago barreled lard." [51] Shippers and consumers seem to have gained lower transport costs. The average margin between Chicago and New York spot prices for barreled lard declined 4 cents in February, and another 5 cents in March, when rate-cutting first took place. The margin declined approximately 7 cents in September and October, when rate-cutting began once again (as shown in the Statistical Appendix to Chapter 4, Table A.3).[52]

There were disruptions in rates on grain as extensive as those on provisions. Board of Trade reports reflected small reductions in the 15-cent official rate in May and June of 1884 but no discounts at all during July and August. Discounting broke out again during September and October, however, while official rates were 25 cents, and Board of Trade rates were reduced approximately to 20 cents per 100 pounds for some weeks. The nonmembers of the pool were blamed: "There has not been, since the railroad war of 1881, so serious a disturbance of the harmony of the trunk-lines, nor of the rates which have been maintained by that harmony, as in the last ten days. . . . At the beginning of this week grain was taken from Chicago to New York at 15 cents and provisions at 20 cents, while the regular rates were 25 and 30 cents . . . The question seems to have been the position of the Grand Trunk with regard to accepting the decisions of

[51] It would be more appropriate to review the effects of rate changes on tonnage shipments and price margins for cattle or dressed beef. This is not possible, however, because the specifications of quality in the Chicago and New York produce reports, for either cattle or beef, are so inexact that the range of prices in each city is greater than price differences because of transport costs between cities. There is information on the particular grade of barreled lard denoted "Chicago" or "western" with a range of spot prices each day not greater than 3 cents per 100 pounds. The Grand Trunk Railway, at the time of the rate-cutting, seems to have gained a larger share of the lard tonnage from Chicago than it received in previous or immediately succeeding years. In February and March, when rate-cutting on provisions began, the Grand Trunk received approximately 40 per cent more than its long-term trend of shares of lard tonnage. This railroad received approximately 100 per cent more than trend in April and May, the period of most extensive cutting (as shown in the Statistical Appendix to Chapter 4, Table A.4).

[52] The average margin between Chicago and New York spot lard prices was not identical with either the official or Board of Trade rate on lard transport between the two cities. Given that there were costs of storage, spoilage, and further transport to warehouses or consuming centers, the margin should have been greater than the rail transport rate. During most of the months in 1884, the margin was from 2 to 5 cents greater than the railroad rate (during the middle of the summer, however, when some lard was transported by steamer, the margin was slightly less than the Board of Trade official rate). The margin was from 6 to 10 cents per 100 pounds greater than listed rates during 1885, and 2 to 5 cents greater than rates during 1886.

the arbitrators . . . and the West Shore claims to be able to deliver freight as promptly as the other roads and demands . . . full rates. . . ." [53]

Discounting continued through the winter season of 1884–1885. There were only six weeks in this season in which the official 25-cent rate was quoted on the Board of Trade, while there were twelve weeks in which rates from 19 to 23 cents per 100 pounds were offered. This rate-cutting was once again the fault of the nonmembers of the pool: ". . . irregularities in eastbound rates . . . [follow] shipments below tariff rates which reach New York or Boston by the West Shore and Lackawanna roads which are brought to them at Buffalo by the Grand Trunk . . . The trouble this year seems to have been exceptionally obstinate though it has never resulted in open war . . . Rates go down for a week or two or three weeks, then are generally restored. . . ." [54]

The breakdown of the Trunkline Executive Committee was formally acknowledged not long after the New Year. All official rate-setting procedures were suspended and all roads were authorized "to meet any cut rates." [55] The last, and killing, act of nonadherence involved a 5-cent discount in westbound rates: "A consequence of contracts by the Lackawanna road, difficult to explain as a deliberate act of sane men." [56]

The cause of the demoralization does not seem to have been "profits from cheating" for any of the railroads taking part. The accused Grand Trunk Railway gained very little additional grain tonnage out of Chicago during the disruptions. The line received slightly more than its assigned share during April, but in the preceding and following months it received less than this share.[57] The rate reductions of the other transporters were sufficiently rapid to prevent any tonnage gain by the Grand Trunk. In fact, the reactions were so rapid that the Pennsylvania received much larger shares of the Chicago tonnage in March and April, and the New York Central was over its regular share in most of the summer months. There was obviously no profit gain for the Grand Trunk: "The eight railroads out of Chicago apparently carried (the grain tonnage) at no profit . . . The 25-cent rate which the railroads tried to get last winter leaves but a narrow margin of profit; the 20-cent rate which they are trying to get now probably only half as much; while the cuts to 17½ and 15 cents which were

[53] "The Trunkline War," *op. cit.*

[54] "Irregularity in Eastbound Rates," the *Railroad Gazette* (January 9, 1885), 26.

[55] "The Freight Pool," the *Railroad Gazette*, *ibid.* This authorization was limited to rate-matching, a technique guaranteed to promote instability (as noted in Chapter 2).

[56] "Reduction in Westbound Rates," the *Railroad Gazette* (January 23, 1885), 56.

[57] The road received 13.4 per cent of total grain tonnage in April when it should have received 11 per cent, according to the apportionment of July 1, 1883. In February and March, however, it received no more than 6 per cent; in May and June it received no more than 7 per cent (as shown in the Statistical Appendix to Chapter 4, Table A.1).

reported at Chicago (in April) must have swept away nearly or quite all the profits that remained." [58]

Whatever the cause, the rate demoralization had a marked effect upon prices in Chicago and New York grain markets. The margin between Chicago spot prices for grain and those in New York during the winter season of 1884–1885 averaged 25.9 cents per 100 pounds, with transport costs of 22 to 24 cents per 100 pounds between the two cities; this was 6 cents more than during the Buffalo–New York rate war of the previous winter, but still 3 cents less than in the "pool" winter of 1882–1883. The margins between these spot prices was more in accord with the Board of Trade rates than with official rates (as shown in Table 4.4, the variance in grain price differences explained by the two rates was approximately the same; but $\hat{\beta}$ for the Board of Trade rate equation was much closer to 1.00 than was $\hat{\beta}$ for the official rate equation). Chicago–New York price differences decreased, during the summer season of 1885 following the breakdown, to 15.5 cents per 100 pounds — the lowest level since the beginning of the new organization in March of 1879. Week-to-week variance in grain price differences was explained more by changes in Board of Trade rates, than by changes in official rates (as shown by the higher values of $\hat{\rho}^2$ for the Board of Trade rate equations in Table 4.4). That is, price spreads declined with the lower rates consequent from cartel breakdown.

Last Attempts to Set and Maintain Rates, 1885–1886

There were some hopes for reestablishing the Committee in the fall or winter of 1885. One of the nonadherent railroads, the West Shore, was removed by merger with the New York Central system on January 1, 1886.[59] There were confident predictions of good results from reorganization of the techniques for rate-setting as well (the new arrangements placed responsibility for rates upon a commissioner for freight business and a second commissioner for passenger traffic, so that changes were to be made more quickly). Most important, the Grand Trunk responded favorably to invitations to join the meetings on reformation.

It was proposed that the railroads allow the commissioners to set new shares of the tonnage and passenger traffic. The commissioners decreed

[58] "Chicago Shipments Eastward," the *Railroad Gazette* (April 24, 1885), 264. The constancy of this railroad's market shares, and the greatly reduced rates, make calculation of "profits from cheating" unnecessary; for $(R_0 - K_2)Q_0 > (R_c - K_2)Q_c$ if $Q_0 \sim Q_c$ and $R_0 > R_c$ (where $R_c \sim R_T$ before rate matching and $R_c \sim R_0^*$ after rate-matching).

[59] Cf. Henry V. Poor, *Manual of Railroads of the United States for 1887* (H. V. and and H. W. Poor: 70 Wall Street, New York, 1888), p. 203.

that the Chicago and Grand Trunk, as the returning prodigal, was to receive a share 3.8 per cent greater than in 1883, the Pennsylvania was to receive 4 per cent more, and the Baltimore and Ohio 1 per cent more. The Erie was given 1 per cent less than its allotment in the 1879–1880 agreement, and the New York Central system was delegated 8 per cent less than its previous assigned share.[60]

The proposal, and the new shares, were accepted by all of the railroads except the Erie. The agreement was put into effect anyway (and the Erie remained outside to form an express company with the Lackawanna to take as much tonnage as it could get).[61]

The rates were set on the basis of 25 cents per 100 pounds for grain, and 30 cents per 100 pounds for provisions, early in 1886. The 25-cent grain rate was cut to 22 cents during two weeks of March, and to 20 to 23 cents during two weeks in April (as indicated by offers on the Chicago Board of Trade). In the two months following the opening of navigation, rates were close to 22 and 25 cents in almost alternating weeks on the Board of Trade, and there were similar periods during September, October, and November. There were only two weeks during the winter of 1886–1887 in which the official grain rate was maintained. Similar discounting took place on provision rates; the average Board of Trade lard rate was less than the official rate each month of the summer season except November (as shown in the Statistical Appendix to Chapter 4, Table A.3).[62] A review of both grain and provision rates for the year 1886 showed that "while the rates of the trunklines and their immediate connections were much better than in 1885, the rates were not really well maintained at any time or for very little of the time. We do not remember any year in the history of the trunkline associations when irregularity seemed so common without resulting in an open break." [63]

The Erie seems to have increased its profits by greatly increasing its tonnage during this rate-cutting. The railroad received 10 to 11 per cent

[60] Cf. "Percentages for the Division of Eastbound Dead Freight from Chicago," File 50114, the Pennsylvania Railroad Archives, Philadelphia, Pennsylvania.

[61] "The Express War," the *Railroad Gazette* (November 26, 1886), 816.

[62] The effect was, as before, reduced transport costs that benefited the shipper. Average margins between Chicago and New York grain prices were less than average official rates; fluctuations in grain prices were in accord with changes in Board of Trade rates, as shown by the regression equations in Table 4.4 for lags in adjustment to Board of Trade rates for the three seasons beginning with the winter season of 1885–1886. The margin between Chicago and New York lard prices also remained considerably below official rates, after discounts were announced on the Chicago Board of Trade in April of 1886. With lower transport costs, shippers in competition on the Chicago and New York exchanges reduced margins between both grain and provision prices in the two cities.

[63] "A Summary of the Year," the *Railroad Gazette* (December 31, 1886), 904.

more of the total than its long-term trend of shares from Chicago in the period from September to the middle of November of 1886 (as shown in the Statistical Appendix to Chapter 4, Table A.2). The lines' profits on such increased tonnage at the cut (Board of Trade) rates were approximately \$98,000 in September, rather than \$59,000 on its "accustomed" share at official rates. In October profits were \$108,000 from the larger share at cut rates, and profits on the trend share of tonnage at official rates were approximately \$69,000 (as shown in Table 4.5). Computed profits at

TABLE 4.5

Profits from Independent Rate-Setting by the Erie Railroad, September to November, 1886

Month	*Year*	*Computed Profits*	*"Trend" Share of Profits at Official Rates*
September	1886	97,766	59,446
October	1886	107,859	68,558
November	1886	37,338	39,657
		\$242,963	\$167,661

Source: "Computed Profits" equal $(R_T - K_2)q$, with R_T the Board of Trade rate, K_2 the average variable costs of 7.9 cents per hundred pounds for shipping from Chicago to New York City (as computed from the sources noted in Chapter 3), q the hundreds of pounds shipped from Chicago by this railroad, as shown in the *Annual Reports* of the Chicago Board of Trade and the weekly summaries of the *Chicago Daily Commercial Bulletin*.

The "Trend Share of Profits" is an estimate of the net revenues from the official rate and per cent share of tonnage shown from the computed trend $\{S = \alpha + \beta T + \gamma X\}$ of Table 4.2. The "trend" profit is $(R_0 - AVC)(S \cdot Q)$, with R_0 the official rate of 25 cents, and Q the total shipments from Chicago each month.

cut rates in November were less than those possible from the trend share that month at official rates — the rate reduction was greater than the 1.2 per cent increase in tonnage share — but the profits from cheating for the entire three-month period seem to have been more than 50 per cent greater than those that would have followed from accustomed tonnage shares at official rates.[64]

This independence of the Erie, and the existence of increased profits from such independence, would seem to have rendered impossible the effective operation of the trunk-line pool. The Executive Committee had been able in 1886 to restore rates from 15 to 20 cents per 100 pounds to

[64] Since there was "harmony" in December — at least to the point that it was possible to agree that official rates were to be increased to 30 cents — the gains and losses from rate-cutting seem to have been limited to three months.

approximately 24 cents per 100 pounds. The restoration was extensive, if not complete; the 25-cent official rate was not maintained without discounts of 2 or 3 cents, as has been seen, and increasing the official rate to 30 cents in the winter only increased the size of the discount to 6 cents. But by the end of the year the trunk-line cartel was evidently at the same point as its predecessor during the winter of 1878–1879 — a point at which it was more profitable for any railroad to offer discounts on official rates, so that the agreement was unstable.[65] The trunk-line association had not been able, in the years since 1879, to find means for maintaining a schedule of official rates.[66]

The Resulting Structure of Through and Local Rates

The many attempts to set long-distance rates during 1884, 1885, and 1886 had been largely unsuccessful. Those rates were in fact lower than during earlier periods of cartel stability. Then, with constancy in local rates set by individual railroads in exclusive or semiexclusive territories there was continued, and increased, discrimination against local shippers.

The railroads had managed to bring official and Board of Trade rates from (approximately) 21 to 22 cents per 100 pounds during the demoralization of 1884–1885, to 24 cents per 100 pounds during 1886 (as reviewed in Table 4.6). This was some improvement for the cartel but no cause for satisfaction since it was 5 cents less than during the 1882–1883 "pooling" years, and 5 to 9 cents below 1880.[67]

These through rates were in contrast with both the level and the stability of announced local rates. In 1880 the Pennsylvania Railroad posted a rate of 18 cents per 100 pounds for fourth-class transport between Philadelphia and points as far as Erie, and these held through the decade (as shown in Table 4.7). This railroad also charged a rate of 8 cents per 100 pounds for shipping grain from Cessna to flour mills in Cumberland, Maryland, from

[65] The agreement was "unstable" in this sense: there was present a profit from nonadherence for each road, so that there was an inherent incentive to cheat. This is assumed to have increased the probability of cheating (over that from cheating out of a mistaken estimate of gains from such, or out of caprice or whim) — as explained in detail in Chapter 2.

[66] The effect of cartel instability on the profits of the railroads is discussed in the concluding sections of the following chapter, by means of comparison of profits with those in the later 1880's and 1890's.

[67] Differences between grain and provision rates had been narrowed, as well: rates on higher priced provisions 10 to 15 cents greater than on grain were discontinued after April of 1885, in favor of a new twelfth-class rate on all provisions equal to or 5 cents higher than the grain rate. Cf. the Interstate Commerce Commission, *Railways in the United States in 1902*, *op. cit.*, Table XLIII, p. 78.

TABLE 4.6

Official Rates, Board of Trade Rates, and Grain Price Differences, 1880 to 1886

Season and Year	Average Official Rate R_0, (Cents per 100 Pounds)	Average Board of Trade Rate, R_T (Cents per 100 Pounds)	Average Grain Price Difference, P_g (Cents per 100 Pounds)	Average Lake-Rail Rate, R_L (Cents per 100 Pounds)
Summer 1880	30.00	30.00	31.67	26.81
Winter 1880–1881	33.80	33.62	37.28	
Summer 1881*	19.10	17.14	17.48	17.07
Winter 1881–1882*	19.63	19.16	19.96	
Summer 1882	25.00	25.00	23.22	18.34
Winter 1882–1883	30.00	30.00	28.06	
Summer 1883	25.30	25.30	20.60	19.64
Winter 1883–1884*	24.86	24.14	19.30	
Summer 1884*	21.89	21.25	18.68	16.82
Winter 1884–1885*	23.61	22.16	25.93	
Summer 1885*	20.52	17.14	15.51	15.24
Winter 1885–1886*	25.00	24.23	22.80	
Summer 1886*	25.00	24.82	17.79	20.00
Winter 1886–1887*	28.33	24.21	27.45	

* Periods in which there were reports of nonadherence to the rates sent by the Trunk-line Executive Committee or the Freight Department of the Central Traffic Association.

February, 1880, to March, 1885, with some discriminations against particular shippers (one shipment was made at 9 cents, one at 12 cents, but one also at 5 cents per 100 pounds as shown in the *Manifest Book* of all shipments from Cessna, Pennsylvania, in Table 4.8). The same railroad increased the rate for shipping grain from Cessna to Bedford, Pennsylvania, in the period November, 1880–September, 1885, from 2.3 cents to

TABLE 4.7

Long- and Short-Distance Local Rates on the Pennsylvania Railroad, 1886

Service	Distance (Miles)	Fourth-Class Rate (Cents per 100 Pounds)
Philadelphia to Altoona, Pa.	230	18
Philadelphia to Pittsburgh, Pa.	325	18
Philadelphia to Williamsport, Pa.	210	18
Philadelphia to Erie, Pa.	415	18

Source: U.S. Interstate Commerce Commission, *Railways in the United States in 1902:* "Part II: A Forty-Year Review of Changes in Freight Tariffs" (Washington, 1903), p. 168.

TABLE 4.8

Local Rates for Transporting Grain from Cessna, Pennsylvania, on the Pennsylvania Railroad

Date: Initial Date and Final Date of Shipment at the Recorded Rate	*Destination*	*Number of Shipments*	*Average Number of Pounds in a Shipment*	*Recorded Rate (Cents per 100 Pounds)*	*"Equivalent" Through Rate, for the Same Tonnage and Distance (Cents per 100 Pounds)*
February–July, 1880	Cumberland, Md.	5	24,120	8.0	1.4
January–January, 1884	Cumberland, Md.	1	36,000	9.0	1.1
January–January, 1884	Cumberland, Md.	1	40,000	8.0	1.1
February–December, 1884	Cumberland, Md.	15	34,400	8.0	1.0
December–December, 1884	Cumberland, Md.	1	30,000	5.0	1.0
January–January, 1885	Cumberland, Md.	1	24,000	12.0	1.0
February–March, 1885	Cumberland, Md.	5	31,452	8.0	1.0
November, 1880–July, 1883	Bedford, Pa.	32	22,645	2.3	0.14
March, 1884–April, 1886	Bedford, Pa.	7	28,806	4.0	0.14
September–September, 1885	Bedford, Pa.	1	28,080	5.0	0.13
March–March, 1880	Baltimore, Md.	3	28,000	18.5	5.7
December, 1880–April, 1881	Baltimore, Md.	12	47,556	14.5	6.5
January–February, 1883	Baltimore, Md.	2	26,670	13.0	6.5
August–October, 1883	Baltimore, Md.	7	30,706	12.0	4.8
December–December, 1883	Baltimore, Md.	1	28,000	13.0	4.7
November–December, 1883	Baltimore, Md.	7	33,408	15.0	4.7
January–January, 1884	Baltimore, Md.	3	46,990	15.0	4.7
January–January, 1884	Baltimore, Md.	1	42,420	10.0	4.7
June–June, 1884	Baltimore, Md.	3	39,943	7.5	4.2

Source: The *Manifest Book* of all shipments from the station at Cessna, Pennsylvania (the Pennsylvania Railroad Archives, Philadelphia, Pennsylvania). The "equivalent rate" is obtained by dividing the average official rate for the same date by the ratio of the Chicago–New York distance to the shorter relevant distance.

4.0 cents, and to 5.0 cents per 100 pounds (as shown in Table 4.8). All of the local rates were higher than the through rates for the same distance; all remained at 1880 levels, while through rates declined sharply during the periods of cartel breakdown.[68]

[68] A particular instance indicating the contrast between exclusive "local" rates and those set by breakdowns in agreements with other lines is found in the rate from Cessna to Bedford as compared to the rate from Cessna to Baltimore, Maryland. Grain producers in the Cessna region, to have shipped at all, had to be able to cover production costs and transport costs into Baltimore with the Baltimore spot grain price. As the Baltimore price declined with reductions in the rates from Chicago, the Cessna-Baltimore rate also had to decline. This rate was recorded at 18.5 cents per 100 pounds in March, 1880, and 14.5 cents per 100 pounds during the winter of 1880–1881. There were further declines during 1883, and large declines in January and June, 1884. The final recorded rate was 7.5 cents per 100 pounds. All of these rates declined while Chicago rates were declining, but all were more than twice the level of Chicago rates. The Chicago-Bedford rate did not decline, and was at more than 20 times the level of Chicago rates (for comparable distance, if not tonnage).

It was apparent that short-distance rates were relatively high and were relatively rigid. The contrast raised questions as to whether "something should be done." From the view of the railroads, local rates were closer to the ideal level than through rates because of imperfections in the cartel organization controlling the long-distance transport. The through rates were a testimony to cartel mismanagement: "Since the first railroad pool was made somewhere about 1875, the prices for the transportation of freight and passengers have been reduced nearly 50 per cent, and as regulators of rates they have recorded a series of unbroken failures." [69] The structure of rates could have been improved by an effective agreement raising, and stabilizing, the long-distance rates. From the view of local and through shippers, the lowest level of through rates on grain was the standard for reasonable rates (accounting for tonnage and distance). Rate differences per ton per mile should have been reduced by reducing local rates. It is with the contrasts in rates, and in standards for the structure of rates, that regulation had to deal.

[69] A comment in the *Chicago Times* of December 22, 1886, as noted in the *Railroad Gazette* (December 24, 1886), 895.

5

Trunk-Line Railroad Rates and the Interstate Commerce Commission, 1887 to 1899

State and national legislative committees, from the Hepburn Committee in the New York State legislature of 1879 to the Senate Select Committee on Interstate Commerce in 1886, investigated the entire range of effects from individual and group rate-setting. The Senate considered such statements as (by an Indiana wholesale grocer receiving goods from the East Coast): "When one shipper through any kind of favoritism secures rebates, he has an unjust advantage over his neighbor resulting very often in the demoralization of trade and resulting in no benefit to the recipient."[1] The effects from rate wars led the Senate Committee of 1886 to conclude that "the secret cutting of rates and the sudden fluctuations that constantly take place are demoralizing to all business except that of a purely speculative character . . . and frequently occasion great injustice and heavy losses."[2] Given that "the great desideratum is to secure equality, so far as practicable, in the facilities for transportation afforded and the rates charged by the instrumentalities of commerce. . . . The burden of complaint is against unfair differences in these particulars as between different places, persons and commodities and its essence is that [present] differences are unjust."[3] The cause for complaint was rate discrimination. The remedy in new legislation was the abolition of secret rates for selected individual shippers and also the removal of "unjust," but publicly acknowledged, differences between rates to various locations.

The remedy would seem to have been realized in trunk-line transport during the late 1880's and early 1890's. There was an increase in the level

[1] Statement of John W. Grubbs, *Report of the Senate Select Committee on Interstate Commerce*, 49th Congress, 1st Session, Report 46, Part 1 (Washington: Government Printing Office, 1886), Appendix, pp. 172–173. It would not seem possible to determine the validity of this statement — whether Mr. Grubbs or other shippers suffered from an unjust disadvantage or an unjust advantage. But, valid or otherwise, the investigators allotted extensive time and space for this allegation in hearings on legislation.

[2] *Report of the Senate Select Committee on Interstate Commerce*, *op. cit.*, p. 181.

[3] *Ibid.*, p. 182.

of through rates, and a decrease in the differences between through and local rates, which conformed to regulations against discrimination. New cartel rate-setting techniques that conformed to the new regulation were successful in maintaining the agreement, and can be credited with increasing the long-distance rates. Subsequently, however, for reasons to do with cartel organization and with ineffective regulation, rates, profits, and grain price margins declined in a general cartel breakdown.

The Act to Regulate Commerce of 1887

The Senate and House of Representatives of the United States Congress enacted the *Act to Regulate Commerce* on February 4, 1887.[4] A great part of the twenty-four sections was concerned with the creation of an Interstate Commerce Commission and its attendant duties (Sections 11 to 21 and 23–24), and with specifying applicability to individuals or corporations (Sections 1, 8, 9, 10, and 22). The remaining sections dealt with what constituted unlawful acts by the railroads. Sections 2, 3, and 4 specified unlawful differences between rates. Sections 5, 6, and 7 outlined the nature of lawful rate-setting procedures — or legal "structural conditions" in the relations between the trunk-line railroads, as contrasted with "rate characteristics" dealt with in Sections 2, 3, and 4.

The rates charged by individual railroads were unlawful if they were unjustly discriminatory. Section 2 stated that "if any common carrier subject to the provisions of this Act shall . . . receive from any person a greater or less compensation for any service rendered . . . than it charges . . . any other person for doing a like and contemporaneous service in the transportation of a like kind of traffic under substantially similar circumstances and conditions, such common carrier shall be deemed guilty of unjust discrimination. . . ."[5] This prohibition of different rates for shippers seeking identical service was extended to shippers seeking comparable service: Section 3 states "that it shall be unlawful for any common carrier subject to the provisions of this Act to make or give any undue or unreasonable preference or advantage to any particular person, company, firm, corporation or locality or any particular description of traffic in any respect whatsoever. . . ."[6] Section 4 dealt with the most extreme form of discrimination in the structure of rates; here it was stated "that it shall be unlawful for any common carrier subject to the provisions of this Act to charge or receive any greater compensation in the aggregate for transpor-

[4] Cf. 24 Stat. 379 or the *Second Annual Report of the Interstate Commerce Commission* (1888), Appendix A, pp. 73 to 79.

[5] *An Act to Regulate Commerce*, *op. cit.*, Section 2.

[6] *Ibid.*, Section 3.

tation . . . under substantially similar circumstances and conditions for a shorter than for a longer distance over the same line in the same direction, the shorter being included within the longer distance. . . ." [7]

The "structural conditions" prohibited members of a rate-setting association from carrying on pooling activities. It was stated that "it shall be unlawful for any common carrier subject to the provisions of this Act to enter into any contract agreement or combination with any other common carrier . . . for the pooling of freights of different and competing railroads, or to divide between them the aggregate or net proceeds of the earnings of such railroads. . . ." [8] The members were denied this particular form of cooperation, but they were permitted to organize the rate-setting process. In fact, Section 7 of the Act specifically considered a lack of interrailroad cooperation in through transport as unlawful; it was stated that no "common carrier . . . [could] prevent . . . the carriage of freights from being continuous from the place of shipment to the place of destination." [9]

The legislation was concerned with making certain that there were no secret rates. Each railroad "subject to the provisions of this Act" was required to "bring and keep for public inspection schedules showing the rates and fares and charges for the transportation of passengers and property . . . which are in force at the time. . . ." [10] Changes in rates could be made only after public notice[11]; an increase in rates required ten days' notice, while an immediate reduction in rates was possible "if the same shall immediately be publicly posted and the changes made shall immediately be made public by printing new schedules. . . ." [12]

[7] *Ibid.* Section 4. There was an exception: "that upon application to the Commission appointed under the provisions of this Act such common carrier may, in special cases, . . . be authorized to charge less for the longer than for the shorter distance for the transportation of passengers or property; and the Commission may from time to time prescribe the extent to which such designated common carrier may be relieved from the operation of this Section of this Act."

[8] *Ibid.*, Section 5. Pooling was generally considered to have been the actual transfer of tonnage or profits so that realized shares for each railroad were made equal to pre-assigned shares.

[9] *Ibid.*, Section 7.

[10] *Ibid.*, Section 6. This was required of both local transport service and (for the trunk-line railroads) through service for export; that is, publication of rates extended to "schedules showing the through rates established and charged by [the] common carrier to all points in the United States beyond the foreign country to which it accepts freight for shipment." (*Ibid.*, Section 7.) Published rates were required for shipments through a foreign country for final receipt in the United States — and if this were not conformed to, the shipments were to be subject to "custom duties as if said freight were of foreign production."

[11] *Ibid.*, Section 6.

[12] *Ibid.*

The administrative agency established to enforce these "rate" and "structural" conditions, the Interstate Commerce Commission, was to consist of five commissioners appointed by the President of the United States. It was to investigate the practices of the railroads in order to detect violations, and was to obtain relief from violations by "cease and desist orders" and by appeal of these orders to the United States Circuit Courts.

The Interstate Commerce Commission's Standards for "Just and Reasonable" Rates

The Commission sought to provide rules by which to determine the legality of rate conditions. The first outline of standards was presented in the agency's first report, in December of 1887 (nine months subsequent to organization). It left the strong impression at once that rate discrimination, within prescribed limits, was legal.

Discrimination in railroad-determined rates was not only possible but favored: "The public interest is best served when the rates are so apportioned as to encourage the largest practicable exchange of products . . . and this can only be done by making value an important consideration and by placing upon the higher classes of freight some share of the burden. . . ."[13] This was to place a stamp of approval on the existing structure of class rates — of higher rates for higher class (of commodities with higher final sales prices).

The Interstate Commerce Commission set narrow limits on the differences between rates on any commodity transported to different locations, however. Rates for transport on different railroads in the same region should have been approximately the same for any given distance: "If the circumstances and conditions under which the traffic is carried by the two roads are substantially the same, the comparison [of rates] would be legitimate and the argument from it of very great force."[14] Applying this pronouncement to the trunk-line railroads, the Commission required the reduction of the difference per mile between long-distance and short-distance rates when "circumstances were the same" — and circumstances were clearly the same unless there were observable differences in the costs of transport.[15]

The *Act* provided the basis for Commission prohibition of short-haul

[13] *The First Annual Report of the Interstate Commerce Commission, December 1, 1887* (Washington: Government Printing Office, 1887), p. 36.

[14] *Ibid.*, p. 39.

[15] As noted in *The First Annual Report of the Interstate Commerce Commission, op. cit.*, p. 39. Rates might have been allowed to differ only by amounts equal to the cost differences as well. This is not specifically stated, however.

rates greater than those for longer distance transport on the same railroad in the same direction, if "under substantially similar circumstances and conditions." [16] The chairman of the Commission attempted to define "similarity of circumstances" in some detail in June of 1887, in the matter of a petition of the Louisville & Nashville Railroad Company for relief from this (fourth) section of *The Act to Regulate Commerce*. This railroad sought relief on the grounds that circumstances for through transport were dissimilar from those for local traffic; Chairman T. M. Cooley stated that application for an exception to the *Act* was not the means for obtaining a Commission decision that the *Act* did not apply. Providing relief on grounds of dissimilar circumstances "in those sections of the country in which the reasons or supposed reasons for exceptional rates are most prevalent, the Commission would in effect be required to act as rate-makers for all the roads, and compelled to adjust the tariffs so as to meet the exigencies of business, while at the same time endeavoring to protect relative rights . . . of rival carriers in rival localities. This in any considerable state would be an enormous task. In a country so large as ours and with so vast a mileage of roads it would be superhuman . . . [a] fact [that] tends strongly to show that such a construction could not have been intended." [17] Rather than seeking a declaration that rates were exceptions, the railroad was to set short-haul rates higher than through rates "on the judgment of its managers . . . at the peril of the consequences." [18] If the resulting rates were for transport in dissimilar circumstances, then there were no adverse consequences because the prohibitions of the *Act* did not apply. The Commission only could reduce the peril of a violation of Section 4 by specifying with some exactness the nature of "dissimilar circumstances."

The Commission stated that a dissimilar circumstance existed when the cost of short-haul transport was considerably greater than that for long-haul service; but that the transporter had to bear the burden of showing that the additional cost of short-haul "would support the greater charge." That is, "to make out the . . . case in which the general rule of the statute may be disregarded on the grounds that the circumstances and conditions are not substantially similar, the difference in cost should itself be exceptional and be capable of proof amounting to practical demonstration." [19]

Differences in the extent of competition made for different "circum-

[16] As stated in Section 4 of *An Act to Regulate Commerce*, *op. cit.*

[17] *In Re Louisville & Nashville Railroad Company*, *Interstate Commerce Commission Reports 1*, 31 (at p. 56). The Interstate Commerce Commission Reports are referred to as *ICC Reports* hereafter.

[18] *Ibid.*, p. 61.

[19] *Ibid.*, p. 64.

stances and conditions" only in one instance. The general rule was stated: "Every railroad ought . . . to so arrange its tariffs that the burden upon freights shall be proportional on all portions of its line . . . ," [20] but, "when there is water competition at leading points, . . . it may be impossible to make some portion of the traffic pay its equal proportion of the whole cost . . . if it can then be made to pay anything toward the cost the railroad ought not in general to be forced to reject it. . . ." [21] Competition among railroads resulting in a reduction of only the through rate generally was not a "dissimilar circumstance" since competition had to begin with one railroad cutting the legal rate. The agency's conclusion was that "the competition with each other of the railroads . . . subject to the federal law can seldom . . . make out a case of dissimilar circumstances and conditions within the meaning of the statute, because it must be seldom that it would be reasonable for their competition at points of contact to be pressed to an extent that would create the disparity of rates on their lines which the statute seeks to prevent." [22]

The Commission's guidelines established that rates on any commodity transported some given distance by each of a group of railroads should have been roughly the same. The rates for that commodity for different distances should have been in accord with differences in the costs of transport; at the least they should not have been so disparate for different distances as to have resulted in higher short-haul charges.

The Interstate Commerce Commission's First Attempts at Enforcement

Enforcement of the rules involved the trunk-line railroads, among others, in hearings before the Commission. In particular, these railroads were

[20] *Ibid.*, p. 79.

[21] *Ibid.*

[22] *Ibid.*, p. 81. The Commission did mention two possible instances of "extraordinary" railroad competition. First, there was a statement that competition of Canadian roads with American roads created dissimilar circumstances. That is, "whenever such roads compete with roads in the United States for business between one point of our country and another, the state of circumstances exists as to such business which justifies American roads in meeting such competition by a corresponding reduction of rates." (*Ibid.*, at p. 80.) The second exception existed in the instance of a particular railroad having been forced to discontinue long-distance service as a result of conforming with the requirements of Section 4. The choice, in such an instance, was between a drastic reduction in local rates (in order to continue carrying through traffic at existing through rates) and no reduction in local rates (with no through traffic). This second instance was a possible, but not established, exception: "whether this position is sound the Commission may determine hereafter . . . it is sufficient to say of the case at this time that it is one in which a strict application of the general rule laid down by the statute must be fatal to competition . . . ." (*Ibid.*, at p. 81.)

involved in a case in 1887 of some importance, since it foreclosed two means of evasion. The northernmost railroad out of New England, the Grand Trunk, was accused by the New York Central of having set lower rates on through traffic from Boston to Detroit than on shipments from Boston to points in Vermont, to the detriment of society and the loss of New York Central revenues. The Interstate Commerce Commission, in *Boston and Albany Railroad Company* vs. *Boston and Lowell Railroad Company*,[23] considered first the accused railroad's claims that the through rates to Buffalo were joint tariffs of a number of connecting roads, and thus were not subject to the *Act to Regulate Commerce* (since they were not tariffs of one line "under a common control" as required by Section 4).[24] The Commission denied this claim: "any common carrier is as much restrained when it unites with one or more others in making the long-haul charge as when it makes such a charge independently." [25] Second, an attempt was made to justify lower long-distance rates on grounds that cost differences made circumstances dissimilar to those for short-distance traffic, and that competition made trunk-line through traffic different from local traffic. Both defenses were denied. Cost differences "fairly justify a difference in rates . . . [but] the calculations put in evidence do not satisfy us that the same kind of freight can be taken from Boston through St. Albans to Detroit at a less cost than from Boston to St. Albans." [26] The Commission did not deny that through traffic involved greater rail competition; in fact, "the evidence is entirely conclusive that the competition which is troublesome to the defendants is that of the trunk-lines" [27] but such competition did not result in "dissimilar conditions." That competition results in dissimilarity "is a claim which could be advanced wherever a route, however circuitous, could be formed for through traffic. . . . we content ourselves with saying that such peculiar facts are not found to exist as will justify the greater charge over the shorter [distance]." [28] The Commission statement, as a precedent, established the right of review over joint tariffs and over through and local rates of any railroad.[29] With respect to the

[23] *ICC Reports, I* (1887), 158.

[24] *Ibid.*, at p. 175.

[25] *Ibid.*, at p. 176.

[26] *Ibid.*, at p. 180. It would seem to have been necessary to demonstrate that total costs of long-distance transport were less or that the marginal transport costs of larger volume and longer distance were negative.

[27] *Ibid.*, at p. 182.

[28] *Ibid.*, at p. 183.

[29] In the case of *E. B. Raymond* vs. the *Chicago, Milwaukee & St. Paul Railway Company* of November, 1887, the Commission reviewed two grain rates, one on transport North of Chicago that was 5 cents higher than one on transport of the same commodity North and West of that city. Previous to the passage of the *Act to Regulate Commerce*, the first rate had been only 2 cents greater than the second; the Commission, noting

trunk-line railroads, this must have made apparent the Commission's right to compare Chicago–New York rates with those for shorter distance transport.

The Commission required a single rate for all grain from Chicago to another (seaboard) city. The standards as applied to Chicago–Boston traffic were outlined in the case of *The Boston Chamber of Commerce* vs. *The Lake Shore & Michigan Southern Railway Company*, *et al.*, of February 15, 1888. There had been a complaint that eastbound rates on grain for export were 5 cents less than rates on grain for local consumption, and that this was an unjust discrimination. This was to be remedied, according to the petition of the Boston Chamber of Commerce, by lowering the rate on domestic grain to the level of the export rate (and incidentally to the level of the Chicago to New York rate on domestic grain).[30] The Interstate Commerce Commission stated that "[each railroad] must treat all alike that are situated alike . . ." [31] but that to require the domestic rate to be reduced was arbitrary.[32]

The question remained as to whether the Boston railroads were required to increase the export rate. The Commission's answer, in the case of the *New York Produce Exchange* vs. *The New York Central & Hudson River Railroad Company* of June, 1889, was that a similar New York export rate had to be raised to the level of the rate on domestic grain: "The rate at seaboard should be specified and should not discriminate against the inland tariff rate unless justifiable conditions exist for a difference. It is not shown that such conditions exist and very clearly they do not exist." [33] The differ-

---

the departure from established differences between the two rates, said that the increase might "divert a large part of the grain trade as to subject Mazeppa to unreasonable disadvantage and give undue preference to Red Wing and Lake City" so that a re-adjustment had to be made to prevent "divert[ing] trade and business to one locality." Cf. *E. B. Raymond* vs. the *Chicago, Milwaukee & St. Paul Railway Company*, *ICC Reports*, *I* (1887), 231.

[30] Cf. *The Boston Chamber of Commerce* vs. *The Lake Shore and Michigan Southern Railway Company*, *et al.*, *ICC Reports*, *I* (at p. 451).

[31] *Ibid.*, p. 452; there was a further requirement, citing the case of *The Board of Trade Union of Farmington et al.*, vs. *The Chicago, Milwaukee and St. Paul Railroad Company*, "that rates should be relatively reasonable when the same carrier transports over different branches of its road to a common market. . . ." *ICC Reports*, *I*, 215, cited at p. 452.

[32] Moreover, it would result in the reduction of the difference between Chicago–Boston and Chicago–New York rates — a difference that should not be disturbed since it is "the result of many years of contention and struggle, in which concessions were necessary to arrive at an adjustment, finally culminating in the creation of a board or tribunal in which all the lines were represented for the settlement of disputes and the maintenance of peace and stability." (*Ibid.*, p. 460.)

[33] *The New York Produce Exchange* vs. *The New York Central and Hudson River River Railroad Company*, *et al.*, *ICC Reports*, *III*, 137 (at p. 184).

ence was unjustifiable because it was a departure from the established structure of rates: "a standard rate . . . uniform both for domestic and foreign consignments, had existed, nominally at least, for more than eleven years and was the agreed result of experience and severe contention as the only practical way by which all the interests involved would be adequately protected." [34]

The two "export rate" decisions provided substantial support for continuing the 1886 structure of official rates on grain to the eastern seaboard. Rates on domestic or export grain from Chicago to any one seaboard city had to be the same; rates on domestic grain had to be in accord with traditional differences for shipment to Boston as compared to New York City.

The railroads were given notice, as well, that discounts on official rates were illegal in most instances because the resulting rate conditions were unduly discriminatory. There was a Commission view against rate wars: "it may be truly said that, while railroad competition is to be protected, wars in railroad rates unrestrained by competitive principles are disturbers in every direction; if the community reaps a temporary advantage it is one whose benefits are unequally distributed. . . ." [35] The agency was prepared to invoke the short-haul clause against those transporters causing breakdowns of official rates: "The Fourth Section of the *Act* has also important possibilities as a restraint upon reckless rate wars. The reduction when such wars are in progress has generally been made chiefly at competitive points a considerable distance apart; and when a reduction of rates at such points involves also the reduction to or from a great number of intermediate points, a resort to a cutting of rates that goes beyond the warrant of legitimate competition becomes unlikely in proportion as it would be injurious to the party inaugurating it." [36] That is, the railroad cutting the official rate on through traffic (either by setting a secret — and thus illegal — lower rate, or an announced lower rate) had to reduce rates on local traffic as well to conform to Section Four of the *Act;* this requirement makes "a rate war, under the present law, a much more serious matter than formerly. . . . Rates between terminals cannot now be lawfully reduced without at the same time requiring large reductions at inter-

[34] *Ibid.*, p. 174. Also, there were allegations that the lower export rate undercut shippers forwarding grain under the domestic rate since those paying export rates secretly sold some of their grain in New York for domestic consumption. The Commission concluded that "substantially the charge of the complaint in respect to discrimination is sustained by the evidence and it was not justified by the circumstances and conditions shown to exist. The discrimination was actual; it was unjust, therefore unlawful." (*Ibid.*, p. 184.)

[35] Cf. *The First Annual Report of the Interstate Commerce Commission, December 1, 1887* (p. 37).

[36] *Ibid.*, p. 35.

mediate points affecting purely local traffic. . . . Reductions often affect many other points than those at first in contemplation, and rates on many other commodities are drawn into the current . . ." [37] with consequent losses for the cheater.

The Interstate Commerce Commission, in its initial decisions, provided the mandate for a single rate on grain to one seaboard city, and for allegiance to established differences in rates on transport from Chicago to different eastern cities. The Commission made a strong appeal for roughly equal rates per 100 pounds per mile. At the least, "a departure from the rule of equal mileage rates as applied to the several branches of a road is not conclusive that such rates are not lawful, but the burden is on the company making such departure to show its rates to be reasonable when disputed." [38] The trunk-line railroads had to set rates on any particular commodity, and the differences between class rates, within the framework of this regulation.

The Trunk Lines' Initial Adjustments to Interstate Commerce Commission Regulation

From the beginning of the hearings until the bill was passed, the Chicago and New York newspapers recorded the daily activities of Congress in the passage of the *Act to Regulate Commerce*. The trunk-line railroads were informed that they were to be faced with regulation in the spring of 1887. The rate-setting associations and the individual railroads undertook adjustments in the rate-setting process and in the structure of rates to comply with the new law and new agency.

The members of the Trunkline Executive Committee met in New York on March 4, 1887, to discuss the required changes. It was realized at the beginning that "the most important question to be considered . . . will be in regard to the action necessary to keep the roads from cutting each other's throats after the Interstate Commerce bill which prohibits pooling becomes a law." [39] The assignment to each railroad of a predesignated share of the tonnage clearly was pooling in form and in substance. This

[37] *The Second Annual Report of the Interstate Commerce Commission, December 1, 1888, op. cit.*, p. 24. The "seriousness" of the matter was clear: while the Justice Department prosecuted the cheater for having departed from published tariffs, the Interstate Commerce Commission prosecuted this road for having set secret rates that were unduly discriminatory — or else the cheater announced his intentions and set lower rates on all traffic.

[38] *P. N. C. Logan, F. D. Babcock and E. M. Parsons, et al.* vs. *Chicago & Northwestern Railway, The Annual Report of the Interstate Commerce Commission* (1889), Vol. III, p. 127. The rule was to hold for transport of any commodity.

[39] Cf. the *Railroad Gazette* (March 4, 1887), 152.

technique had to be abandoned, and new ones devised for maintaining loyalty to agreed rates. It was obvious that changes were required in the levels of through relative to local rates. The structure of rates was in violation of the (expected) requirements for "just and reasonable rates" and even of the long haul–short haul clause of the new legislation, as a result of the breakdown of control of through rates from 1884 to 1886.

The rate structure was dealt with preliminary to solving the problem of how to enforce loyalty to the agreement. There were two drastic means by which the differences between short- and long-distance rates could have been made "just and reasonable": all short-distance rates (per ton-mile) could have been lowered to equality with the Chicago–New York equivalent class rates, or the Chicago–New York rates could have been increased to a level equal to the short-distance rates. Each had an important disadvantage. Reducing local rates required the railroads to forego profits where they were most secure — where each of the individual roads provided exclusive or semiexclusive transport service. Increasing through rates required the Executive Committee to have great confidence in the rate agreements (that is, cartel operation had to expect to duplicate the control enjoyed by any one line on its own local traffic in order to set comparable through rates). The trunk lines avoided these disadvantages in part by lowering a number of local rates and increasing through rates only enough to establish rough comparability with the average local rate.[40]

[40] The disadvantages of the extreme adjustments faced all of the long-distance railroads in the country; nowhere were they more important, however, than for the lines west of Chicago where cartel management of rates had been lacking for some time. [The last attempt at cartel organization had been terminated in August, 1884, with the withdrawal of the Chicago and Northwestern Railroad from the existing pool. Cf. J. Grodinsky, *The Iowa Pool: A Study in Railroad Competition 1870–1884* (The University of Chicago Press, 1950), p. 162.] One of these railroads, the Chicago, Burlington and Quincy, had decided to set the higher through rate and discontinue carrying this tonnage entirely, rather than to decrease local rates: "the situation is this: on the Mississippi River at Hannibal, Quincy, Keokuk, and Burlington we have not been doing much business since the Interstate Law took effect, because the rates made by the competing lines from the eastward have been lower than we felt willing to accept, on account of its effect on our local rates in Illinois. . . ." (Memorandum from the New York general freight agent of the Chicago, Burlington, and Quincy to Vice-President C. E. Perkins of that railway, dated October 11, 1887; the Archives of the Chicago, Burlington and Quincy Railroad, the Newberry Library, Chicago, Illinois.) This railroad considered the alternative of lowering local rates: "by voluntarily making a reduction of 15 or 20 per cent, the chances are that we shall be obliged to face a loss on local business of somewhere from $200,000 to $250,000, against which we shall have, however, what we gain by going again into the competitive traffic on the Mississippi River and at interior points in Illinois. What this gain would amount to it is impossible to estimate with any accuracy. The direct gain at the various competing points Mr. Ripley estimates at about $100,000 gross earnings, only a portion of which would be net." (*Ibid.*) The other possibility considered was "that we might, by risking a decision

For through rates, the twelve classes for eastbound traffic constructed by the Joint Executive Committee in 1879–1880 were canceled and replaced by a six-class "New Official Classification." This classification was to apply on both eastbound and westbound traffic.[41] The new class rates increased some of the charges for through transport. Westbound rates from New York to Chicago had been 75 cents, 60 cents, 45 cents, and 35 cents for the four classes and 25 cents for "special" class from November 18, 1885, to April 1, 1887[42]; the New Official Classification was on rates of 75 cents, 65 cents, 50 cents, 35 cents, 30 cents, and 25 cents for the six classes. For nine large volume westbound commodities — lumber, salt, iron products, hops, tobacco, tea, coffee, sugar, and dry goods — there were three rate decreases, one rate increase, and five rates the same.[43] Eastbound higher class rates were reduced considerably, particularly those on first, second, third, and fourth class[44]; but the rates on a number of the important bulk commodities were increased. The trunk-line rate on grain from Chicago to New York was set at 25 cents per 100 pounds, the same as during the summer of the previous year, and the rate on live cattle was also the same as in the summer of 1886 (35 cents per 100 pounds); but the live hog rate was increased 5 cents per 100 pounds (from 30 cents to 35 cents)[45] and

---

against us under the Interstate Law, reduce our rates so as to meet the competition at Mississippi River towns and at interior common points in Illinois, leaving our local rates say 15 or 20 per cent higher than would be necessary to comply with the long and short-haul regulation. The probability is that the Commission would decide against us. . . . The Interstate Commerce Commission seems to have taken the ground in the *Vermont Central* case that competition by rail or by rail and water did not constitute a difference of circumstances under the law." (*Ibid.*)

[41] As noted in *The New York Times*, March 3, 1887, p. 2; and "Fares and Freight Rates Under the Interstate Commerce Law," the *Railroad Gazette* (March 4, 1887), 152.

[42] The Interstate Commerce Commission, *Railways in the United States in 1902*, Vol. II, Table XXIX, p. 44.

[43] Cf. the Interstate Commerce Commission, *Railways in the United States in 1902*, Vol. II, Table XIX, "Comparison of Official Classification and Rates Effective During the Years 1886, 1887, 1890, *et al.* in the Territory Now Covered by the Central Traffic and Trunkline Associations," pp. 24 to 28. Increases followed both from higher rates on the same class, and from shifting the commodities to higher classes.

[44] The rates in force as of December 27, 1886, were 100 cents, 85 cents, 70 cents, 60 cents, 50 cents, 45 cents, 40 cents, 35 cents, 35 cents, 30 cents, 30 cents, and 35 cents for twelve classes; the "New Official Classification" of April 1, 1887, was the same for eastbound as noted above for the westbound traffic. As a result, the first-class rate after April 1 was lower than either the first- or second-class rate at the end of 1886; the second-class rate was halfway between the third- and fourth-class rate under the old classification. None of these rates had very much effect, however, since eastbound tonnage consisted of shipments of grain, cattle, provisions, and various other agricultural products ranging from beans to hay, all of which were not in the first four classes.

[45] Cf. the Interstate Commerce Commission, *Railways in the United States in 1902*, Vol. II, p. 86.

rates on the important categories of the dressed meat and provision traffic increased 10 cents per 100 pounds.[46]

Since selective increases were all that was achieved in through commodity rates, a number of short-distance rates were lowered to render the structure of rates in accord with the new law. The Pennsylvania Railroad announced rates of 17 cents per 100 pounds from Philadelphia to points on its main line up to 210 miles distant, and 19 cents per 100 pounds to points from 230 to 415 miles away (as shown in Table 5.1). The rate to Pittsburgh

TABLE 5.1

Long- and Short-Distance Rates on the Pennsylvania Railroad in the Official Classification of 1887

Service	*Distance (Miles)*	*Fourth-Class Rate, (Cents per 100 Pounds)*	*Fourth-Class Rate (Cents per 100 Pounds, as of 1886)*
Philadelphia to Altoona, Pa.	230	19	18
Philadelphia to Pittsburgh, Pa.	325	19	18
Philadelphia to Williamsport, Pa.	210	17	18
Philadelphia to Erie, Pa.	415	19	18

Source: U.S. Interstate Commerce Commission, *Railways in the United States in 1902*, Vol. II, p. 168.

had been 18 cents, the same as that to Williamsport, so that the correction was made at the time of the New Official Classification by raising the longer distance rates by 1 cent and decreasing the shorter distance rate by 1 cent. The Penn decreased local grain rates from Cessna, Pennsylvania, to render them comparable to through rates. The grain rate from Cessna to Cumberland, Maryland, was decreased from 9 cents per 100 pounds to 7 cents some time between March, 1887, and February, 1888 (as shown in Table 5.2). This reduced the difference between the recorded local rate and the Chicago–New York official grain rate for the same distance to the smallest amount since April of 1876.[47]

Other trunk lines followed the example of the Pennsylvania road. For one, local rates were reduced on the Lake Shore and Michigan Southern Railway from Buffalo to the closest local stations, from Toledo to the closest stations en route to Chicago, and from Chicago to stations a short distance away in southern Michigan, so that higher local rates were exclu-

[46] The charge for shipping 100 pounds of dressed hog in refrigerated cars increased from 55 cents to 65 cents while the charge on the same commodity in common cars was increased from 50 cents to 60 cents. (*Ibid.*)

[47] As can be seen from comparing the last two columns of Table 3.13 and the last two of Table 4.8 with the last two columns of Table 5.2.

TABLE 5.2

Grain Rates on the Pennsylvania Railroad from Cessna, Pennsylvania, to Cumberland, Maryland, 1887 to 1889

Date of Shipment (Month and Year)	*Volume Shipped (Hundreds of Pounds)*	*Recorded Rate (Cents per 100 Pounds)*	*The Rate for Comparable Distance, Using the Chicago–New York Official Grain Rates as Bases*
January, 1887	360	9.0	1.29
March, 1887	245	9.0	1.29
February, 1888	309	7.0	1.29
March, 1889	351	7.0	1.29
April, 1889	32	7.5	1.29

Source: The *Manifest Book* for Cessna, Pennsylvania, of the Pennsylvania Railroad; the Archives of the Pennsylvania Railroad, Philadelphia, Pennsylvania. Note: the last named rate is a "less than carload" rate.

sively for longer distance transport.[48] The changes were according to a set of "official discounts" on the Chicago–New York rate for shorter distance traffic devised by the Trunkline Executive Committee (termed "Joint Committee Circular No. 805," March 3, 1887); the discounts were part of an agreement among the trunk-line roads on March 4, 1887, that "where through rates are quoted by agreement to any point on any railroad, no bill of lading at a higher rate can be issued to any shorter intermediate point on any line." [49]

While the new classifications and schedules were being worked out, there were also discussions of new techniques for securing the loyalty of all of the members to the new official charges. There was to be no money pooling, nor were there to be transfers of tonnage to some railroads from others. The proposed mechanism for guaranteeing shares of the tonnage was the price mechanism: the railroad with less than its share was to announce a discount on official rates sufficient to recover the deficit tonnage. That is, "the procedure is that if a railroad is actually found to be losing traffic it will be allowed to quote a differential; but if the action results in depriving the other roads of their traffic, it will not be continued on the same scale." [50] The Executive Committee was to supervise closely this

[48] Cf. the Interstate Commerce Commission, *Railways in the United States in 1902*, Vol. II, Tables LIII, LIV, and LV, pp. 170–171.

[49] As quoted from the Trunkline Executive Committee agreement by *The New York Times*, March 4, 1887, p. 2.

[50] "The Trunkline Presidents' Apportionment Scheme," the *Railroad Gazette* (December 23, 1887), 830. Since the setting of shares would have been considered illegal pooling, a railroad's "traffic" was its "usual share." There were difficulties in determining the "usual share."

posting of discounts.[51] There had to be certification of a decrease in the tonnage, and agreement that the rate reduction was no more than necessary to regain the usual tonnage. The committee did not have to prove "premeditated" disloyalty but only the existence of unusual shares.[52]

The new proposal was accepted as the means for rate adjustments. This was thought to have been sufficient to gain the loyalty of all the railroads, since the new law greatly penalized disloyalty. Regulation under the *Act to Regulate Commerce* reduced the profits from cheating on official rates. Under circumstances in which disloyalty did not violate the law, cheating resulted in smaller profits for a shorter period of time under regulation for two reasons. First, any rate cut that was "just and reasonable" had to be extended to all traffic — so that increased profits from cutting rates on grain to New York had to be accompanied by lower profits from the required cutting of local rates.[53] Second, the discount had to be announced, and announcement — if followed by reduction of the official rate — reduced the length of time in which the disloyal firm received larger profits.[54]

[51] "Proceedings of the Trunkline Board of Presidents," Circular No. 639, as reported in *The New York Times*, December 21, 1887, p. 2. These new procedures were partially a result of the anticipated regulation. But they were not "new"; as was noted at the time, "it is simply a repetition of the familiar practice of the Southern Railway and Steamship Association in which percentages were equalized not by diversion of freight but by increase of rates on the routes which were running ahead. If, as we presume to be the fact, there is an implied agreement that the differentials will be revised if they are found to give the weaker roads more than their share of the traffic, the resemblance is complete." The *Railroad Gazette* (January 13, 1888).

[52] The setting of differentials obviously made for legal discounts to deficit railroads greater than any original secret discount of any disloyal member, however. In the analysis of Chapter 2, the optimal reaction to cheating is to set a new official rate R_0^* below the cheater's rate R_c. The new agreement set $R_0^* < R_c$ simply to make up lost shares. The expectation of such an R_0^* should have added some measure of cartel stability, because the decline in cheaters' expected profits provided a disincentive for nonadherence.

[53] In terms of the demand and cost relations assumed in Chapter 2, cheating is profitable when the sum of profits π_c (from the exclusive offer of a cut price) and profits π_c^* (from the cartel undercutting the lower price after its presence is known) are greater than profits π_s (from adhering to the agreed price for the entire period). A regulatory requirement of the trunk-line railroad that long- and short-distance rates be proportional to distance should reduce π_c: the consolidated demand for all transport at a uniform per-mile rate has a larger value of β with given γ and δ from cheating on long-distance rates, and these relative values reduce the profits from cheating as compared to smaller β, given γ and δ, from cheating on the rate on Chicago–New York traffic alone. Cf. the Appendix to Chapter 2, for the comparison of profits from some values of β, given γ and δ.

[54] The time period in which π_c was realized was decreased by the regulatory requirement of public announcement of the discount, since cheating "detected" was cheating immediately acted upon. The sum total effect of regulation should have been to decrease the size of profits from cheating each week and also the number of weeks of any profit gain.

Under circumstances in which the cheater was acting illegally (by setting a secret rate or by discriminatory cutting of through rates), the revenue gains from such activity might have been the same as those received before the *Interstate Commerce Act*. But the disloyal railroad would have had to deduct the "costs" of prosecution in United States Court for violation of the Criminal Code. Profits from cheating and a criminal record together should have made disloyalty less attractive than were the profits from cheating in the years before the arrival of the Interstate Commerce Commission.[55]

The new agreement and the new regulatory penalties against cheating must have been expected to add to cartel control of Chicago–New York freight rates. The shippers expected as much: "There was much inclination to attribute the unusual volume of business by the railroads in March [of 1887] to . . . shippers anticipat[ing] the effects of the new rates and classifications. . . . Such shippers as had special rates availed themselves of them to the greatest extent possible before the law prohibiting special rates and rebates was effective." [56] Expectations were realized to an appreciable extent.

Cartel Control of Long-Distance Rates, 1887 to 1890

The Interstate Commerce law went into effect with the accompanying adjustments in trunk-line transport rates. *Official Rate Schedule No. 1* was posted in April of 1887 and remained in effect until the new year. Fourth-, fifth-, and sixth-class rates were increased during January of 1888 and remained at the higher levels while the Great Lakes remained frozen over. There was a readjustment to the previous summer's level on March 5, 1888; there was a sharp reduction in all class rates on November 12, 1888, but a return to the original schedule on December 18, 1888, which lasted until the spring of 1892.[57]

[55] In the early years of regulation, the "costs" of illegality were high because the statute was interpreted liberally by the courts; in later years, prosecution was far less likely. In the early years, then, illegal cheating was not as likely.

[56] "Trunkline Through-Shipments for the First Quarter of 1887," the *Railroad Gazette* (April 29, 1887), 289. It was mentioned that "the general prosperity of the country has had a very material interest in the continued large business of the last eight months too, of course." *Ibid.*, p. 289.

[57] None of these adjustments involved differences between eastbound and westbound long-distance rates. Local or short-distance rates were "in line" with the official through rate schedules of 1887 and 1890; at least local rates on the Pennsylvania Railroad and the Baltimore and Ohio in 1890 bore the same relation to through rates in 1890 as to *Official Rate Schedule No. 1* in 1887. [Cf. the Interstate Commerce Commission, *Railways in the United States in 1902*, Vol. II, pp. 167, 169.]

The arrangement for guaranteed shares of the tonnage was also put into effect less than a month after it was announced. On January 13, 1888, "a new trunkline tariff [was] published which requires the New York Central and the Pennsylvania to allow other roads a differential varying from 5 cents per hundred on first-class freight to one cent on fifth and sixth class. . . . Presumably there has been some shift to higher percentage traffic for the Pennsylvania and perhaps the New York Central." [58] The "premium" and "discount" railroads operated with this set of differentials until April, when sharp reductions of a particular commodity rate "eroded" some differentials.[59] The Baltimore and Ohio Railroad was promoted from a "discount" to "a premium line" in 1889,[60] and the 5-cent differential remained in force, and then was removed, for periods of varying length until April 1, 1895.[61]

The official rates and the sharing arrangement for tonnage were adhered to in the first year of regulation. The railroad newspapers noted that "when the *Act* first went into effect there was a remarkably successful effort to abolish special rates distinct from the published tariff . . . taking the railroads of the country as a whole it is pretty certain that a larger proportion of their business was done at published rates in the latter part of 1887 than had been the case for many years previous." [62]

This was realized by the customers of the trunk-line railroads as well: in correspondence with his New York freight agent, the president of the Chicago, Burlington and Quincy Railroad inquired whether the railroads East of Chicago and St. Louis had been able to put the 25-cent official rate into effect (rather than the 20 to 24 cent rate of 1886)[63] and received the answer that "this question can be answered in two ways. More people are

[58] The *Railroad Gazette* (June 13, 1888).

[59] There was an attempt to create further erosion in November when "the New York Central announced a reduction of all rail rates from New York to Chicago on the basis of 50 cents first class, one-third less than the tariff which has been in force . . . the reason is said to be [the] tariffs [of] the so-called weaker lines." [The *Railroad Gazette* (November 16, 1888).] In the instance of the general reduction, the "Erie, West Shore, Lackawanna, and Lehigh Valley met the cuts of the New York Central by issuing tariffs on the basis of 45 cents to Chicago." [The *Railroad Gazette* (November 23, 1888).] No further action on the part of the New York Central was forthcoming, so that the differentials remained in force. The changes in the one commodity rate — that on dressed beef — are discussed in detail below.

[60] "The New Trunkline Agreement," the *Railroad Gazette* (February 24, 1889), 124.

[61] Cf. the Interstate Commerce Commission, *Railways in the United States in 1902*, Vol. II, p. 43; the *Railroad Gazette* (July 12, 1889), 43; the *Railroad Gazette* (November 25, 1892), *et al.*

[62] "Some Impressions of the Act," the *Railroad Gazette* (January 4, 1889), 9.

[63] That is, whether shippers to the seaboard from points on the Burlington were paying more for transport on that proportion of mileage beyond this railroad's terminus, to the detriment of the demand for Burlington transport.

paying lower rates than are paying higher ones — but I think more merchandise tonnage is paying higher rates than those current prior to April 5, especially in the territory between the Mississippi and Missouri Rivers and also in Illinois. The through or competitive rates are substantially the same as those nominally in effect prior to April 5 and the only advance has been the amount of concession either by open or secret rates which formerly existed and which has been almost wholly done away with. . . ." [64] This reply seemed clear to the president: "I have yours of the twenty-second, with answers to some of my questions. If the roads East of Chicago and St. Louis are maintaining tariff rates now, while, before the law, they did not maintain tariff rates, then the country West of Chicago and St. Louis, taken as a whole, is paying more for transportation between these points and the seaboard than it did formerly. I do not see how this conclusion can be avoided. It may be, as you say, that there is more equality among shippers, and that large shippers like Marshall Field and others are not getting lower rates than smaller shippers, as they formerly did; nevertheless in the competition to secure business I have no doubt that a very considerable part of the total drawbacks paid by the eastern lines prior to the law found their way eventually into the pockets of the consumers." [65]

In addition, the trunk-line railroads themselves had few complaints with the first months of the operation of the new agreement. There was the statement in May: "It is claimed in Chicago that the Baltimore and Ohio, and Chicago and Grand Trunk, are making low rates on export freight via Baltimore and Montreal respectively. . . . It is alleged that provisions are taken to Glasgow at 28 cents per hundred pounds while the regular rate from Chicago to the seaboard is 30 cents. . . ." [66] No further mention was made of this matter, nor were discounts in rates recorded in the *Chicago Daily Commercial Bulletin*. A single statement was made in August that: "the Lake Shore has taken shipments from Minneapolis to trans-Atlantic ports at about 34 and 35 cents, which deducting $7\frac{1}{2}$ cents for the road west of Chicago and 11, 12, or 13 for the ocean charges, leaves only about 15 cents for the roads east of Chicago." [67] But this was not proven, nor were there allegations of repetitions. Rather, "tariff rates of 25 cents

[64] Letter from E. P. Ripley to Mr. C. E. Perkins, President, the Chicago, Burlington and Quincy Railroad Company, dated September 22, 1887; the Archives of the Chicago, Burlington and Quincy Railroad Company, the Newberry Library, Chicago, Illinois.

[65] Letter of Mr. C. E. Perkins to E. P. Ripley, dated September 23, 1887; the Archives of the Chicago, Burlington and Quincy Railroad, *op. cit.* Mr. Perkins's impression that drawbacks and rebates led to a decline of producer and wholesaler margins on commodity prices is substantiated, perhaps, by findings here of declines in Chicago–New York grain price differences in periods of railroad wars.

[66] "Changes in Rates," the *Railroad Gazette* (May 20, 1887), 344.

[67] "Grain Rates and Shipments from the Midwest," the *Railroad Gazette* (August 26, 1887), 560.

have, on the whole, been well maintained since the first of April and at least some of the credit for the calmness must be accorded to the Interstate Law." [68]

The Executive Committee's control of rates continued. In the course of setting and maintaining rates over the three years 1888 to 1890, there were, however, four departures from or breakdowns of official rates.

Two of these incidents were considerably different from cheating as observed in previous years. Instead of beginning with a secret reduction by one railroad, and continuing with declines in official rates, trouble began with a public warning by one railroad that it was going to reduce rates and continued with the cartel's announcement that the independent rate would be matched as soon as it went in force. The issue was clearly not the gains from undetected cheating, but dissatisfaction with some aspect of the agreement. Public rate-cutting was resorted to in order to change the agreement.

Two other incidents, however, followed the earlier pattern of secret discounts and cartel breakdown upon the detection of such discounts. These were short-lived and had little effect on rates or on prices in the relevant commodity markets.

The first incident after 1887 involved "cheating." There were reports of "unsettled conditions" in rates in the Chicago Board of Trade from April 7 to April 13, 1888, when "agents of the leading trunklines generally reported tariff rates adhered to, but shippers intimated that reductions of $2\frac{1}{2}$ to 3 cents per hundred pounds could be obtained on round lots through to the seaboard points." [69] During the week April 14 to April 20, "Agents generally insist that tariff rates are adhered to but shippers intimate that lower prices can be obtained." [70] This dispute on the actual level of rates continued into the first week of May, when a report that "a very unsettled feeling prevailed in freights during the past week, and there is little doubt but concessions in rail rates have been made . . . of 5 cents per hundred pounds . . . and that 20 cents on flour, grain and feed to New York has been accepted with like reductions to other points." [71] There would appear to have been cheating; but a 20-cent rate was never matched by the cartel, nor was an individual railroad identified as a rate-cutter. During the remaining weeks of May, the official rate of 25 cents remained posted, and, finally, the Board of Trade grain rate returned to the tariff level. Thus cheating — now illegal rate-cutting — was limited to six weeks and did not lead to the deterioration of the level of agreed rates.

The second incident of rate disruption was in contrast with these events of April and early May and seems rather to have been "dissatisfaction"

[68] *Ibid.*

[69] The *Chicago Daily Commercial Bulletin*, April 13, 1888, column entitled "Freights."

[70] The *Chicago Daily Commercial Bulletin*, April 20, 1888.

[71] The *Chicago Daily Commercial Bulletin*, May 4, 1888.

with the agreement. There had been some controversy early in November of 1887 over a differential rate on the more roundabout shipments of dressed beef from Chicago to the eastern seaboard. The differential was necessary because "One large shipper refuses to send goods by the Grand Trunk Railway unless he receives a lower rate by that line than by any other. The Grand Trunk cannot refuse to make the differential without losing its business, the other railroads cannot allow such a differential to be made without running the risk of losing their business." [72] The controversy was not settled so that "the Chicago and Grand Trunk [announced] a new tariff from Chicago to seaboard ports several times during the past week. Each day a reduction of from 3 to 5 cents has been made and the latest quoted rates are on the basis of 34½ cents per hundred on dressed meats in refrigerator cars. . . ." [73] The rate reached a low of 31 cents by the third week of November. With agreement on a 4-cent differential in favor of the Canadian route, this rate was increased to 65 cents per 100 pounds in December of 1887.[74]

In June of 1888 "disagreement" broke out once again upon the Pennsylvania Railroad "notifying shippers that on and after Monday [provisions and beef] rates would be 55 cents to New York and 61 cents to Boston, a uniform reduction of 10 cents. Michigan Central and Lake Shore companies issued a similar notice to their patrons, the only difference being that they made the rate to Boston 60 cents rather than 61." [75] On July 3, 1888, there was more of the same: "Today's developments were as predicted. The rate on dressed beef fell to 23 cents per hundred pounds Chicago to New York, which is the lowest rate ever made in this article of freight by any railroad even in times of war. When the inner-belt railroads received advance information that the Erie would make the cut to 23 cents they were so prompt to meet the reduction that they actually quoted the new

[72] "The War in Dressed Beef Rates," the *Railroad Gazette* (December 2, 1887), 782.

[73] *Ibid.* The Grand Trunk apparently issued a "commodity tariff" which lowered rates on dressed beef, but did not disturb the six class rates. This conformed with the long haul–short haul provisions of the Interstate Commerce Law, but was not in strict accordance with "just and reasonable" differences between rates on the various bulk commodities. But all of the roads saw the effects of lowering the entire official rate schedule: "the [dressed beef] reduction had best be arranged by what seem . . . to be known as 'Commodity Tariffs' — or otherwise our other business would be too seriously affected." (From a letter to the New York Central's General Traffic Manager, Mr. N. Guilford, from Mr. Arthur Mills of the Lake Shore and Michigan Southern, dated November 14, 1887, the Archives of the New York Central Railroad, Cleveland, Ohio.)

[74] The increase of official rates on dressed beef was from 31 cents per 100 pounds to 65 cents per 100 pounds announced on November 26, 1887; cf. the Interstate Commerce Commission, *Railways in the United States in 1902*, Vol. II, p. 86; and the discussion of tariff differentials in effect on January 9, 1888, the *Railroad Gazette* (January 13, 1888).

[75] "Rate War on Eastbound Freights," *The New York Times*, June 30, 1888, p. 2.

rate before the official notification was received by the Erie's western connection, the Chicago and Atlantic, to be put into effect on that line."[76] The Pennsylvania initiated this new series of open reductions, but it was the Erie that pressed the "argument," by demanding the Grand Trunk's differential for itself.

There were further open rate reductions during the first week of July even though "some of the railroad men . . . are beginning to hope that dressed beef rates will go no lower. They are waiting to see if the Erie continues to claim its differential. If it does the cutting will be continued."[77] The last reduction was not until the last week of July when "cattle rates have touched bottom . . . the Pennsylvania went as low as 10 cents to New York and then stopped . . . the Erie, Grand Trunk and other roads went out of the cattle business some time ago. . . ."[78] The basis for terminating the rate quarrel was "the Erie resumed the slaughter yesterday morning by making the rate 7 cents Chicago to New York . . . [but this] road is making no special efforts to secure the business at the prevailing low rates and rather seems to discourage shipments by its line."[79] The roads' reluctance to accept freight was evidence of its willingness to discuss the quarrel[80]; meetings were held which were successful in bringing an end to cutting rates in order to make a point. Rates were increased from 7 cents to 22½ cents per hundred pounds on August 3, 1888.[81]

[76] "The War Still Rages," *The New York Times*, July 4, 1888, p. 2.

[77] *Ibid.*

[78] The *Railroad Gazette* (July 20, 1888).

[79] *Ibid.*

[80] *The New York Times* also documents the conditions under which the rounds of rate-cutting were brought to an end: "The Erie, and Chicago and Atlantic, made their usual cuts today in the rate on dressed beef, dropping from 7 to 6 cents per hundred pounds from Chicago to New York. For the first time since the war began the Vanderbilt and Pennsylvania lines failed to meet the cut. A consultation of the freight officials of the Michigan Central, Lake Shore, Panhandle, and Fort Wayne roads resulted in an agreement to stand by the rates in effect on Saturday and ignore the Erie's cut for one day at least. What brought the Vanderbilt and Pennsylvania lines to this halt was the discovery that they were getting all the dressed beef business. Inasmuch as every carload handled at existing rates means a loss to the roads, this state of things was more than they had bargained for. The agents on these lines say that the Chicago and Atlantic has taken a very peculiar position. That road, fortunately for itself, has been carrying no dressed beef out of Chicago for several days, the reason being that notice was given to shippers last week that, in consequence of present low rates, these shipments would have to take chances with ordinary freight instead of being forwarded by special fast freight lines." From "Trying to Escape Business: A Peculiar Situation on the Eastbound Freight," *The New York Times*, July 17, 1888, p. 2. The Erie, in effect, by reducing the quality of service at the 7-cent rate had shifted the tonnage to other lines. These lines correctly considered the reduction in quality as equivalent to an increase in the 7-cent rate — or as a discontinuance of the downward movement of rates.

[81] Cf. the Interstate Commerce Commission, *Railways in the United States in 1902*, Vol. II, p. 86. It is interesting to inquire of the effect on local or short-distance rates on

With the resumption of "remunerative" rates, there was introduced an official differential in favor of the Erie. It would seem that this railroad had been able to make its point. Moreover, the railroad had gained more than a reputation for stubbornness: the larger share of cattle tonnage at differential rates might have compensated for the period of rate warfare. This was not "profits from cheating," [82] but (if present) was "profits from reform of the agreement." [83]

The third incident of a disruption of rates was a new variation on "cheating" on the agreement. There were rumors of discounts and rebates on official grain rates during the early weeks of August, 1888.[84] These were followed by allegations that the Baltimore and Ohio Railroad, and the Chicago branch of the Erie Railroad, were shading rates from Chicago to points on Lake Erie.[85] But this time "the reduction only averages one-half to three cents per hundred pounds," [86] and nothing was done by the Executive Committee; there was "a disposition on the part of the railroad managers to work in harmony [on the question of] cuts made in rates . . . west of Pittsburgh." [87] Harmony was discontinued the second week of September: the Pennsylvania announced that it was reducing grain rates 5 cents per hundred pounds and both the B & O and the New York Central followed this example,[88] (although the Baltimore railroad increased its rate for a short period "in which the demand for cars was greater than its

---

dressed beef to ascertain whether, for one, the short haul–long haul clause was violated; unfortunately no tariffs or bills of lading survive for this commodity for this year.

[82] Cf. There could not have been profits from cheating because there was no lag between the weekly reports of the *Chicago Daily Commercial Bulletin*, which gives actual rates, and the official rates listed in the Interstate Commerce Commission, *Railways in the United States in 1902;* Vol. II, p. 86 each week.

[83] Nor was there any suggestion that this highly publicized series of changes was designed to divert attention from cheating on rates on other bulk commodities. To the contrary, the Chicago Board of Trade reported each week that "a further material reduction in dressed beef, livestock, and provisions rates has been submitted, but no . . . announcement has been made of a cut in flour and grain charges. . . ." The *Chicago Daily Commercial Bulletin*, July 6, 1888. When the dressed beef rate was at the lowest level of the year, it was stated that "rates have been reduced materially on livestock and dressed beef, but shippers of flour and grain have been unable to obtain any concessions." The *Chicago Daily Commercial Bulletin*, July 13, 1888, and July 20, 1888. The *Railroad Gazette* noted "grain rates are still maintained at 25 cents per hundred pounds" on July 20, 1888.

[84] Cf. the *Chicago Daily Commercial Bulletin*, "The Weekly Review of Freights," August 10, 1888 and August 17, 1888.

[85] Cf. the *Chicago Daily Commercial Bulletin*, "The Weekly Review of Freights," August 24, 1888.

[86] *Ibid.*

[87] The *Chicago Daily Commercial Bulletin*, "The Weekly Review of Freights," August 31, 1888.

[88] The *Railroad Gazette*, "Changes in Trunkline Through Rates" (September 12, 1888).

immediate supply" [89]). Finally on October 8 the cartel organization gave up attempts to restore the tariff of 25 cents per 100 pounds and announced a "special commodity rate" for grain of 20 cents per 100 pounds.[90]

This sequence of events was that of a breakdown in an agreement: reports of a secret rate followed by open rate-cutting in defense. The Baltimore and Ohio Railroad or the Erie Railroad was left with the blame for rate cuts, and perhaps with increased profits from cheating. There were noticeable differences between this difficulty with grain rates and the usual evidence of cartel instability, however. The reports of cut rates, no more than scattered rumors, were followed quickly by open reductions of rates. The alleged cut in rates, "one-half to three cents," while comparable to rebates and drawbacks in 1886, was less than earlier cuts. Most important, one railroad's disloyalty did not lead to a breakdown of all through rates or even of those on bulk commodities moving in the same direction — as it did in 1884, 1885, and 1886.

But the aftereffects from this incident were out of proportion to its briefness. A committee of three members of the Interstate Commerce Commission set up a special inquiry in Baltimore early in 1889 to determine whether there had been secret reductions in rates. Evidence was to lead to the prosecution of the railroad officials that cut official rates. The Pennsylvania repeated its accusations against the Baltimore and Ohio, but could not provide evidence: Mr. W. H. Joyce of the Pennsylvania declared that his company had decided to reduce rates "in consequence of the discovery that large sales of grain were being made in Nebraska for the Baltimore market higher than market quotations . . . [suggesting] that a cut was made in railroad rates presumably by the Baltimore and Ohio." [91] The inference was denied by the general freight manager of the Baltimore and Ohio and by "the [grain shipper] alleged to have been favored." [92] Baltimore shippers were also heard by the Commission, but their statements were open to a variety of interpretations — there were complaints

[89] The *Chicago Daily Commercial Bulletin* notes that "the Baltimore and Ohio, which gave notice of a reduction to the same figure, has rescinded its action and will restore rates to twenty-five cents after the expiration of the usual time of notification" in its "Weekly Review of Freights" for September 21, 1888.

[90] Cf. the *Railroad Gazette*, "Trunkline Freight Rates" (October 15, 1888). This special rate was in effect until December 15, when it was allowed to lapse in favor of the regular sixth class rate, except for the change which was part of the New York Central Railroad's reduction of all of its rates "to the basis of fifty cents first class, one-third less than the tariff which has been in force" on November 16, 1888. The reason given for this action, as noted above, was the Central's dissatisfaction with the results of allowing the rate differential on provisions in October. Cf. the *Railroad Gazette* (November 16, 1888). It did not have the effect of reducing the differential, and was short-lived, since the previous level was restored in the middle of December.

[91] As reported in the *Railroad Gazette* (February 22, 1889).

[92] *Ibid.*

that "various merchants [found] that their competitors [over]bid them 2 cents on grain bought in Nebraska . . . [and] that when they got reasonable rates there was a scarcity of cars and when cars became plenty they did not get suitable rates." [93] The Commission was not able to obtain information sufficient for a Justice Department complaint, but the officers of the trunk-line railroads were on notice that secret rates invited investigations of criminal violations of Section 6 of the *Interstate Commerce Act*.[94]

The last of the four incidents of rate disruption began in June of 1890 and continued at least until November 14 of that year.[95] The time had come for a review of differential rates on dressed beef traffic, and the review

[93] *Ibid.*

[94] Two prosecutions were undertaken in 1889 against trunk-line railroads for alleged shading of announced rates so that the roads were on notice that not all investigations were going to be fruitless. Cf. *United States* vs. *Michigan Central Railroad Company*, 43 Fed. 26 (1890) and *Interstate Commerce Commission* vs. *Baltimore and Ohio Railway Company*, 43 Fed. 37 (1890). Perhaps a second, if more indirect, legacy of the incident of the fall of 1888 was some change in the Interstate Commerce Law. The abruptness of the Pennsylvania's decrease in rates, and the supposed behavior of the Baltimore and Ohio, seem to have provided information pertinent to two amendments: (1) making it illegal for any carrier participating in a joint through rate to accept compensation below his announced share of the official joint rate; (2) making it necessary for three days' notice to be given of a rate reduction as well as ten days' notice of a rate increase. The first (in summary form) is an amendment to Section 7 and the second to Section 6 of the Act as of March 2, 1889 (25 Stat. 885).

[95] There was a relatively minor incident in July of 1889, as well, that involved publicly announced discounts of rates. From all appearances the Baltimore and Ohio had been receiving less than its usual share of tonnage into the east coast seaports during the months of April to June — the first months after having become a "premium" line (as can be seen below in the Appendix table of tonnage shares). This railroad announced a reduction of the rate on grain by 5 cents per 100 pounds the first week of July, and the Pennsylvania matched the reduction on July 6. It is not certain that all of the other railroads followed: "the officers of the other roads being inclined to allow the B & O to continue its reduced rates alone, believing that the traffic affected by it will be chiefly that from the more southern ports. The Grand Trunk, however, announces that if the Pennsylvania makes a reduction it will take the same action in rates to all New England points. The Executive Committee of the trunklines and their western connections met in New York on Wednesday but took no definite action." ["Eastbound Grain Rates," the *Railroad Gazette* (July 12, 1889), 470.] The rate of each railroad was between 20 cents and 25 cents per 100 pounds, with some railroads offering transport for all grain at 25 cents while others transported wheat and oats for that rate and corn for 20 cents. This continued until August 1 (when rates of all of the trunk lines were announced as 20 cents for corn and 25 cents for wheat and oats) only because of the required waiting periods before rate changes could go into effect: "At this time the rates by different roads are in a very mixed condition, notices of reductions to take place three days in the future and of advances to take place ten days in the future having been intermingled in a way to cause considerable uncertainty." ["Eastbound Grain Rates," the *Railroad Gazette* (July 19, 1889), 486.] The Baltimore and Ohio seems to have become convinced quickly that a differential on grain rates set at the lower level of 20 cents per 100 pounds was not an advantage, and the argument was not continued.

began with the reopening of the old controversy. The Central held that the differential should be eliminated; the Grand Trunk reaffirmed the necessity for the existing 3-cent difference in its favor. Each held to its own view in conference and in the war of rates that followed. Discussion took place in June until the Central announced a reduction in its dressed beef rate to that of the Canadian railroad, and this railroad in turn reduced its rate to maintain the 3-cent differential. The pattern of challenge and response was repeated: "The Lake Shore people gave notice that they would further reduce the rate on dressed beef to 39 cents per hundred pounds . . . to meet the action of the Chicago and Grand Trunk which has filed a new tariff making a rate of 39 cents to preserve its differential. The Lake Shore is still determined to abolish the differential and will continue to meet the Grand Trunk's rates." [96]

There were further rounds of rate decreases, as well as statements of firm intent to continue: "The Chicago and Grand Trunk has not given up the fight on dressed-beef rates . . . this company has given notice of its intention to adopt a rate of 36 cents per hundred pounds beginning June 20. This is the date on which the 39 cent rate [of the Lake Shore–New York Central] becomes effective. As soon as the Lake Shore people received this intelligence they called a special meeting of the Chicago committee and announced that they would meet the 36 cent rate to take effect on June 26. Immediately . . . the Grand Trunk sent a notice to the Interstate Commerce Commission announcing a further reduction to 33 cents taking effect June 23." [97] On the last round, during the first week of July, all of the trunk-line railroads announced and adhered to a 30-cent rate.[98]

Reconciliation at higher rates was most difficult. No one of the railroads was willing at this point to compromise on the size of the rate differential — in other words, "the western roads would rather fight than make money" [99] — so that the 30-cent rate prevailed throughout the fall and early winter. In November the Grand Trunk was allowed a 2-cent differential and the right to pay shippers $\frac{1}{4}$ cent per mile more than other lines for the use of the shippers' refrigerator cars. Finally the rate on dressed beef transported from Chicago to New York was increased from 30 cents to 45 cents per 100 pounds on November 24, 1890.

This incident clearly occurred for the same reason as the rate disruption

[96] *The New York Times*, June 14, 1888, p. 5. These rate changes again all involved "commodity tariffs" so as not to disturb the class rates.

[97] "More Rate Cutting," *The New York Times*, June 19, 1890, p. 4.

[98] The trouble had been reasonably well limited to the differential on dressed beef — in fact, "the assertion has been made that the only point at issue is whether the Grand Trunk Railway shall be allowed a 3-cent differential on dressed beef. . . ." ("The East Bound Freight Question," *The New York Times*, November 6, 1890, p. 2.)

[99] As quoted by *The New York Times* from "an impatient exclamation of a trunkline president," November 6, 1890, p. 2.

during the summer of 1888. All rates were publicly announced and submitted to the Interstate Commerce Commission according to legal requirements. The reductions were part of efforts either to maintain or to destroy differentials in official rates. Using the level of rates as an "argument" was the chosen, if expensive, means by which to change the content of an agreement on tonnage shares.

The two disagreements on rate differentials and the two cases of cheating each had some effect on rates and on prices in grain and beef markets. There were reductions in Board of Trade rates in all instances. As a matter of course, the lower rates were accompanied by lower margins between midwest and east coast prices for provisions. But there was in fact little effect upon differences between Chicago and New York grain prices from the two short breakdowns in grain rates. Grain price margins seem to have shown the results of the operation of a newly successful cartel.

There is indirect evidence that reductions in provisions rates had substantial effects upon markets for meat products, from the average differences between Chicago and New York lard prices in these periods. Regulation began in April of 1887 with a 35-cent rate and with an average lard price difference of 33 cents per 100 pounds (as shown in Table 5.3). The 35-cent rate, and slightly higher lard price margins, prevailed until July of 1888, when the "dressed beef war" reduced Board of Trade rates and average lard price differences to 25 cents.[100] The rate on the Board of Trade was "restored" during the month of September, and the lard price difference increased to 35 cents. This margin continued at levels greater than or equal to 35 cents until May of 1890, when the second "dressed beef war" broke out. Then the reductions of rates to 25 and 23 cents were accompanied by reductions in price margins to between 24 and 30 cents per 100 pounds, the lowest recorded in the 4-year period.

After regulation was installed, at the beginning of the summer season of 1887, differences between spot prices in the Chicago and New York grain markets were close to the 25-cent to 28-cent official rates on grain transport. The Board of Trade rate equaled the official rate of 25 cents for the entire summer season of 1887, and also equaled the official rate for each of the slight increases in the winter season of 1887–1888. These rates were matched by seasonal average grain price differences (as shown in Table 5.4; the summer average price difference was within 1 cent of the average lake-rail rate, while the winter average price difference was within

[100] The official rates remained at 35 cents not because of "cheating" on official rates, since these changes were listed in the newspapers and in the *Railroad Gazette*. They were not listed in the Interstate Commerce Commission volume, *Railways in the United States in 1902*, presumably because they did not involve a change in the official rate schedule but rather in a particular commodity rate (and the commodity rate for lard in the year 1888 was not given).

TABLE 5.3

Railroad Rates and Lard Price Differences: 1887 to 1890

Month	*Official Rate* (*Cents per 100 Pounds*)	*Board of Trade Rate* (*Cents per 100 Pounds*)	*Chicago–New York Western Lard Price Mean Difference* (*Cents per 100 Pounds*)
January, 1887	35	34	32
February	35	29	26
March	35	27	22
April	35	35	33
May	35	35	35
June	35	35	34
July	35	35	35
August	35	35	35
September	35	35	37
October	35	35	38
November	35	35	45
December	35	35	34
January, 1888	32.5	32.5	30
February	32.5	32.5	32
March	35	35	40
April	35	33	39
May	35	32	33
June	35	35	30
July	35	25*	23
August	35	31*	33
September	35	35	35
October	35	35	42
November	30	30	42
December	35	35	36
January, 1889	35	35	47
February	35	35	42
March	35	35	40
April	35	35	36
May	35	35	39
June	35	35	37
July	35	35	37
August	35	35	38
September	35	35	39
October	35	35	48
November	35	35	45
December	35	35	36
January, 1890	35	35	37
February	35	35	39
March	35	35	37
April	35	35	36
May	35	34	31
June	25	25	27
July	23	23	30
August	23	23	26
September	23	23	25
October	23	23	24
November	23	23	26
December	30	30	42

Source: The official rates were the fourth-class rates listed in the Interstate Commerce Commission, *Railways in the United States in 1902*, Vol. II, p. 78, unless the Board of Trade *Daily Commercial Bulletin* noted a change in class and listed the new class rate. The Board of Trade rate is from the "Weekly Review of Freights," the *Daily Commercial Bulletins*, 1887 to 1890. The western lard price difference is the average of the differences of the daily spot prices for that grade listed in the monthly summaries of the Chicago *Daily Commercial Bulletin* and the *Annual Reports* of the New York Produce Exchange.

* Based upon Board of Trade announcements of changes in general provision rates.

TABLE 5.4

Official Rail Rates, Board of Trade Rail Rates, and Differences Between Chicago and New York Grain Prices, 1887 to 1890

$$R_T = \alpha + \beta R_0 + \nu \quad (1)$$

R_T, the rail rate, in cents per hundred pounds Chicago to New York, from the Board of Trade *Daily Commercial Bulletin;* R_0, the rail rate from compilations of posted tariffs and from announcements of the various traffic associations.

$$P_g^* = \alpha + \beta R_0 + \gamma R_L + \nu; \qquad P_g^* = \alpha + \beta R_T + \gamma R_L + \nu \quad (2)$$

P_g^*, the weekly average of the daily differences between Chicago and New York spot prices, in cents per hundred pounds, for the number 2 "western" grade of corn and oats; the official rate R_0 or the Board of Trade rate R_T; the lake-rail rate R_L from Chicago to New York via steamship to Buffalo and rail from Buffalo to New York as quoted on the Board of Trade.

$$P_{g,t} = P_{g,t-1} + \epsilon(P_g^* - P_{g,t-1}) \quad (3)$$

$P_{g,t}$, the average of the daily Chicago–New York grain price differences during week t and $P_{g,t-1}$, the average for the previous week; $P_g^* = \alpha + \beta R + \gamma R_L$ for the summer season and $P_g^* = \alpha + \beta R$ for the winter season from an equilibrium adjustment to changes in transport rates.

The standard error of each coefficient is shown in parentheses below the coefficient.

SUMMER SEASON, 1887:

(31 weeks); $R_T = R_0 = 25.00$ cents per 100 pounds for the entire period

$\bar{R}_L = 20.933$ cents per 100 pounds

$\bar{P}_g = 19.867$; standard deviation $SP_g = 2.228$ cents per 100 pounds

(30 weeks); $P_g = 15.572 + \underset{(0.000)}{0.000}R_0 + \underset{(0.209)}{0.205}R_L$; $\hat{\rho}^2 = .033$

(28 weeks); $P_{g,t} = \underset{(0.131)}{0.749}P_{g,t-1} + 0.251(11.325 + 0.429R_L)$; $\hat{\rho}^2 = .134$

WINTER SEASON, 1887–1888:

(21 weeks); $\bar{R}_T = \bar{R}_0 = 26.143$ cents per 100 pounds

$\bar{P}_g = 25.705$ cents per 100 pounds; $SP_g = 1.897$ cents per 100 pounds

(21 weeks); $P_g = 25.238 + \underset{(0.352)}{0.016}R_0$; $\hat{\rho}^2 = .0007$

(15 weeks); $P_{g,t} = \underset{(0.212)}{0.605}P_{g,t-1} + 0.395(13.484 + 0.083R_0)$; $\hat{\rho}^2 = .196$

SUMMER SEASON, 1888:

(33 weeks); $R_T = 3.833 + \underset{(0.128)}{0.808}R_0$; $\hat{\rho}^2 = .563$

$\bar{R}_0 = 23.636$; $\bar{R}_T = 22.939$; $\bar{R}_L = 17.897$

$\bar{P}_g = 17.732$; $SP_g = 4.417$

(32 weeks); $P_g = -17.874 + \underset{(0.307)}{0.381}R_0 + \underset{(0.387)}{1.484}R_L$; $\hat{\rho}^2 = .337$

(32 weeks); $P_g = -31.223 + \underset{(0.297)}{0.702}R_T + \underset{(0.404)}{1.832}R_L$; $\hat{\rho}^2 = .415$

TABLE 5.4 (Continued)

SUMMER SEASON, 1888: (Continued)

(32 weeks); $P_{g,t} = 0.845P_{g,t-1} + 0.155(28.119 - 0.561R_0 + 0.146R_L)$; $\hat{\rho}^2 = .089$
$\qquad (0.132)$

(32 weeks); $P_{g,t} = 0.782P_{g,t-1} + 0.218(-22.687 + 0.803R_T + 1.218R_L)$; $\hat{\rho}^2 = .103$
$\qquad (0.136)$

WINTER SEASON, 1888–1889:

(16 weeks); $R_T = R_0 = 25.000$ for the entire period
$\bar{P}_g = 22.219$; $SP_g = 1.400$

SUMMER SEASON, 1889:

(33 weeks); $\bar{R}_0 = 23.000$; $\bar{R}_T = 22.867$; $\bar{R}_L = 14.067$
two weeks in which $R_0 = 25.000$ and $R_T = 20.00$, otherwise $R_0 = R_T$
$\bar{P}_g = 18.111$; $SP_g = 1.858$

(32 weeks); $P_g = -0.749 + 0.196R_0 + 1.001R_L$; $\hat{\rho}^2 = .332$
$\qquad (0.216) \quad (0.270)$

(32 weeks); $P_g = -7.031 + 0.454R_T + 1.035R_L$; $\hat{\rho}^2 = .473$
$\qquad (0.155) \quad (0.218)$

(13 weeks); $P_{g,t} = 0.794P_{g,t-1} + 0.206(18.963 - 0.685R_0 + 1.072R_L)$; $\hat{\rho}^2 = .185$
$\qquad (0.124)$

(13 weeks); $P_{g,t} = 0.715P_{g,t-1} + 0.285(6.203 + 0.239R_T + 1.344R_L)$; $\hat{\rho}^2 = .166$
$\qquad (0.142)$

WINTER SEASON, 1889–1890:

(18 weeks); $R_0 = R_T = 22.000$ for the entire period
(18 weeks); $\bar{P}_g = 21.284$; $SP_g = 1.867$

SUMMER SEASON, 1890:

(34 weeks); $\bar{R}_0 = 21.382$; $\bar{R}_T = 21.324$; $\bar{R}_L = 15.076$
$R_0 = R_T$ each week except for two when $R_0 = 22.00$, $R_T = 21.00$, and $R_0 = 21.00$, $R_T = 20.00$
$\bar{P}_g = 16.350$; $SP_g = 2.443$

(33 weeks); $P_g = 1.838 + 0.843R_0 - 0.240R_L$; $\hat{\rho}^2 = .067$
$\qquad (0.804) \quad (0.586)$

(33 weeks); $P_g = 7.3286 + 0.712R_0 - 0.421R_L$; $\hat{\rho}^2 = .062$
$\qquad (0.707) \quad (0.550)$

(33 weeks); $P_{g,t} = 0.729P_{g,t-1} + 0.271(-81.363 + 2.929R_0 + 2.404R_L)$; $\hat{\rho}^2 = .236$
$\qquad (0.142)$

(33 weeks); $P_{g,t} = 0.719P_{g,t-1} + 0.281(-35.001 + 1.312R_T + 1.600R_L)$; $\hat{\rho}^2 = .201$
$\qquad (0.146)$

.4 cent of the average official–Board of Trade rail rate). As a result, the average grain price difference was 2 cents per 100 pounds higher than in the previous summer, even though official rates were the same for both of these seasons. The average grain price difference was 4 cents higher in the winter of 1887–1888 than in the winter of 1885–1886, again with the same official rates for both seasons.[101] Variations in week-to-week grain price differences this first year of regulation did not entirely conform with changes in rates, however; changes in rates explained no more than 20 per cent of the variance in price margins.[102] But changes in rates were few and of small size, so that price margin adjustments to rate changes were necessarily of minor importance.

The discounts on official grain rates in the summer of 1888 somewhat disrupted the conformity of grain price differences to cartel-set rates. Reductions in Board of Trade rates in April and again in September[103] explained a greater proportion of the variance in the grain price differences than did changes in official rates, given either a simultaneous adjustment or a lag in adjustment to rate change (as shown by the values of the coefficients of multiple determination for the relevant computed equations in Table 5.4; the addition to $\hat{\rho}^2$ from replacing R_0 with R_T in the lag equation is 1.4 per cent of total variance, while the addition for the simultaneous adjustment equation is 7.8 per cent). The average margin between Chicago and New York grain prices declined as well, by more than 2 cents per hundred pounds. The average Board of Trade rate decreased by more than 2 cents per 100 pounds while the official rate declined only 1.36 cents (as compared to rates the previous winter). The impression is that the disruption of rates decreased average grain price margins by 2 cents, and by somewhat more than 2 cents for short periods.

Grain prices in the subsequent seasons during 1888, 1889, and 1890 were not affected by any deviations from official rates. There were very few changes in official rates and Board of Trade rates during the winter of 1888–1889 or the summer of 1889 (Board of Trade rates equaled official rates, except for the two-week period in July when the Baltimore and Ohio was seeking independently to return to a differential rate). Board of Trade and official rates were from 22 cents to 25 cents per 100 pounds, while

[101] The winter of 1887–1888 would seem more comparable to that of 1885–1886 than to that of 1886–1887. The winter of 1886–1887 was a period of *transition* to markets soon to be regulated and soon to have an entirely new scheme for group adjustments of rates. Rates in the winter of 1885–1886 were not regulated and not under the influence of a new agreement. Those in the winter of 1887–1888 were under the influence of both, so that contrast with 1885–1886 of the effects from cartel and regulatory activities seems possible.

[102] Also, any week's grain price difference was roughly equal to 66 per cent of the previous week's price difference plus 33 per cent, an adjustment to "an equilibrium grain price difference." The "equilibrium difference" increased or decreased by as much as 32 per cent of any change in official rates and by as little as 8 per cent (as shown by the computed equations for $P_{g,t} = (1 - \epsilon)P_{g,t-1} + \epsilon(\alpha + \beta R + \gamma R_L)$ in Table 5.4).

[103] These were sufficiently extensive that R_0 explained only 56.3 per cent of the variance in R_T, given that the coefficient of correlation $\hat{\rho}^2 = .563$ for the computed equation $R_T = 3.833 + 0.808R_0$ for this season as shown in Table 5.4.

lake-rail rates were 14 cents. The average seasonal grain price difference was 2 cents below the announced rate during the winter[104] and halfway between the rail and the lake-rail rates during the summer. There were no changes in the official-Board of Trade rate of 25 cents in the winter, and very little fluctuation in the grain price difference (the standard deviation of P_g was the lowest since regulation began, as shown from comparing the values of SP_g in Table 5.4). Variations in the official rate during the summer explained 18 per cent of the variance in the lagged grain price difference, and 33 per cent of the variance in nonlagged grain price differences (while the Board of Trade rates, including the independent rate of the Baltimore and Ohio in July, explained slightly more variance in the nonlagged equation, and less in the lagged equation, as shown in Table 5.4). In the winter of 1889–1890 all rates were 22 cents, and the average grain price margin was 21.3 cents; in the summer of 1890 there were very slight reductions in rates, the weekly grain price differences averaged slightly more than the lake-rail rates, and fluctuated in accord with changes in official rates.[105]

Comparison of the fluctuations in the price margins on lard and those on grain would seem to indicate more effect from rate disruptions in the lard markets. Most of the time, however, in both commodities the margins reflected strict adherence to official rates and to the regulations of the trunk-line cartel associations. Moreover each of the disruptions in official rates, as short-lived as they were, may have helped in attaining cartel stability. The railroads cutting the official grain rates would appear to have made no great gains in shares of grain tonnage from Chicago, or in tonnage into the east coast seaports; the small tonnage increases did not compensate for the reductions in rates so that the disloyal members realized decreased profits from cheating. The lesson demonstrated was that disloyalty did not pay.

Shares of grain tonnage from Chicago, during the first four years of regulation, were relatively more stable than during earlier periods (as in Table 5.5). There were no periods of three months or more in which a railroad was considerably above the long-term trend of its shares, while other railroads were considerably below; rather, the months in which any

[104] There is one possible explanation, at least. There was a paucity of shipments of oats to New York and a listing of a "nominal" oats price in that city, so that the grain price difference may not have reflected transactions prices based upon the costs of transport.

[105] The explained variance in the lagged adjustment model is particularly high, as shown in Table 5.3. The explanation for the extensive decline in lake-rail rates and grain price differences is that the war in dressed beef rail rates had shifted transport of provisions away from lake transport, so that the supply of steamer tonnage available for the grain traffic was much larger than in previous years. This made the lake-rail rate on grain much lower and consequently the average grain price difference much lower in conformity with this particular rate.

TABLE 5.5

Grain Shipments from Chicago, 1887 to 1890

The trend of the percentage share of grain tonnage of each railroad has been defined as

$$s = \alpha + \beta T + \gamma X + \nu$$

with s = a railroad's percentage of total tonnage of Chicago shipments; T = month of shipments, numbered from 1 to 156 for the period 1887 to 1899; X = "1" for the winter months in which the lakes were closed to traffic, "0" for the summer months. The estimates of trend, as calculated by the method of least squares, are as shown in the Statistical Appendix to Chapter 5.

The "trend" share for any month is obtained from inserting the relevant values of T and X in the computed equations. The actual shares are shown in the table, along with the difference between actual and trend shares for that season.

	The New York Central Railroad		*The Pennsylvania Railroad*		*The Baltimore and Ohio Railroad*		*The Erie Railroad*		*The Grand Trunk Railway*	
Season and Year	*Actual Share*	*Actual Share Minus "Trend" Share*	*Actual Share*	*Actual Share Minus "Trend" Share*	*Actual Share*	*Actual Share Minus "Trend" Share*	*Actual Share*	*Actual Share Minus "Trend" Share*	*Actual Share*	*Actual Share Minus "Trend" Share*
Winter 1886–1887	27.6	−5.9	21.2	+5.8*	5.0	−0.4	13.8	+3.5	11.0	−4.4
Summer 1887	30.1	−1.0	20.5	+6.6*	7.9	+2.8	8.9	+3.1	9.4	−1.4
Winter 1887–1888	29.4	−3.9	17.7	+1.9	8.3	+2.8	7.3	−3.1	18.3	+3.3
Summer 1888	26.6	−4.6	17.0	+2.8	5.0	−0.3	11.8	−0.3	9.3	−1.0
Winter 1888–1889	31.2	−2.6	17.8	−0.4	6.9	+1.2	11.9	+1.4	21.3	+6.8*
Summer 1889	33.9	+3.0	14.5	−0.2	6.9	+1.4	13.5	+1.3	9.1	+0.8
Winter 1889–1890	37.5	+4.5	15.9	−1.0	10.7	+4.8*	10.7	+0.1	12.0	−2.0
Summer 1890	35.0	+4.2	12.7	−2.3	4.7	−1.1	14.1	+1.8	11.8	+2.4

* Greater than the relevant standard error of estimate shown in the Statistical Appendix to Chapter 5, Table A.1.

Source: Weekly shipments by railroad and lake steamers of wheat, corn, oats, and flour are listed in the *Chicago Daily Commercial Bulletin*, 1880 to 1899.

railroad's share was off the trend line of its shares[106] were widely spaced in the four-year period.[107] When there were changes in shares, grain tonnage did not shift to the rate-cutter as a reward for breaking the agreement.[108] In the first instance of grain rate disruption, during April and May of 1888, the cheater was never identified.[109] If the Grand Trunk carried more than its usual tonnage share in April,[110] and somewhat less than its usual share in May[111] as a result of rate-cutting, then gains resulted from April's larger

[106] It is assumed that a significant deviation from trend is a value of the residual from the trend line greater than the standard error of estimate; such values (as asterisked in Table 5.5) are expected to occur in approximately 30 per cent of the observations given that the residuals are normally distributed with zero mean. There seems no *a priori* reason, other than cheating, for them to occur in sequence of three or more months.

[107] The marked stability of shares of grain shipments from Chicago is matched by the stability of the tonnage shares into the east coast seaports shown in the Appendix to Chapter 5, Table A.2. This was the experience in shipments into the East Coast, except for a number of successive months in which the Lehigh Valley Railroad experienced less than its trend share while initiating service into New York and for some months of New York Central deficits and Pennsylvania surpluses in 1890. There is no simple explanation for the last instance. Shipments from Chicago were distributed as in previous months, while shipments into Philadelphia were so much greater that this city must have received more than its usual share of lake-rail shipments. However, at the same time, the differential in favor of "the weaker lines" was denounced by the New York Central; and the ocean rates to Europe from New York were relatively higher than from the more southern ports (cf. *New York Produce Exchange* vs. *Baltimore and Ohio Railroad*, *ICC Reports*, *VIII* (1898), 614, at p. 629, which shows the New York rate the *same* this year as from Baltimore while it had previously been 1 to 3 cents less). The reduction in the size of the differential in favor of the weaker lines should have favored the New York Central and Pennsylvania. The reduction in ocean freights from southern ports should have favored the Pennsylvania. The second of the favors might have been conclusive.

[108] The trend value of shares in Table 5.5 is taken to be the value "expected" by all members of the associations for any one of the railroads. This trend value is a sophisticated version of the expectation that each of the lines was to receive "its usual share." As mentioned above, there was no policy of formally allotting shares in violation of the antipooling clause of the Interstate Commerce Act.

[109] If the cheater was the New York Central Railroad, the Baltimore and Ohio Railroad, or the Erie Railroad, then nothing whatsoever was gained: each of these lines received shares less than or equal to expected shares. If the Pennsylvania Railroad initiated the rate decreases, then tonnage was gained too late since the gain took place in May (after the general reconciliation) but not in April.

[110] This line also received much more than its expected share in March, however, which would suggest either that cheating went on completely undetected during the month of March to the advantage of this railroad, or that the March-April shift was a random occurrence.

[111] The Grand Trunk might have been the receiver of larger shipments into the East Coast as well. Shipments into five major east coast cities were quite a bit less than expected for the Baltimore and Ohio, and only slightly more than expected for the Erie and Pennsylvania in the month of April, 1888. This makes it possible, but not necessary, to argue that the Montreal road (for which there are no tonnage figures on east coast receipts) was the recipient of increased tonnage.

share but losses followed from the low rates of May and June. The estimated profits under cheating for three months were $93,000, while the estimated profits from the usual tonnage share at official rates were $116,000. The figures suggest that the Grand Trunk, if guilty of cheating, was also guilty of initiating a highly unprofitable venture (as in Table 5.6).

TABLE 5.6

Profits from Independent Rate-Setting, April–May, 1888

Month	*Year*	*Railroad*	*Computed Profits (in Dollars)*	*"Trend" Share of Profits at Official Rates (in Dollars)*
April	1888	Grand Trunk	67,443	65,977
May	1888		17,087	27,339
June	1888		9,281	22,760
			93,811	116,076

Source: "Computed Profits" equal $(R_T - K_2)q$, with "R_T" the Board of Trade rate in Table 5.4, "K_2" the average variable costs of 10.02 cents per hundred pounds for shipping from Chicago to New York City (as computed from the sources noted in Chapter 3), "q" the hundreds of pounds shipped from Chicago by this railroad, as shown in the *Annual Reports* of the Chicago Board of Trade.

The "Trend Share of Profits" is an estimate of the net revenues from the official rate on the tonnage share shown from the computed $\{S = \alpha + \beta T + \gamma X\}$ of Table 5.5. The profits equal $(R_0 - K_2)(S \cdot Q)$, with Q the total shipments from Chicago each month.

The second breakdown in grain rates seems to have been even more disastrous for the instigators. The Grand Trunk Railway and the Erie Railroad were accused of having offered 5-cent discounts on grain rates in August, 1888; the other railroads matched the 20-cent rate and revised the differentials in September. Both railroads received no more than their accustomed tonnage shares out of Chicago in August and September so that there were no increases in profits at the lower rate[112]; for the two months, Baltimore and Ohio profits on the grain traffic would have been $12,000 more and Erie profits would have been $26,000 more[113] if official rates had been maintained at 25 cents per 100 pounds.[114]

[112] The Baltimore and Ohio received -0.4 per cent from its trend share, while the Erie received $+2.4$ per cent from its trend, as shown in Appendix A.1. Both received approximately 2 per cent less than expected of receipts into the East Coast, as shown in Table A.2 of the Statistical Appendix to Chapter 5.

[113] As calculated from multiplying the actual grain tonnage of the railroads by 5 cents per 100 pounds. With no significant increase in tonnage, the "computed profits from cheating" equal $(R_T - K_2)q$, in contrast to $(R_0 - K_2)(S \cdot Q)$. In this instance $\{S \cdot Q \simeq q\}$ so that the revenue loss is $(R_0 - R_T)q$.

[114] The losses from having initiated the extended wars in provisions rates were far

The breakdowns were unprofitable not only for the disloyal member of the Executive Committee, but also for the loyal firms since they received the same tonnage shares at lower rates as well. It should have been apparent that cheating did not pay, given the revised rules of the cartel under regulation. The "costs" of rate-cutting in a disagreement on tonnage shares had been shown to be great. The results should have suggested that adherence to the cartel agreement was the best policy.

Strengthened Regulation and Strict Cartel Control of Rates, 1891 to 1893

The peace and quiet[115] continued, with the years 1891, 1892, and 1893 characterized by strong cartel control of rates. This followed from the activities of the regulator and of the Trunkline Executive Committee.

The Interstate Commerce Commission sought to prohibit a good part of the rate pattern from rate wars. The Commission stated the view that

---

greater than those in grain, but were part and parcel of the initiator's argument from the beginning. They did not demonstrate the ill effects from cheating, since they were not the result from cheating. They may have indicated the "high cost" of arguments carried out by rate reductions, however. Shares of the traffic in lard show that there was remarkably little shifting of tonnage at the beginning of the more extended periods of warfare, and that the shifting that did take place did not increase profits of the roads taking the initiative. The Grand Trunk Railway and the Erie Railroad began the dispute over the allowed discount on the provisions rate in the months of June, July, and August of 1888. The Grand Trunk received approximately 5 per cent more than its accustomed share of the lard traffic in the month of June, approximately .7 per cent more than expected in July, and approximately 3.5 per cent more than expected in August. The gain in tonnage did not come close to compensating for the 30 per cent decrease in lard rates in June and July (as shown in Table 5.3). The Erie, having taken up the struggle for the differential in July, received approximately 9 per cent more of the total tonnage in that month and approximately 8 per cent less in the succeeding month (as shown in Table A.3 of the Statistical Appendix to Chapter 5). The Erie's initial percentage gain was greater than the percentage reduction in rates on lard, but the gain was clearly not sufficient to compensate for the decline in August.

The Grand Trunk appears to have done no better in the war in provisions rates from June to November of 1890. The rates on lard transport to New York dropped from 35 cents per 100 pounds to 25 cents in the month of June, and to 23 cents in July. The 23-cent rate remained in force until December. The Grand Trunk received no surplus above its expected tonnage share of the lard shipments in either June or July. In August and September this road experienced an appreciable increase in share, when the 23-cent rate was in effect. The average tonnage gain throughout the entire period was less than 3 per cent while the average decrease in rates was more than 30 per cent. This exchange of lower rates for a larger tonnage share was clearly not profitable.

[115] The summers of 1888 and 1890 having been exceptions to general harmony for reasons mentioned above.

changes in official rates were "reasonable" only if permanent — so that a decrease in grain rates in a breakdown of rates had to be extended permanently. Changes in rates were reasonable when cost-justified and spaced over a considerable period, but 25 per cent to 50 per cent temporary discounts of rates were not reasonable: "a railroad company by putting in force a rate furnishes evidence that the rate is profitable, which is more convincing when such rate is long maintained." [116] When the rate was a departure from the existing geographical structure of rates, the Commission suspected that the rate was unreasonable: "Transportation charges are required to be relatively reasonable as well as reasonable in themselves, to prevent unjust discrimination between localities. A locality not widely dissimilar in situation and in respect to the transportation services of the same carrier to other localities where lower rates are given is entitled to rates that bear a just relation to the lower charges made." [117] The burden of proof was the railroad's, and it required comparison of rates over time and between localities to establish a rate reduction as legal. An independent rate reduction had become a matter of long-term policy, for each of the trunk-line railroads, rather than short-term profit.[118]

The Interstate Commerce Commission's emphasis upon maintaining an existing structure of official rates was impressed upon the trunk-line railroads in four important cases decided in 1890 and 1891. The first considered whether recent grain rates were "just and reasonable." The other three considered whether changes in an official commodity rate during a rate war were "just and reasonable" without reference to other commodity and class rates. The decisions in all four made it more difficult and costly for these railroads to induce, or take part in, a departure from agreed rates.

The Commission investigated freight rates on grain because the United States Senate passed a resolution in February of 1890 requiring them to do so. The issue was whether rates on food products were "too high"; the Interstate Commerce Commission argued that prices long maintained must have been no more than sufficient to cover costs, and prices higher than usual resulted in greater than normal profits. That is, "the frequency with which all grain including wheat and flour has been carried from Chicago at less than 20 cents . . . warrants the belief that some lower rate than 25 cents will be fairly remunerative. . . . 23 cents from Chicago is a reasonable limit over which rates on wheat and flour should not be

[116] Cf. *Coxe Brothers and Company* vs. *Lehigh Valley Railroad Company*, *ICC Reports*, *IV* (1891), 536.

[117] *The Manufacturers' and Jobbers' Union of Mankato, Minnesota* vs. *The Minneapolis and St. Louis Railway Company et al.*, *ICC Reports*, *IV* (1890), 79.

[118] Cf. also *Independent Refiners Association of Titusville et al.* vs. *Western New York and Pennsylvania Railroad Company et al.*, *ICC Reports*, *V* (1892), 415; *Union Pacific Railway Company* vs. *Goodrich*, 149 U.S. 680 (1893); *Foster* vs. *C. C. C. and St. L. Railway Company*, 56 Fed. 434 (1893).

made."[119] Since 20-cent rates were established during the short breakdown of rates in the summer of 1888, it turned out that war rates were treated as "just and reasonable" rates.[120]

In the first of three cases concerned with the structure of rates, the Commission reviewed the difference between the Pennsylvania Railroad rate on corn and that on corn products from Indianapolis to the eastern seaboard.[121] The two rates had been the same until 1888 when the summer reductions in grain rates were not accompanied by reductions in grain product rates. After a complaint by a shipper of corn products, the Commission found that the rate difference was not "reasonable" because there had been no such differences previously; upon a petition for rehearing, the Commission was told that there had been previous rate differences, as a result of secret discounts on the corn rate, so that this disparity was part of an "established pattern." The Commission dismissed the argument because secret reductions were not made on the basis of "some rule of uniformity as applied to differing times or conditions."[122] The agency made it clear that highly selective reductions in rates on competing commodities were prohibited in favor of maintaining a long-established structure.

The decision in the second of the three cases, the "commodity rate case" of 1891,[123] did allow the railroads to set some commodity rates, but only on the basis of comparability with given class rates. The Commission considered the Grand Trunk Railway's policy of setting "commodity rates" so as to cut some rates on westbound traffic in May of 1890.[124] The Com-

[119] "In the Matter of Alleged Excessive Freight Rates and Charges on Food Products," *ICC Reports, IV* (1890), 48. At the time the investigation was made, the official rate on wheat from Chicago was 25 cents per 100 pounds, while the rates on corn and oats were between 20 and 22½ cents per 100 pounds. These rates were all set at 25 cents per 100 pounds on December 29, 1890, and the common rate remained in effect throughout 1891 and the spring of 1892. Cf. the Interstate Commerce Commission, *Railways in the United States in 1902*, Vol. II, p. 79. Thus the Commission's conclusions were not a threat to declare any rate greater than 23 cents on grain from Chicago as not "just and reasonable."

[120] In the terminology of Chapter 2, the imposition of this rule required that the prospective cheater consider not the profits before cheating was detected as compared to the losses for the few months after detection, but rather the profits before detection and permanent losses. This should have provided a disincentive for the cheater to initiate a unilateral rate change — a disincentive not distinct from those imposed at the initiation of regulation, but of greater extent than earlier general requirements for all distance and commodity rates.

[121] *H. Bates et al.* vs. *The Pennsylvania Railroad Company, ICC Reports, IV* (1890), 281.

[122] *Ibid.*, p. 291.

[123] *The New York Board of Trade and Transportation et al.* vs. *The Pennsylvania Railroad Company, ICC Reports, IV* (1891), 447.

[124] Commodity rates reduced some Montreal to Chicago rates from 22 cents to 13 cents per 100 pounds; the changes in rates were filed with the Interstate Commerce

mission questioned whether, if sixth-class rates had been "just and reasonable" up to that time, temporary discounts on sixth class could have been "just and reasonable"; but the suspicion of illegality was not sufficient: "Manifestly the purpose . . . [is] to prevent a preference being given in rates or otherwise to a particular class of freight against other freight of the same class. . . . Like other rates they are lawful when they are just and reasonable and perform a lawful office. . . ." [125] Those rates which were large and temporary departures from the existing structure of rates were defensible if not "preferential."

In the "Squire and Company case" later in 1891, the Commission stated that there was a just and reasonable relation between the rate on live animals and that on dressed meat from Chicago to Boston. One of the characteristics of such a relation was that it "protected the business interest of the individuals and prevented such discriminations as would operate to the injury of one of a class of persons engaged in the same business with others in whose favor such discriminating rates would operate." [126] Another characteristic was that the rates "should be arrived at so far as practicable upon permanently continuing fixed factors and conditions . . . [immediate] commercial considerations alone [do] not furnish a sufficiently stable and fixed rule for guidance in making a rate which ought to remain substantially permanent. . . ." [127] This did not require the railroads to keep particular firms in business but to set differences between rates "determined from the cost of service" and from "differences which had long been established." [128]

These cases on the structure of rates and on the level of grain rates set simple standards: changes in rates were approved which were to be in effect for long periods of time and were because of changes in the costs of transport. Such standards made "war" rates either illegal or permanent.

Changes in the structure of rates wrought by the cartel at the same time made breakdowns more unlikely than ever.

The Trunkline Executive Committee's contribution was to devise, in 1891 and the early part of 1892, uniform rates for all of the lines on any class of traffic for any given distance. This was not achieved in 1887, as noted by an official of the Pennsylvania: "My impression is that the bases on which the tariffs of the several roads were prepared were not at all uniform; some of the roads undertook to add an amount to represent the

---

Commission according to the required procedure and was also submitted to the Trunkline Executive Committee. *Ibid.*, p. 526.

[125] *Ibid.*, p. 527 and pp. 531–532.

[126] *John P. Squire and Company* vs. *Michigan Central Railroad Company et al.*, *ICC Reports*, *IV* (1891), 611, at p. 622.

[127] *Ibid.*, p. 625.

[128] *Ibid.*, p. 627.

loading and unloading charge; others did not, and while rates between the principal commercial centers of the trunk lines were upon a fixed basis — that is, a percentage of the through rates to the seaboard — the rates between [local] points were not upon any well-defined basis." [129] The reform took place at this time because of the necessity to adjust rates to new state laws while not violating the Interstate Commerce Law. It happened that "in the year 1891 the Indiana legislature enacted laws which heavily increased the taxes of the railroads in that state, in consequence of which some of the important roads felt there was a necessity for increasing railroad earnings in Indiana state traffic in order to recoup the losses which they would sustain under the unjust tax laws." [130] It was questionable, however, whether higher rates in Indiana would have been "just and reasonable" when compared to through rates to the seaboard. In the course of meetings in 1892, "it was found that it was not practicable to readjust rates within the state of Indiana alone and that in order to adjust rates on that traffic it would be necessary to also readjust rates on traffic between points in Indiana and points in adjoining states . . . after several meetings the conclusion was reached that it would be impracticable to accomplish any substantial results unless there was a general revision of state and interstate rates throughout Central Freight Association territory and at this time the question of uniform mileage scales was discussed." [131]

A round of meetings in March of 1893 produced a schedule of rates said to have been "a compromise between the several bases of rates then in effect, and . . . based upon experience, observation, and intuition of the traffic officers, the commercial necessities, and observance of the law. . . ." [132] The upper limit on all rates for distances up to 500 miles was set at the Chicago to Pittsburgh rate for that class (so as not to violate the long haul–short haul clause of the *Act to Regulate Commerce*). The rates for shorter distance were designed to compensate for the high taxes in Indiana, but to be within the maximum rate law of the state of Ohio (as shown in Table 5.7). The result was a schedule very roughly in accord with distance, and not obviously in violation of any set of regulatory rules.[133]

[129] From a letter written by Mr. D. T. McCabe, Fourth Vice-President of the P. C. C. and St. L., to Mr. J. B. Hill, general freight agent, P. C. C. and St. L., Pittsburgh, Pennsylvania, May 15, 1907, as quoted in detail in *Historical Memorandum Concerning Class Rates in Central Freight Association Territory* by E. A. Tullis, special agent, Freight Rate Department, Pennsylvania Railroad Company, dated 1918. The memorandum is in the historical files of the Pennsylvania Railroad, Chicago, Illinois.

[130] *Ibid.*

[131] *Ibid.*

[132] E. A. Tullis, *op. cit.*, p. 5.

[133] The schedule also dictated lower rates for the Pennsylvania Railroad in central Pennsylvania: the rate for fourth-class tonnage from Philadelphia to Altoona was 19

TABLE 5.7

Central Traffic Association Scale of Uniform Class and Mileage Rates for All Railroads, Circa 1893

	Classes — Governed by Official Classification Rates in Cents per 100 Pounds					
Miles	*1*	2	3	4	5	6
5	7.5*	7.5*	7.5*	6	4	3
10	7.5*	7.5*	7.5*	6†	4.5	3†
15	7.5*	7.5*	7.5*	7	5	3.5
20	7.5*	7.5*	7.5*	7†	5†	4†
25	7.5*	7.5*	7.5*	7†	5.5	4.5
30	7.5*	7.5*	7.5*	7†	6	5
35	8.5*	8.5*	8	7.5	6.5	5.5
40	9.5*	9.5*	9	8	7	6
45	10.5*	10.5*	10	8†	7.5	6.5
50	12.0*	11.5	10.5	8.5	7.5†	6.5†
55	13.0*	12.5	11.5	9	7.5†	6.5†
60	14.5*	13.0	12.0	10.0	7.5†	6.5†
65	15.5*	14.0	13.0	10.0†	7.5†	7
70	17.0*	15.0	13.5	10.0†	8	7†
75	18.0*	16	15	10.5	8†	7†
80	19.5*	18.5	17	11	8.5	7.5
85	21*	19	17†	11.5	8.5†	7.5†
90	22*	20	17†	12	9	8
95	23*	22†	18	12†	9†	8†
100	24*	22†	19	12.5	9†	8†
110	24.5	22†	19.5	12.5†	9†	8†
120	25	22†	19.5†	12.5†	9.5	8†
130	26	23	19.5†	13	10	8.5
140	27.5	24	20	13†	10†	8.5†
150	28.5	25	20†	13.5	10.5	8.5†
160	30	26	21	13.5†	11	9
170	31	26.5	21.5	14	11†	9†
180	31.5	27	21.5†	14†	11.5	9†
190	32	28	22	14.5	11.5†	9.5
200	33	28.5	22†	15	12	9.5†
210	34	29.5	22.5	15†	12.5	10
220	35	30	22.5†	15†	13	10†
230	35.5	30.5	23	15.5	13†	10.5
240	36	31	23†	16	13†	10.5†
250	37	32	23.5	16†	13.5	10.5†
275	38.5	33	24.5	16.5	14	11
300	40	34	25	17	14.5	11.5
325	41	35	26	18	15	12
350	42	36	27	18.5	15.5	13
375	43	36.5	27.5	19	16.5	13.5
400	44	37.5	28.5	19.5	17	14
425	44.5	38.5	29	20.5	17.5	14.5
450	45	39	30	21	18	15

* Rates represent depressions imposed by Ohio maximum rate law.
† Rates represent failure to logically progress rates.
Source: As reproduced in full from the memorandum of Mr. E. B. Tullis, the Archives of the Pennsylvania Railroad, *op. cit.* The footnotes are from Mr. Tullis's manuscript, as is the note, "information taken from Central Traffic Association circular No. 2369."

The "rationalization" of local rates made discounts at some locations all too obvious. Thus the New Official Classification placed responsibility on any potential rate-cutter on a particular commodity or class of traffic to justify the single change or reduce the schedule of rates. The first was more difficult, given the recent Commission decisions; the second was more expensive, given the extent of the new schedule.

There were, in fact, no noticeable disruptions in through rates on grain and provisions in these three years. There was a controversy on dressed beef rate differentials, again, but rather than utilizing rate reductions as part of the argument, the whole matter was referred to arbitration. This "arbitration . . . has resulted in a decision that the lines north of Lake Ontario should charge 2½ cents per hundred pounds less than the lines south of this lake. . . . The differential awarded the Grand Trunk in the dressed beef question has been accepted by this line . . . and the tariff rates continue at 45 cents to both New York and Boston on an average of probably 100 cars a day out of Chicago."[134] Also, a plan for more flexible differentials on other types of transport was put into effect: "so as to apportion traffic. . . . The commissioners [are to] know accurately from monthly reports how freight has gone before they decide how it ought to go, and, after they decide that, rates are reduced approximately 5 per cent [for the lines with a deficit in market share]. . . . The commissioners will endeavor to fix the rates about the twentieth of each month which means that any [deficit] in December shall be corrected by a change in rates in February."[135] It was also agreed that there should be "a joint agency at the New York Produce Exchange to revise and adjust the rates on grain as necessary. An endeavor will be made to apportion the tonnage about in proportions that the roads have secured in the past three years."[136] All of these changes were not in reaction to cheating on the existing agreement, or to publicly announced rate changes during a disagreement over differentials. They were attempts to improve sharing arrangements considered quite adequate, so as to correct for random changes in shares of tonnage.

Steadfast loyalty to official rates resulted in rigid margins between prices in the Chicago and New York grain markets. During each of the summer seasons, the average price difference was approximately 5 to 6 cents below the average official all-rail rate, and 3 to 6 cents greater than the announced

---

cents in 1887 (as shown in Table 5.1), while the 230-mile rate in the New Classification was 15.5 cents (as shown in Table 5.7); the rate from Cessna to Cumberland, Maryland, was halved (as can be seen from comparing the 9.0 cent rate of Table 5.2 with the 20-mile rate for the new sixth class in Table 5.7).

[134] The "Railroad Rates" column of the *Railroad Gazette* (February 20, 1891).

[135] "Trunkline Pooling," the *Railroad Gazette* (November 25, 1892), 884.

[136] "A Joint Agency at the New York Produce Exchange," the *Railroad Gazette* (March 31, 1893), p. 244.

lake-rail rate.[137] The average official rates for each of the winters of 1890 to 1893 were close to 25 cents per 100 pounds, as were the average Board of Trade rates for these seasons. The average Chicago–New York grain price difference was as low as 22 cents in the winter of 1890–1891, and as high as 29 cents in the winter of 1891–1892, but the general impression is that the average differences were close to 25 cents per 100 pounds. There was very little variation in grain price differences, and only three changes in all-rail rates. Small variance in the week-to-week grain price differences was explained in part by the very few changes in official rates and the slightly more frequent changes in lake-rail rates (given an adjustment period before the grain price difference responded in full to a change of rates, then changes in rates and the adjustment process together explained from 20 to 40 per cent of the variance in the grain price difference, as shown in Table 5.8).[138]

Price variations in the provision markets — as represented by New York–Chicago price differences on barreled western lard — were also dictated by official rates (if not as closely as were grain price differences).[139] There were some differences between Board of Trade rates and official rates on lard to New York: in May and June of 1891, the Board of Trade listed provisions rates at 25 cents to 27½ cents per 100 pounds while the official rate was 30 cents per 100 pounds; during June of 1892, the Board of Trade listed a rate of 27 cents with official rates based on 30 cents per 100 pounds; during the last five months of 1893, there were no official changes in rates, but the Board of Trade noted discounts of 1 cent per 100 pounds for transporting lard to New York.[140] Otherwise, Board of Trade rates and official rates

[137] Except for the summer of 1891, when a particularly low lake-rail rate seems to have had little effect upon the average grain price difference, as shown in Table 5.8.

[138] In these seasons there was a two- or three-week adjustment period and a tendency for a change in the grain price difference to have been from 16 per cent to 190 per cent of the change in rates — with the more frequent change in the price difference having been from 60 to 90 per cent of the rate change. The percentage changes are indicated by the values of computed $\hat{\beta}$ in Table 5.8.

[139] Board of Trade rates could not have explained a greater proportion of the variation in the week-to-week grain price difference, because these rates were the same as official rates in all except one week during the summer of 1891, as listed in the Interstate Commerce Commission, *Railways in the United States in 1902*, *op. cit.*

[140] This was a discount on "a published rate of 50 cents per 100 pounds" when the usual fifth-class lard rate was 30 cents. It would appear that the discount from August to December of 1893 followed after a shift of lard from fifth to third class. No official report of such a shift in class has been found, however.

These few differences between Board of Trade and official rates were not accompanied by allegations or evidence of nonadherence to the agreement. The first instance contained week-by-week listings of what the Board of Trade called "changes in official rates," so that there may have been no more than an error of omission in the 1902 Interstate Commerce Commission listings of previous official rates. The second instance involved a Board of Trade announcement of a "change in official rates" for two weeks previous

TABLE 5.8

Official Rail Rates, Board of Trade Rail Rates, and Differences Between Chicago and New York Grain Prices, 1890 to 1893

$$R_T = \alpha + \beta R_0 + \nu \quad (1)$$

R_T, the rail rate, in cents per hundred pounds Chicago to New York, from the Board of Trade *Daily Commercial Bulletin;* R_0, the official rate from compilations of posted tariffs and from announcements of the various traffic associations.

$$P_g{}^* = \alpha + \beta R_0 + \gamma R_L + \nu; \qquad P_g{}^* = \alpha + \beta R_T + \gamma R_L + \nu \quad (2)$$

$P_g{}^*$, the weekly average of the differences between Chicago and New York daily high and low spot prices, in cents per hundred pounds, for the number 2 "western" grade of corn and oats; the official rate R_0 or the Board of Trade rate R_T; the lake-rail rate R_L from Chicago to New York via steamship to Buffalo and rail from Buffalo to New York as quoted on the Chicago Board of Trade.

$$P_{g,t} = P_{g,t-1} + \epsilon(P_g{}^* - P_{g,t-1}) \quad (3)$$

$P_{g,t}$, the average of the daily Chicago–New York grain price differences during week t and $P_{g,t-1}$, the average for the previous week; $P_g{}^* = \alpha + \beta R + \gamma R_L$ for the summer season and $P_g{}^* = \alpha + \beta R$ for the winter season from equilibrium adjustment to changes in transport rates.

The standard error of each coefficient is shown in parentheses below the coefficient.

WINTER SEASON, 1890–1891:

(17 weeks); $\overline{R}_0 = \overline{R}_T = 24.529$ and $R_0 = R_T$ each week

$\overline{P}_g = 21.852$; $SP_g = 0.932$

(15 weeks); $P_g = 12.779 + \underset{(0.316)}{0.359}R$; $\hat{\rho}^2 = .090$

(13 weeks); $P_{g,t} = 0.640P_{g,t-1} + 0.360(1.651 + \underset{(0.218)}{0.957}R)$; $\hat{\rho}^2 = .304$

SUMMER SEASON, 1891:

(35 weeks); $R_0 = 25.00$ each week; $\overline{R}_T = 24.428$; $\overline{R}_L = 11.000$

$\overline{P}_g = 18.226$; $SP_g = 3.774$

(15 weeks); $P_g = -8.121 + \underset{(0.908)}{2.410}R_L$; $\hat{\rho}^2 = .351$

(15 weeks); $P_{g,t} = 0.472P_{g,t-1} + 0.528(-3.293 + \underset{(0.304)}{1.970}R_L)$; $\hat{\rho}^2 = .219$

WINTER SEASON, 1891–1892:

(19 weeks); $R_0 = R_T = 25.000$ for the entire period

(14 weeks); $\overline{P}_g = 29.549$; $SP_g = 1.775$

SUMMER SEASON, 1892:

(34 weeks); $\overline{R}_0 = \overline{R}_T = 23.412$ and $R_0 = R_T$ each week; $\overline{R}_L = 12.813$

$\overline{P}_g = 15.203$; $SP_g = 1.429$

(34 weeks); $P_g = 11.604 + \underset{(0.308)}{0.152}R - \underset{(.0267)}{0.001}R_L$; $\hat{\rho}^2 = .019$

(30 weeks); $P_{g,t} = 0.454P_{g,t-1} + 0.546(13.289 - \underset{(0.136)}{0.0001}R + 0.149R_L)$; $\hat{\rho}^2 = .394$

WINTER SEASON, 1892–1893:

(18 weeks); $R_0 = R_T = 25.000$ for the entire period

(16 weeks); $\overline{P}_g = 23.534$; $SP_g = 1.929$

TABLE 5.8 (Continued)

SUMMER SEASON, 1893:

(34 weeks); $\bar{R}_0 = 24.853$; $\bar{R}_T = 24.441$; $\bar{R}_L = 13.939$

$\bar{P}_g = 19.008$; $SP_g = 2.505$

(31 weeks); $P_g = 8.680 + \underset{(0.351)}{0.396}R_0 + \underset{(0.469)}{0.047}R_L$; $\hat{\rho}^2 = .052$

(31 weeks); $P_{g,t} = \underset{(0.147)}{0.584}P_{g,t-1} + 0.414(6.554 + 0.414R_0 + 0.162R_L)$; $\hat{\rho}^2 = .229$

were the same. The average lard price differences, during the summer seasons of 1891 to 1893, were similar to official and Board of Trade rates when the rates were equal, and closer to the Board of Trade rates when the rates differed (when "similar" is considered to be a price margin approximately 4 cents greater than the rates, as in Table 5.9). During the winter seasons, lard price differences were mostly within 3 to 6 cents of listed rates.[141] The lard markets indicate shipments at agreed rates close to 30 cents per 100 pounds during much of the three-year period.

Regulation and Cartelization: The Weakening of Regulation and the Disruption of Rates in 1894

At this point of great success in the control of rates, a number of changes were in the process which were to render further control impossible. Court decisions in 1892 and 1893 weakened Interstate Commerce Commission regulation after 1893. Questions arose as to the extent of regulatory jurisdiction over certain rates in the trunk-line territory, and the answers were not forthcoming or it was found that there was a lack of jurisdiction. Court decisions in 1896 and 1897 greatly decreased regulatory control further. The weakening of national authority nullified all of the gains of regulation and cartelization over the structure and level of rates.

One of the western railroads, during proceedings leading to a Supreme Court decision in 1892, questioned whether through rates over connecting

to the Interstate Commerce Commission dating of this change; this again may have been an error in listing, since there were no accompanying reports of cheating. The third instance seems to have resulted from incomplete reports of rates by both the Board of Trade and the rate-setting associations — or a lack of any reports whatsoever.

[141] The largest of these differences occurred at the same time that the Board of Trade announced "a blockade of transport." A blockade lasted for some weeks in the winter of 1891–1892, and again in November and December of 1892; the costs at these times for transporting to the seacoast included substantial (implicit) charges for waiting for transport, as well as the rail rate — so that the lard price differences may have been close to "the real transport charge."

TABLE 5.9

Railroad Rates and Lard Price Differences: 1891 to 1893

Month	*Official Rate (Cents per 100 Pounds)*	*Board of Trade Rate (Cents per 100 Pounds)*	*Average of the Daily Chicago–New York Western Lard Price Differences (Cents per 100 Pounds)*
January, 1891	30	30	36
February	30	30	36
March	30	30	33
April	30	30	25
May	30	28	28
June	30	25	28
July	30	30	24
August	30	30	32
September	30	30	34
October	30	30	33
November	30	30	34
December	30	30*	41
January, 1892	30	30*	37
February	30	30*	34
March	30	30	33
April	30	30	33
May	30	30	29
June	30	27	29
July	25	25	29
August	25	25	31
September	25	25	30
October	30	30	34
November	30	30*	56
December	30	30	—
January, 1893	30	30	36
February	30	30	32
March	30	30	29
April	30	30	30
May	30	30	32
June	30	30	28
July	30	30	30
August	50†	50†	50
September	50†	50†	47
October	50†	50†	50
November	50†	49	52
December	50†	40	46

* Blockade of transport, as noted in the Chicago *Daily Commercial Bulletin.*
† Suspected shift of lard from fifth class to third class.
Source: As in Table 5.3.

lines and local rates were quoted under "like circumstances and conditions," and thus whether the structure of rates was subject to scrutiny under the long haul–short haul clause of the Interstate Commerce Law. The Interstate Commerce Commission's view had been that railroads utilizing one continuous road bed formed "one line," whether or not connecting railroads were independent companies. The Commission also had followed the policy of comparing the joint rates of "one line" with the local rates of any of the railroads involved.[142] This policy was considered by the United States Court and found wanting; instead, each line was "distinct" when setting a local rate and when becoming a party to a joint tariff so that "a through tariff on a joint line is not the standard by which the separate tariff of either company is to be measured or condemned." [143]

The *Osborne* decision was a setback for the Interstate Commerce Commission's policy of requiring comparability between through and local rates. The agency itself stated that "[this] case originating in the Iowa Federal Court was carried to the Circuit Court of Appeals and a construction of the section entirely at variance with the meaning announced by the Commission in 1887, and long accepted and believed in by the great majority of carriers, was announced. . . . The decision of the Court of Appeals holds that there are as many lines as there are carriers, and that each is wholly independent of any other as regards the legality of rates under the fourth section. Having settled upon this view it then proceeds to say in substance that a rate over one railroad may, for anything in the fourth section, be higher than the rate over that with another connecting road or over that with any number of connecting roads. . . . Congress never intended that the fourth section should permit rates over connecting roads which would present such extravagant inequalities as related to distance. . . . This ruling of the Appellate Court gave character and force to the opposition against this section." [144] The public view was that the decision allowed the roads to set rates without regard to distance: "[given this decision] neither company is bound to adjust its own local tariff to suit the other . . . if they make a joint tariff it is not a basis by which the reasonableness of a local tariff of either line is determined." [145]

[142] As has been noted above in the discussion of the first four years of regulation according to the *Act to Regulate Commerce.*

[143] *Osborne* vs. *Chicago and Northwestern Railroad Company*, 52 Fed. 912 (1892), at 915.

[144] *The Seventh Annual Report of the Interstate Commerce Commission, December 1, 1893* (Washington: Government Printing Office, 1893), pp. 33–34.

[145] "The Long and Short Haul Controversy," the *Railroad Gazette* (October 28, 1892), 806. The trunk-line railroads had opportunities to divide through rates, between such "independent" railroads as the Lake Shore and Michigan Southern, the Nickel Plate, the New York Central, and the Boston and Albany — all controlled and operated by

The lack of comparability between through and local rates was confirmed by the Circuit Court for the Northern District of Georgia in the summer of 1893. This court found that the James N. Mayer Buggy Company of Cincinnati, Ohio, had no complaint against a local railroad in Georgia because the through rates from Cincinnati to Atlanta, Georgia (a distance of 474 miles) and to Augusta, Georgia (a distance of 645 miles) were the same, while the joint through-local rate to Social Circle, Georgia (approximately 526 miles) was 30 cents higher.[146] The railroad asserted that it had merely added its local tariff for service from the main line of a long-distance transporter to the through tariff from the main line to Cincinnati, and that it was not a party to "an arrangement for continuous carriage or shipment." The Court accepted these assertions in light of the *Osborne* decision, and concluded that the rate to Social Circle was a combination of two separate tariffs and was not comparable with the joint through tariffs to Atlanta and Augusta.[147] From the point of view of the Interstate Commerce Commission, "the result of this is to exempt this railroad, and all others similarly situated, from the provisions of the *Act to Regulate Commerce* on all interstate traffic upon which such a road chooses to exact its local rates as its share of the total charge. . . . It will thus be seen that in addition to the embarrassments imposed by the original 'line' decision [the *Osborne* decision], the very jurisdiction of the law itself is invaded by the extension . . . indulged in by the Georgia federal court." [148]

Two changes in regulatory rules during 1893 and 1894 complemented the *Osborne* and *Social Circle* decisions in weakening Interstate Commerce Commission control of the structure of rates. The first change, in fact, followed from the *Osborne* and *Social Circle* "separate and independent line" doctrine. There had been some controversy among lake steamer companies, railroads, and the Interstate Commerce Commission as to whether the lake-rail rates from Chicago to the eastern seaboard were subject to regulation. As the Interstate Commerce Commission noted, "from the date that *Act to Regulate Commerce* became affective [sic] until it was amended in March 1889 its provisions, as far as they require the publi-

---

the managers of the New York Central. If the managers set a lower rate on through traffic to the eastern seaboard than the 450-mile rate in the New Official Classification, then the Interstate Commerce Commission, proceeding on grounds of "unjust and unreasonable" rates, might have encountered the argument that the local classification was "distinct" from the through rates.

[146] Social Circle, Georgia, was served only by the one railroad while the other two cities were on the main lines of a number of railroads.

[147] Cf. *Cincinnati, New Orleans and Texas Pacific Railroad Company* vs. *Interstate Commerce Commission*, 162 U.S. 184 (1896), "Statement of the Case," pp. 185–187; cf. also "The Social Circle Rate Cases," The *Railroad Gazette* (July 7, 1893), 507.

[148] *The Seventh Annual Report of the Interstate Commerce Commission, December 1, 1893* (Washington: Government Printing Office, 1893), p. 35.

cation of tariffs and notice of advances and reduction in rates, were not considered to apply to this traffic. When however, the *Act* was so amended as to include joint tariffs, a serious question arose as to whether it included the rates made by the regular lines of lake carriers operating in connection with rail lines." [149] The railroads agreed in 1889 to submit and publish the rail rate from Lake Erie to the eastern seaboard, but the steamer companies did not agree to list the lake rate from Chicago to Lake Erie. During the early 1890's the railroads followed the procedure of adding to the official rail rates from Lake Erie to the eastern seaboard, the daily lake-rail rates quoted on the Board of Trade.[150] But the new court decisions changed the necessity for this procedure; the New York Central Railroad, for one, was in a position to claim that its Buffalo to New York City rate was an intrastate rate, since the railroad did not anywhere cross the New York state borders with this particular transport tonnage, and that this rate also was "separate and independent" from all-rail Chicago to New York rates. The Pennsylvania Railroad was able to claim the same on rates for transport from Erie, Pennsylvania, to Philadelphia. Filing of any part of the lake-rail rates, and justification of changes in these rates, were not required of these railroads.[151]

There was another change in the regulatory rules which subtracted from the enforceability of given standards for "just and reasonable" rates. The Commission had heretofore taken testimony from railroad officials on rates for various purposes and had discovered in some instances that secret discounts or rebates had been made available. The discovery of such vio-

[149] *The Seventh Annual Report of the Interstate Commerce Commission, December 1, 1893* (Washington: Government Printing Office, 1893), Appendix E: "Traffic and Rates Upon Through Lines Operating Via the Great Lakes," p. 239.

[150] As a matter of fact, one (unnamed) railroad filed through tariffs of lake-rail rates under the same procedures as for all-rail rates from 1889 to 1893. As the Interstate Commerce Commission noted, the company, in a letter of April 17, 1889, stated its policy: "We publish a tariff of rates for through grain from Chicago to the seaboard and other eastern points and file the same with the commissioner as directed by law. We do not advance our rates without ten days previous notice or reduce them without three days previous notice." The Interstate Commerce Commission also stated that this company "specifically disclaimed any intention to assent to the propositions that it is impracticable to publish through rates on grain and bulk carriage via lake and rail." (*The Tenth Annual Report of the Interstate Commerce Commission*, *op. cit.*, p. 240.)

[151] The Interstate Commerce Commission believed that intrastate rates were cut secretly, but legally, in the fall of 1893. Board of Trade listings of lake-rail rates indicated that there were discounts on official rates on grain and provisions to the eastern seaboard and to Europe. The Commission found that "in regard to export shipments . . . no attempt to meet the legal requirements governing the publication and filing of tariffs has been made." (*The Tenth Annual Report of the Interstate Commerce Commission*, *op. cit.*, pp. 255–256.) The "usual legal requirements" were not enforced in these unusual circumstances, because it was possible for some trunk lines to set rates on shipments from Chicago to New York outside of the jurisdiction of national regulation.

lations may have been incidental to findings of "just and reasonable" rates, but it had been customary for the Commission to report the testimony on secret cuts to the Justice Department for criminal prosecution. In 1890 a shipper refused to disclose the actual rate he had received, because this might have constituted self-incrimination contrary to the Fifth Amendment of the Constitution. In 1892 the Supreme Court of the United States found that this shipper had the legal right to deny rate information to the Interstate Commerce Commission, even though this prevented the Commission from proceeding to examine the level and structure of existing rates relative to the requirements of Sections 2 to 4 of the *Act to Regulate Commerce.*[152] This thwarted regulatory investigations of rates set in the fall of 1893: "Several indictments for unjust discrimination were ready for trial on November 21 [1893] in the United States Circuit Court sitting in Chicago. . . . The manager of the Lackawanna Fast Freight Line, the chief clerk in the freight office of the Burlington system, the Western freight manager at Buffalo of the Delaware, Lackawanna and Western, the general freight agent of the New York, Chicago and St. Louis road, an agent of the Nickel Plate Fast Freight Line located at Chicago, and the manager of the Traders Dispatch Fast Freight Line refused to testify [because of] rebates alleged to have been paid to one of the large packing firms of Chicago, and the secretary of that firm and four other persons employed by it also declined to answer questions propounded in the same matter. Each of the witnesses set up his constitutional privilege as the basis of his refusal to testify. Under this condition of affairs it was deemed useless by the District Attorney and the Court to try the indictments found . . ."[153] This decision was not important because it prevented Interstate Commerce Commission detection of departures from the established tariff — such departures were, as the Commission pointed out, "crimes [which] do not differ from other misdemeanors . . . the ordinary machinery of the criminal law must be employed in enforcing the penal provisions of the *Act to Regulate Commerce* and there is no other way by which these provisions can be enforced . . ."[154] — but rather because Commission investigations no longer provided sufficient information to judge the "reasonableness" of whatever rate was actually in effect, and to require readjustment of departures from a given structure of rates.[155]

[152] Cf. *Counselman* vs. *Hitchcock*, 142 U.S. 547 (1892).

[153] *The Sixth Annual Report of the Interstate Commerce Commission, December 1, 1892* (Washington: Government Printing Office, 1892), pp. 28–29.

[154] *The Seventh Annual Report of the Interstate Commerce Commission, December 1, 1893* (Washington: Government Printing Office, 1893), p. 7.

[155] There was hope for improvement of this situation in 1893 with the enactment of the *Compulsory Testimony Act of February 11, 1893* (27 Stat. 443), since this granted immunity from prosecution for any witness providing testimony in hearings on "just and reasonable" rates; but there was no assurance that the grant of immunity was

After a point, all of the changes in regulation were in effect. The "separate and independent line" doctrine became a matter of concern in the summer of 1892 and a cause for despair in the Interstate Commerce Commission after the Circuit Court decision in the *Social Circle* case during the summer of 1893. The controversy in lake rates was publicly acknowledged in a Commission report at the end of 1893. The setback in investigatory procedures took place with the *Counselman* decision in 1892; there was renewed hope, but much uncertainty, after the passage of the *Compulsory Testimony Act in 1893*. The year 1894 was then the year for changes in regulation to have affected trunk-line rate-setting.

The effects were first realized a bit earlier. During the late fall and winter of 1893 there was a fairly general disruption of official rates. The first notice appeared in the Chicago Board of Trade on November 10 when "rumors of cut rates were numerous, reductions of 2–3 cents were said to have been made, although the nominal rate to New York was 25 cents per 100 pounds for flour and grain and 30 cents for provisions." [156] The next week it was said that "provision rates are unsettled, the tariff rate being 30 cents but it was cut and there was a reduction of 5 cents per 100 pounds on provisions to the South." [157] Small discounts of both grain and provision rates continued to be offered until the first week of December, when all eastbound rates were "demoralized with open cuts of 5 cents per 100 pounds on flour and grain and 7½ to 8 cents per 100 pounds on provisions to New York." [158] Moderate rate-shading had deteriorated into a war in eastbound freight.

The Trunkline Executive Committee proceeded to reduce official rates.[159] The Erie Railroad filed lower rates with the Interstate Commerce Commission during the second week of December, at a level it thought to be equal to that of the secret rates. At this time, "all of the eastern lines and

---

sufficient to protect the witness from self-incrimination, until the Supreme Court favorably reviewed the legislation of 1893 in *Brown* vs. *Walker* [161 U.S. 591 (1896)] during the 1896 term. The Commission was still complaining in 1894: "the powers [of the *Compulsory Testimony Act*] apparently sufficient but in reality extremely difficult to successfully exercise [have] been resisted by carriers and unwilling witnesses in various ways. . . ." [*The Eighth Annual Report of the Interstate Commerce Commission, December 1, 1894* (Washington: Government Printing Office, 1894), p. 13.]

[156] "Railroad Freights," the *Chicago Daily Commercial Bulletin*, November 10, 1893.

[157] "Railroad Freights," *op. cit.*, November 17, 1893.

[158] "Railroad Freights," *op. cit.*, December 8, 1893.

[159] There was also a threat of prosecution for violation of the criminal code: "It was reported from Chicago yesterday that there are likely to be results from the present railroad war in eastbound rates little expected by the railroad officials responsible for it. The Interstate Commerce Commission is understood to have agents at work in that city collecting all the evidence obtainable as to the cuts and those responsible for making them." ("Railroad News and Notes," *The New York Times*, December 9, 1893, p. 3.) But nothing more was heard of this.

their western connections have issued eastbound tariffs to meet the open cut of the Erie last week of sixth-class rates, including rates on grain and dairy products. These reductions are now in force between New York and Chicago. . . . The Erie people insist that they made the reduced rate on eastbound freight only because they had proofs that the other lines were cutting rates secretly." [160] On December 17, the official rate was decreased from 20 cents to 17½ cents on grain to New York; on December 25 this rate was reduced to 15 cents. The rate for transport of live hogs was also reduced from 30 cents to 22½ cents on December 11, and to 20 cents on December 18. The official lard rate, in contrast, was not changed even though widely announced discounts of 2 cents per 100 pounds throughout November widened to 5 cents in December.[161]

The disruption of agreed rates this winter continued in January, February, and March. An attempt was made to reestablish the 25-cent rate on January 1; the Board of Trade listed immediate cuts, however, of 5 cents per 100 pounds. By January 15 discounting was less extensive, and both official and Board of Trade rates were raised to 25 cents per 100 pounds for the remainder of the month. Discounting reappeared during both February and March: February discounts of 2 to 5 cents were followed by a reduction of the official rate from 25 cents to 20 cents per 100 pounds, while March discounting took the Board of Trade rate down to 15 cents per 100 pounds for the greater part of three weeks.

A number of railroads each accused the other of having caused this 40 per cent decrease in rates. The Erie's officers stated that they had evidence of cheating in November, and implied that the New York Central Railroad was the party responsible.[162] A New York Central Railroad official, in turn, indicted all of the railroads with the rate differential (including the Erie): "The situation which confronts us is very plain. We entered into agreements which were signed by the executive officers of all the companies which were members of the Trunkline Executive Committee and of the Central Traffic Association. These agreements have been secretly violated. The five so-called weaker lines were given such differentials as were supposed to be sufficient . . . in consideration of these concessions these lines agreed not to pay commissions . . . it now appears that one of them has made a secret contract with a steamship line which is to run the whole of this year to allow commissions of from one-half to one-third the total fare from Chicago to New York." [163] There were no investigations to

[160] The "Freight Rate Situation," *The New York Times*, December 15, 1893, p. 2.

[161] Cf. the Interstate Commerce Commission, *Railways in the United States in 1902*, *op. cit.*, pp. 79 and 86. For lard rates, cf. the discussions in the *Chicago Daily Commercial Bulletin* for November and December of 1896 in the "Weekly Freights" column.

[162] *The New York Times*, December 5, 1893, *op. cit.*, p. 2.

[163] "Mr. Depew on Rate Cutting," *The New York Times*, March 25, 1894, p. 3.

establish which of the accusations was correct; there was general acknowledgment that a number of lines had offered discounts in a number of separate instances. Given the extent of secrecy, it was probably not feasible to collect sufficient evidence to prove the guilt of the cheater in each of the many instances.[164]

The rate demoralization was accompanied by marked declines in margins between commodity prices in midwest and east coast markets. The average official rate was 22 cents per 100 pounds, as compared to the average Board of Trade rate of 18.7 cents; both were 25 cents the previous winter.[165] The average difference between Chicago and New York spot grain prices was 16.8 cents, as compared to 23.5 cents the previous winter. The lower, and independent, Board of Trade rates explained more than 26 per cent of the concurrent variance in the grain price differences; the official rate explained little more than 3 per cent of this variance (given the computed values of the coefficient of determination $\hat{\rho}^2$ in, respectively, the least-squares estimates of $P_g = \alpha + \beta R_T$ and $P_g = \alpha + \beta R_0$ shown in Table 5.10). Assuming that the grain price difference depended upon the previous weeks' grain price differences and an adjustment to an equilibrium difference, changes in Board of Trade rates explained 46 per cent of the variance in observed grain price differences, while changes in official rates

[164] The Erie Railroad, if guilty of cheating in October and November of 1893, realized an appreciable gain in its share of Chicago grain tonnage as a result (as shown in the Statistical Appendix, Table A.1) but no gain in its share of tonnage into the eastern seaboard (as shown in the Statistical Appendix, Table A.2). The Erie realized a slight gain in its share of the lard tonnage out of Chicago (as shown in the Statistical Appendix, Table A.3).

The New York Central Railroad, if the major party guilty of cheating, gained appreciably from such activity in December in grain traffic out of Chicago, and in January in grain traffic into the eastern seaboard. This line was not able to hold its own in the traffic in lard, however.

Perhaps both lines were unjustly accused, and the Pennsylvania Railroad was chiefly responsible for the disruption in rates. This line received slightly more than its (trend) share in December of 1893, and more than twice its share in January of 1894, of grain tonnage from Chicago. The Pennsylvania was able to obtain far more than its long-term trend of shares of lard tonnage from Chicago from August through December of 1893.

The point is that in the absence of evidence from the cartel supporting an allegation that any of three railroads reduced its rate, only surmise is possible. Contrary to the experience in earlier instances of general breakdowns of rates, the secrecy seems to have been sufficient in the 1893–1894 instances to have prevented the naming of the cheater *independent* of examining the shift in market shares resulting from cheating. Consequently no conclusions can be reached on the actual profits that followed from cheating in grain or lard tonnage.

[165] Board of Trade rates were independent of official rate changes given that less than 36 per cent of the variance in Board of Trade rates was explained by announced decreases in official rates, as shown by the computed value of the coefficient of determination $\hat{\rho}^2$, for $R_T = \alpha + \beta R_0$ in Table 5.10.

TABLE 5.10

Official Rail Rates, Board of Trade Rail Rates, and Differences Between Chicago and New York Grain Prices, 1893 to 1895

$$R_T = \alpha + \beta R_0 + \nu \qquad (1)$$

R_T, the rail rate, in cents per hundred pounds Chicago to New York, from the Board of Trade *Daily Commercial Bulletin;* R_0, the rail rate from compilations of posted tariffs and from announcements of the various traffic associations.

$$P_g{}^* = \alpha + \beta R_0 + \gamma R_L + \nu; \qquad P_g{}^* = \alpha + \beta R_T + \gamma R_L + \nu \qquad (2)$$

$P_g{}^*$, the weekly average of the daily differences between Chicago and New York spot prices, in cents per hundred pounds, for the number 2 "western" grade of corn and oats; the official rate R_0 or the Board of Trade rate R_T; the lake-rail rate R_L from Chicago to New York via steamship to Buffalo and rail from Buffalo to New York.

$$P_{g,t} = P_{g,t-1} + \epsilon(P_g{}^* - P_{g,t-1}) \qquad (3)$$

$P_{g,t}$, the average of the daily Chicago–New York grain price differences during week t and $P_{g,t-1}$, the average for the previous week; $P_g{}^* = \alpha + \beta R + \gamma R_L$ for the summer season and $P_g{}^* = \alpha + \beta R$ for the winter season from equilibrium adjustment to changes in transport rates.

The standard error of each coefficient is shown in parentheses below the coefficient.

WINTER SEASON, 1893–1894:

(15 weeks); $R_T = -1.993 + \underset{(0.366)}{0.979} R_0; \quad \hat{\rho}^2 = .355$

$\overline{R}_0 = 22.063; \quad \overline{R}_T = 18.733; \quad \overline{P}_g = 16.813; \quad SP_g = 2.034$

(15 weeks); $P_g = 14.279 + \underset{(0.172)}{0.114} R_0; \quad \hat{\rho}^2 = .032$

(15 weeks); $P_g = 13.871 + \underset{(0.072)}{0.157} R_T; \quad \hat{\rho}^2 = .266$

(15 weeks); $P_{g,t} = 0.854 P_{g,t-1} + \underset{(0.157)}{0.146}(-17.845 + 1.551 R_0); \quad \hat{\rho}^2 = .381$

(15 weeks); $P_{g,t} = 0.715 P_{g,t-1} + \underset{(0.145)}{0.285}(8.619 + 0.432 R_T); \quad \hat{\rho}^2 = .460$

SUMMER SEASON, 1894:

(34 weeks); $R_T = 5.104 + \underset{(0.168)}{0.729} R_0; \quad \hat{\rho}^2 = .365$

(34 weeks); $\overline{R}_0 = 20.429; \quad \overline{R}_T = 20.000; \quad R_L = 12.000$ (a constant)

$\overline{P}_g = 13.919; \quad SP_g = 2.717$

(34 weeks); $P_g = 11.221 + \underset{(0.338)}{0.132} R_0 + 0.000 R_L; \quad \hat{\rho}^2 = .005$

(34 weeks); $P_g = 7.202 + \underset{(0.320)}{0.333} R_T + 0.000 R_L; \quad \hat{\rho}^2 = .032$

(34 weeks); $P_{g,t} = 0.578 P_{g,t-1} + \underset{(0.148)}{0.422}(3.432 + 0.509 R_0 + 0.000 R_L); \quad \hat{\rho}^2 = .227$

(34 weeks); $P_{g,t} = 0.558 P_{g,t-1} + \underset{(0.148)}{0.442}(2.513 + 0.568 R_T + 0.000 R_L); \quad \hat{\rho}^2 = .234$

TABLE 5.10 (Continued)

WINTER SEASON, 1894–1895:

(17 weeks); $R_T = -11.778 + 1.411R_0$; $\hat{\rho}^2 = .607$
(0.293)

$\overline{R}_0 = 22.353$; $\overline{R}_T = 19.764$; $\overline{P}_g = 13.964$; $SP_g = 0.861$

(17 weeks); $P_g = 15.289 - 0.059R_0$; $\hat{\rho}^2 = .026$
(0.091)

(17 weeks); $P_g = 14.927 - 0.049R_T$; $\hat{\rho}^2 = .066$
(0.049)

(16 weeks); $P_{g,t} = 0.421P_{g,t-1} + 0.579(17.098 - 0.145R_0)$; $\hat{\rho}^2 = .460$
(0.193)

(16 weeks); $P_{g,t} = 0.409P_{g,t-1} + 0.591(15.719 - 0.094R_T)$; $\hat{\rho}^2 = .479$
(0.242)

SUMMER SEASON, 1895:

(39 weeks); $R_T = 18.036 + 0.036R_0$; $\hat{\rho}^2 = .001$
(0.181)

$\overline{R}_0 = 19.102$; $\overline{R}_T = 18.718$; $\overline{R}_L = 11.795$; $\overline{P}_g = 13.219$; $SP_g = 2.618$

(38 weeks); $P_g = -3.022 + 0.139R_0 + 1.149R_L$; $\hat{\rho}^2 = .329$
(0.188) (0.277)

(38 weeks); $P_g = -7.410 + 0.278R_T + 1.304R_L$; $\hat{\rho}^2 = .409$
(0.181) (0.266)

(38 weeks); $P_{g,t} = 0.861P_{g,t-1} + 0.139(8.272 - 0.843R_0 + 1.813R_L)$; $\hat{\rho}^2 = .102$
(0.106)

(38 weeks); $P_{g,t} = 0.830P_{g,t-1} + 0.170(-3.384 + 0.018R_T + 1.403R_L)$; $\hat{\rho}^2 = .074$
(0.109)

explained only 38 per cent of the variance.[166] The disruptions in rates, as measured by the Board of Trade rates, caused a greater part of the fluctuations in the weekly grain price differences, and decreased the average of such price differences.

There is less that can be said concerning the effect of rate disruptions on provision markets. The recorded prices for "Chicago barreled lard" in Chicago and New York suggest that the price margin declined with the cuts in transport rates very slightly in November and substantially in December of 1892. The margin had been 50 cents or more during the autumn months. During November and December, there were Board of

[166] Also, an average of four weeks was required to adjust to a change in Board of Trade rates, while adjustment to changes in official rates took seven weeks as seen from the computed values for ϵ in the two lagged equations in Table 5.10. In these equations, for each decrease in a Board of Trade rate there was an eventual decrease in the weekly grain price difference 43 per cent as large; for each decline in the official rate, there was a 155 per cent decline in grain price differences. This suggests that the Board of Trade rates may have "led" the grain price differences, while the official rates "lagged" these differences.

Trade reports of rate discounts of 1 to 4 cents per 100 pounds; in the former month, the average price margin declined by one cent and in the latter month there was a 10-cent decline. Rising Chicago prices and falling New York prices were in accord with speculators shipping substantial volumes at the lower Board of Trade rates.

The season-long decline in official rates, Board of Trade rates, and grain and lard price differences, was the first such experience for the Interstate Commerce Commission. The reaction of both the regulatory authorities and the cartel was to seek reforms which would reestablish the level of rates and prevent further disruptions.

Regulation and Cartelization: Evasion of Regulation and the Rate Demoralizations of 1894–1895

There were extremely modest reforms in regulatory procedures. The Interstate Commerce Commission, under the *Compulsory Testimony Act of 1893*, began new investigations of discriminatory rates, but these were temporarily halted by a federal judge in Chicago.[167] The Commission also appealed the *Social Circle* case to a higher federal court, and appealed to Congress to extend the scope of regulation so that joint through rates could be considered comparable to local rates in the same region.[168] But there were promulgated no new regulatory rules.

Reform of the rate-setting procedures was attempted by the Trunkline Committee. The need was evident; as President Depew of the New York

[167] The Interstate Commerce Commission describes the first court treatment as follows: "During the investigation held by the Federal Grand Jury at Chicago, some months [after the passage of the *Compulsory Testimony Act*], Judge Grosscup decided in the case of James, who had pleaded his privilege and refused to testify, that it is beyond the power of Congress to pass a law which would afford protection to a witness as broad as the immunity provided in the Constitution, and the chief basis of the decision was that while the witness might be freed by such a law from the legal consequences of his testimony, the government could not by an enactment save him from the disgrace and taint upon his character which a disclosure of his connection with crime might entail." (*The Eighth Annual Report of the Interstate Commerce Commission, December 1, 1894*, op. cit., p. 11.) The proceeding at least embarrassed the Commission: "As no appeal could be taken from this ruling, it will not be possible to obtain the decision of the Supreme Court upon this point until the question can be raised in another jurisdiction, an order obtained directing the witness to answer, and an appeal taken by the witness to the proper appellate court. A law providing for appeals by the government . . . would remove many of the embarrassments which now surround prosecutions for criminal offenses." (*Ibid.*) The Commission proceeded with hopes of an eventual favorable ruling by the Supreme Court, and this ruling came finally in 1896.

[168] Cf. *The Eighth Annual Report of the Interstate Commerce Commission, op. cit.*, pp. 58–59.

Central stated, "the present demoralization has probably reduced us to a point where we will not again receive more than 20 cents on our grain business . . . if we have a war to the death . . . when the roads get together again the Commission may insist that war rates shall be pool rates. That will be the cost of the war to the railroad companies." [169] The trunk-line committee informally agreed the last of March that the rate commissioner had authority "to order any line ahead of its equitable proportion of traffic to temporarily go out of such business until the equilibrium has been restored"[170]; it was proposed that pooling be reintroduced, in substance if not in form, by the adoption of a $10,000 penalty for any refusal to make up deviations from accustomed proportions of the traffic.[171] The presidents of some of the trunk lines also proposed setting up a staff at the New York Produce Exchange to calculate each railroad's share of New York receipts, and to expedite "voluntary" transfers of tonnage when railroads had surpluses or deficits from usual shares.[172]

These proposals were not all adopted, and those that were accepted were not greatly effective as means for controlling rates. The penalty contract was not put into effect — at least there were no records of payments from any "commisioner's fund." The shifting of tonnage was attempted but abandoned during the extensive disruptions of official rates in the summer of 1894 and the winter of 1894–1895. Most important, rates were not maintained.

Disruptions of summer rates — and concurrent disruptions of grain and lard price differences — followed which were quite different from those during previous rate wars. Rates were the lowest since the summer of 1881, but there was little evidence of cheating on the official all-rail rate: there were two periods of two weeks each in which Board of Trade rates were approximately 15 cents while official rates were 20 cents, but during the

[169] "Mr. Depew on Rate Cutting," *The New York Times*, March 25, 1894, p. 3. There was hope of reform since there was not yet proof that the agreement was unstable — that the disloyal railroad profited from nonadherence, as during the later 1870's and during 1886. The secret discounts were thought to be "mistakes" for the cheater as well as the loyal railroads: in the words of Mr. Depew, "No matter where rates go, everybody adopts them and we each get the same tonnage as before. . . . What one does the others meet and all lines are soon on the same basis and each one of us recognizes to himself, when he communes with himself, that he would like to commune with a person who was not forced to be so big an idiot." (*Ibid.*)

[170] "Eastbound Freight Rates," the *Railroad Gazette* (March 30, 1894), 238.

[171] Cf. "Eastbound Freight Rates," the *Railroad Gazette* (May 4, 1894), 330; or *The New York Times*, May 19, 1894, "Railroad News and Notes," for note that the penalty contract had been signed by all of the trunk-line railroads. The contract risked violation of the antipooling clause of the *Act to Regulate Commerce* for the sake of a more binding agreement.

[172] A review of allocation of tonnage shares is found in "Competition Between Particular Trunk-line Railroads," the *Railroad Gazette* (May 10, 1895), 298.

remainder of the season both official rates and Board of Trade rates were 20 cents (from February until early November) or 25 cents (for the last two weeks of November). Grain price differences averaged less than 14 cents in each season, and did not vary in accord with changes in official or Board of Trade rates. If simultaneous adjustments of rates and grain price differences are assumed, the listed changes in rates explained no more than 3 per cent of the variance in the price margin (as shown by the computed values of the coefficient of determination $\hat{\rho}^2$ in the simultaneous adjustment equations of Table 5.10)[173]; if a lag in the adjustment of grain prices to rate changes is assumed, the proportion of "explained" variance in the price margin was only 23 per cent (for the computed value of $\hat{\rho}^2$ from $\{P_{g,t} = (1 - \epsilon)P_{g,t-1} + \epsilon(\alpha + \beta R_T + \gamma R_L) + \nu\}$ as shown in Table 5.10). The disruption in rates did not take the form of downward movements of Board of Trade rates, followed by reductions in grain price differences and ultimately in official rates.

There was instead "a condition bordering on demoralization" in east-bound lake-rail rates. Official rates had declined, and there were further discounts: "From the standard rate of 20 cents . . . the official rate has worked down to $12\frac{1}{2}$ cents, [while] . . . the actual rate is said to have been less than 12 cents." [174] Even though "repeated efforts have been made to place them on a more satisfactory basis . . ." [175] there were no quotations greater than 12 cents and a breakdown of this rate in the late summer and early fall. Notice was taken in August and September of a one-cent discount[176] of the steamer rates which was followed by the 5-cent discount in the all-rail rates for most of three weeks in October. It was not until the close of navigation in November that harmony among the railroads in the Central Traffic Association was achieved and rates were re-adjusted to the official level.[177]

The discounts in lake-rail rates were not attributed to one particular

[173] The lake-rail rate listed on the Board of Trade was of no assistance in explaining the variability of grain price differences, since it remained at 12 cents per 100 pounds for the entire season.

[174] "The Lehigh Valley Yields," *The New York Times*, May 11, 1894, p. 2, in reference to "another matter discussed at yesterday's meeting [of the Executive Committee of the Trunkline Association]."

[175] "Adjusting Lake and Rail Rates," *The New York Times*, May 24, 1894, p. 2.

[176] The "Weekly Freights" columns in the *Chicago Daily Commercial Bulletin* in October listed the announced rate and, for the first time, statements that there probably were discounts.

[177] Cf. the "Weekly Freights" in the *Chicago Daily Commercial Bulletin*, October 1–November 12, 1894, for the disruption and reestablishment of all-rail grain and provision rates. The complaints of demoralization, and the attempt to establish a lake-rail tonnage pool, are discussed in *The New York Times*, September 3, 1894, p. 2. The re-establishment of rates in November is noted in "The Central Traffic Association Meeting," the *Railroad Gazette* (November 2, 1894), 766.

railroad. During the August-October period of discounting, however, there were changes in tonnage shares of grain shipped from Chicago and of grain received at the eastern seaboard consequent from cheating by the New York Central and the Pennsylvania. The New York Central Railroad, from August to January, experienced the largest decrease in tonnage shares out of Chicago in its experience as a member of the Central Traffic Association. The company received 10 per cent to 16 per cent less than the trend of its percentage shares of the total grain tonnage (as shown in the Statistical Appendix to Chapter 5, Table A.1; the average decrease over the winter season was 7.3 per cent, as shown in Table 5.11). The Pennsylvania Railroad in these same months received from 4 per cent to 8 per cent less than the trend of its shares of the total Chicago tonnage. Both railroads, however, increased their shares of grain tonnage into the eastern seaboard cities. The New York Central transported from 2 to 5 per cent more than its trend share of the total eastern grain tonnage during August, September, October, and December of 1894. The Pennsylvania Railroad transported up to 5 per cent more of the total tonnage during the period from July, 1894, to January, 1895 (except for the month of December, when it received 5 per cent less of the total than usual, as shown in the Statistical Appendix to Chapter 5, Table A.2, *Grain Shipments into the Eastern Seaboard*, 1891 to 1899). The shipments from Chicago of the two large railroads decreased, but these railroads gained back Chicago deficits with increased transport from the lakes to the seaboard.[178] The intrastate rates from the Great Lakes to the seaboard of these two particular railroads were not subject to Interstate Commerce Commission regulation. By offering secret and legal discounts on the rail portion (from Buffalo and Erie) of the lake-rail rates from Chicago to the eastern seaboard, the two railroads shifted tonnage in Chicago to the lakes and to themselves[179] on the lake-to-seaboard portion of the traffic to the East Coast.[180]

[178] The tonnage gain for the two lines on the eastern seaboard was greater than the Chicago loss, given that a 10 to 15 per cent loss in Chicago (of from one-quarter to one-third of the New York tonnage) was less than a 5 per cent gain in New York (of the larger New York tonnage). The gain must have been at the expense of the Erie Railroad — another line with deficits in Chicago.

[179] The contrast between Chicago and eastern seaboard tonnages is not proof of cheating but rather makes cheating plausible, after the fact, for these two lines. If there were accusations by the members of the trunk-line association against the New York Central and/or the Pennsylvania Railroad and there was also information on secret lake-rail rates, it could be assumed that these two lines set independent, and disloyal, rates. Given such an assumption it would then be possible to consider the more interesting question as to whether the two lines profited from cheating.

[180] Disruptions in all-rail rates did appear during 1894, however, in the transport of provisions. The reports of the Chicago Board of Trade indicate discounts in rates for transporting lard in September and further discounts in October, but rates were restored late in November to 30 cents. The differences between spot prices for lard in Chicago

TABLE 5.11

Grain Shipments from Chicago, 1894 to 1896

The trend of the percentage share of grain tonnage of each railroad has been defined as

$$s = \alpha + \beta T + \gamma X + \nu$$

with s = a railroad's percentage of total tonnage of Chicago shipments; T = month of shipments, numbered from 1 to 156 for the period 1887 to 1899; X = "0" for the winter months in which the lakes were closed to traffic, "1" for the summer months. The estimates of trend, as calculated by the method of least squares, are given in the Statistical Appendix to Chapter 5, Table A.1.

The "trend" share for any month is calculated from inserting the relevant values of T and X in the computed equation. The actual shares are shown in the table, along with the difference between actual and trend shares for that season.

Season and Year	The New York Central Railroad		The Pennsylvania Railroad		The Baltimore and Ohio Railroad		The Erie Railroad		The Grand Trunk Railway	
	Actual Share	Actual Share Minus "Trend" Share	Actual Share	Actual Share Minus "Trend" Share	Actual Share	Actual Share Minus "Trend" Share	Actual Share	Actual Share Minus "Trend" Share	Actual Share	Actual Share Minus "Trend" Share
Winter 1893–1894	32.5	+0.2	21.2	+3.1	4.2	−2.0	9.0	−2.1	10.7	−1.4
Summer 1894	21.9	−8.2*	11.3	−5.4	4.2	−2.1	12.9	+0.2	12.4	+5.5*
Winter 1894–1895	24.8	−7.3*	16.1	−2.4	5.5	−1.2	11.1	−0.1	11.0	−0.7
Summer 1895	24.0	−6.0	15.8	−1.2	5.2	−1.3	10.7	−2.2	11.9	+4.8*
Winter 1895–1896	27.8	−4.2	14.5	−4.3	5.1	−1.7	16.1	+4.8*	13.7	+2.5
Summer 1896	27.6	−2.2	13.5	−3.9	9.1	+2.4	12.4	−0.6	6.9	+0.3

* Greater than the relevant standard error of estimate S_u.
Source: As noted in the Statistical Appendix to Chapter 5, Table A.1.

The demoralization in lake-rail rates this summer affected grain transport all of the following winter. The New York Central and Pennsylvania continued to carry far less than their usual tonnage shares of grain in December; but the New York Central, at least, received its usual tonnage share on the eastern seaboard that month.[181] The average difference between Chicago and New York grain prices did not increase with the closing of navigation; in fact, the average of the daily grain price differences for December was 14.10 cents per 100 pounds (0.2 cent greater than the summer season's average). The variance in the week-to-week grain price differences was not explained by changes in either official or Board of Trade all-rail rates.[182] Computations assuming simultaneous adjustment, and those assuming lagged adjustment, indicate that grain price differences increased as rates decreased (the computed values $\hat{\beta}$ were negative, as shown in Table 5.10). Shipments seem not to have taken place at the listed rates but at unlisted rates from Buffalo from the large volumes stored there before the lakes closed.[183] It was not until February that shipments were made from Chicago[184] and that spot price movements reflected the winter transport costs (the Chicago–New York grain price difference increased that month from 14 cents to close to 18 cents). At this time there were downward movements in Board of Trade rates and official rates:

---

and New York adjusted to the cut and to the restoration of rates. The decline in rates and the average decline in the lard price differences were 5 and 7 cents, respectively, in September, 1 and 2 cents in October. The November increases were 4 and 8 cents, respectively.

[181] The Pennsylvania received 5 per cent less during December than expected from the trend of its tonnage to the eastern seaboard. It "made this up," however, by receiving 5 per cent more than its usual tonnage share in January, and 4 per cent more than usual in February (as shown in the Statistical Appendix to Chapter 5, Table A.2.)

[182] During the winter season, the equations for lagged adjustment to equilibrium include $\hat{\rho}^2 = .460$ for R_0 and $\hat{\rho}^2 = .479$ for R_T, but the values of $\hat{\beta} < 0$ for these two equations.

[183] Cf. the *Chicago Daily Commercial Bulletin*, the "Daily Freights" column, for the month of December, 1894. The "Weekly Receipts and Shipments" columns indicate that roughly half as much corn and wheat was being received as in the previous year, but less than 1 per cent as much was being transported out of Chicago in the middle two weeks of December. During the last week of December, "eastbound shipments of flour, grain and mill stuff from Chicago by rail for the week were 10,447 tons. . . . Total shipments of all commodities by rail were 30,787 tons against 40,486 tons the previous week and 113,516 tons for the corresponding time last year" (the *Chicago Daily Commercial Bulletin* for Monday, December 31, 1894, p. 1).

[184] New York stocks of grain were so low that it was now profitable to transport the longer distance for New York spot sale. The *Chicago Daily Commercial Bulletin* for March 11, for example, shows 6.8 million bushels of wheat in New York, as compared to 12.6 the previous year, .3 million bushels of corn rather than .6 million the previous year, .5 million bushels of oats as compared to .5 million the previous year. Cf. the *Chicago Daily Commercial Bulletin*, March 11, 1894, p. 2.

both were 25 cents in January, both declined to 20 cents in February, and the Board of Trade rate declined to 11 to 16 cents in March and April. As grain price differences were increasing to reflect longer distance transport rates, the rail rates were being cut. Grain prices were the legacy of the previous summer's cutting in lake-rail rates, and of new disruptions in all-rail rates.

The breakdown of all-rail rates in March and April of 1895 was more extensive than the disruptions previously experienced under regulation. There were ominous reports early in February: "The 25-cent basis has not been maintained for some weeks except by possibly one or two roads; the other roads have been taking all the grain in sight at lower rates, as low as 17½ cents having been reported. It is now said that all the roads avail themselves of the privilege to bill all the grain they could at cut rates even below 20 cents before the new rates became effective. Of course, none of the roads have published these low rates; consequently they were willfully violating the Interstate Commerce Law if they took the freight as alleged. . . . The provision and dressed beef rates are said to be even worse than the grain rates but no serious attempt has been made to regulate these. The much advertised Chicago eastbound freight agreement does not seem to amount to any more than the host of preceding agreements that have been made only to be broken."[185] The Board of Trade listed cut rates averaging 26 cents per 100 pounds in January and 25 cents per 100 pounds in February, rather than official rates of 30 cents in both months, for lard and other provisions. The Board of Trade reported reductions in grain rates from 25 cents to 16 cents in January, and from the new 20-cent official rate to 16 cents in late February. There was little improvement in March: quoting from the Chicago *Inter-Ocean*, the *Chicago Daily Commercial Bulletin* said, during the first week of March, "After about one week of fairly well maintained freight rates the eastbound lines are back to their usual practice of securing the most business at the lowest rates. Grain rates are from 5 to 8 cents below tariff, and provision rates are being slashed 7 to 10 cents per hundred pounds; in fact provision rates have not been maintained at tariff rates by all lines for six months; they have never been completely restored and probably never will be as long as the shippers have anything to ship."[186] The Board of Trade reporter commented "it is a wise shipper who can guess the rates of freights east for two consecutive days."[187] Meetings were held late in March, the "official rate was restored," but there were reports within one week that a railroad was offering a 3-cent discount and within two weeks that "proof has been obtained that two eastbound lines have already broken the rate agreement and it will be

[185] "Eastbound Freight Rates," the *Railroad Gazette* (February 1, 1895), 76.
[186] Cf. the *Chicago Daily Commercial Bulletin*, March 4, 1895.
[187] *Ibid.*

submitted to the Interstate Commerce Commission."[188] There was a complete breakdown of official rates the first week of April after the filing at the Interstate Commerce Commission of a 12-cent grain rate by the Grand Trunk Railway, and the subsequent independent filings of 12- to 15-cent rates by the other trunk-line railroads.

The beginning of the summer season of 1895 found official rates still at the 20-cent level established in February, and Board of Trade listings of "actual" rates close to 20 cents. The lake-rail rates had opened at 11 cents per 100 pounds Chicago to New York — the lowest level ever announced on the Board of Trade. All of these rates continued into the first week in June, when the Board of Trade announced that "eastbound rail rates are being cut, 3 cents on grain and 5 cents on provisions as the agreement has about collapsed."[189] The collapse was timed to follow the termination date of the existing agreement: "Shipments last week consisted mainly of freight which roads ahead in the tonnage pool had requested shippers to hold back until June 1, the time when the contract became ineffective,"[190] and, according to the Board of Trade, was not confined to Chicago on a few commodities, but spread throughout the Midwest on all commodities: "not for a long time has the rate situation in St. Louis been so deplorably and hopelessly bad as it is at present. In all directions upon all classes and commodities and over all roads shippers can get about any rate they want. Tariff sheets have been laid aside and freight men are engaged in a wild struggle to see who can get the most business at the least profitable rate."[191] After four weeks of these rates, a new agreement adjusted the schedule of 20 cents per 100 pounds.[192] Official and Board of Trade rates remained at 20 cents per 100 pounds, with only intermittent changes, until the cessation of summer traffic in December.

Loyalty to the official rail rate after July was little more than a matter of form, however. As in the previous summer, there were appreciable discounts of the Buffalo to New York City intrastate rate, beginning with

[188] The *Chicago Daily Commercial Bulletin*, March 25, 1895. The Pennsylvania Railroad was accused: "The Pennsylvania Railroad refuses to permit an examination to disclose whether or not it has been cutting rates by means of mileage book trades with scalpers." (The *Chicago Daily Commercial Bulletin*, March 20, 1895.) The accusation was not pressed, nor does it seem consistent with shares of shipments from Chicago (which show a disproportionate increase in the Pennsylvania's grain tonnage in February, and a disproportionate decrease in March) or with shipments into New York. Cf. the Statistical Appendix to Chapter 5, Tables A.1 and A.2.

[189] The *Chicago Daily Commercial Bulletin*, June 8, 1895.

[190] "Eastbound Freight Rates," the *Railroad Gazette* (June 14, 1895), 388.

[191] The *Chicago Daily Commercial Bulletin*, "Items of General Interest," June 10, 1895.

[192] The *Railroad Gazette* noted that "The restoration was decided at a meeting in New York June 27. . . . After the meeting the presidents dined with Mr. J. P. Morgan and the gossips therefore called this another gentlemen's agreement." ["The Trunkline Situation," the *Railroad Gazette* (July 5, 1895), 450.]

1 cent on 11 cents during the first week of May.[193] A 1-cent discount was said to have had a disrupting effect: "The agreement at Buffalo for carrying traffic to New York has broken down. The grain rates have been reduced to below 3 cents a bushel, which the boatmen say is as low as corn or wheat can be carried by canal without loss. The New York Central and the Erie are accused of taking most of the grain that has reached Buffalo by lake this season, and the boatmen say that wheat has been taken as low as $2\frac{3}{8}$ cents per bushel which is equal to 3.96 cents per hundred pounds. . . . It is to be noticed that traffic from New York to Buffalo is not interstate and therefore not subject to the *Act to Regulate Commerce;* the roads therefore have been free to adjust rates in any way they pleased. . . ." [194] The 11-cent rate was restored in June, but then cut during August and the first three weeks of September.[195] For the remainder of the season the announced 11- to 13-cent rates were maintained.

The cuts in lake-rail rates were reflected in lower price margins between Chicago and New York grain markets.[196] The average Chicago–New York grain price difference this summer was approximately 13.2 cents per 100 pounds, slightly less than the previous summer and winter averages of 13.9 cents. The average was quite close to the average lake-rail rate of 11.8 cents, and 6 cents below the all-rail rate. The variance in the week-to-week grain price differences was explained, in good part, by variations in the

[193] The *Chicago Daily Commercial Bulletin*, "Weekly Trade Review," for May 10, 1895. This is the first specific reference of the Board of Trade to intrastate rates from Buffalo to New York City, as contrasted with weekly references to lake-rail rates from Chicago via all Lake Erie ports to New York.

[194] "Competition Between Particular Trunkline Railroads," the *Railroad Gazette* (May 10, 1895), 298. The *Chicago Daily Commercial Bulletin* reports the same condition: "Trunklines from Buffalo have been making such low rates on grain as to drive even the canal men out of the business. The latter say they cannot do business at less than 3 cents a bushel. Few of them have obtained even that figure. This rate has been continuously cut as low as $2\frac{3}{8}$ cents." (The *Chicago Daily Commercial Bulletin*, May 10, 1895.)

[195] Cf. "Weekly Trade Review," of the *Chicago Daily Commercial Bulletin* for August and September of 1895. For this season, for the first time, there are listings of specific lake and rail rates on grain to New York City via Buffalo each week; these listings for the two months are, with infrequent exceptions, 6 or $6\frac{1}{4}$ cents per bushel on corn.

[196] There were effects upon provision markets from these rate discounts, as well. The lack of large volume lake-rail shipment of provisions made for less effect from lake-rail discounts. But the general disruption during June of 1895 of rates on all-rail transport of lard had substantial effects. The "demoralization" in Chicago and (particularly) St. Louis provision rates reduced the average Board of Trade rate to 20 cents per 100 pounds that month. The average Chicago–New York lard price difference declined from 27 and 30 cents in the previous two months to 19 cents per 100 pounds. Since this left no margin for costs other than transport, the suspicion is that some spreaders were obtaining rates considerably below 20 cents per 100 pounds. The general impression is that discounts reduced the Chicago–New York lard price differences perceptibly. Cf. the Statistical Appendix to Chapter 5, Table A.4.

lake-rail rates via Buffalo. Assuming that the price difference adjusted to rate changes announced the same week, any change in the lake-rail rate was accompanied by an equal-sized change in the grain price difference, while any change in the all-rail rate brought about a change in the price difference only one-fifth as great (as seen from the contrasting computed values for $\hat{\beta}$ and $\hat{\gamma}$ in $P_g = \alpha + \beta R + \gamma R_L + \nu$ of Table 5.10). Assuming a lag in the adjustment of grain price differences to changes in transport rates, any change in the lake-rail rate eventually brought about a similar change in the grain price difference, while there were no observable changes in price differences following from changes in all-rail rates (as shown from the contrasting computed values for $\hat{\beta}$ and $\hat{\gamma}$ coefficients in either of the lagged adjustment equations in Table 5.10).[197]

The rate discounts on transport via Buffalo in May, June, August, and September were accompanied by shifts of tonnage to the New York state roads. Both the New York Central Railroad and the Erie Railroad received far less than their "trend" shares of Chicago tonnage during the middle months of the summer: the New York Central received 7 per cent less in May and 11 per cent less than its trend share of total tonnage in June; the Erie received 4 per cent less in July and 7 per cent less than its trend in August; and the New York Central received 16 per cent less than its trend in August and 12 per cent less in September. But the two railroads received far more than their trend shares of the total east coast receipts of grain. The New York Central received 6 per cent more than trend in May, and from 7 to 11 per cent more each month from August through December (cf. the Statistical Appendix to Chapter 5, Table A.2; the average deviation from trend for the New York Central was +7 per cent, for the Erie +6 per cent, for the season, as shown in Table 5.12). The Erie received from 5 to 8 per cent more than its expected shares of the total grain tonnage each month from May to November. The reduction of shares of these two roads in Chicago, and the expansion of shares in New York, are consonant with cuts in the lake-rail rate increasing steamer traffic out of Chicago and rail traffic out of Buffalo.

If the reason for the cuts in the lake-rail rate was to increase tonnage shares for these two railroads, then the results were sufficient to prove the cartel agreement unstable. There were increased profits for the rate-discounter from the increase in tonnage at the discounted lake-rail rate. As

[197] It might be noticed that Board of Trade rates, more than official rates, affected price differences. In the simultaneous adjustment model, the coefficient of determination $\hat{\rho}^2$ for the Board of Trade rate equation is .409, while the coefficient in the official rate equation is .329. In the lagged adjustment model, the values of $\hat{\rho}^2$ are comparable but the computed coefficient for the official rate is negative (so that an increase in the official rate was accompanied by a lagged decrease in the grain price difference) while the coefficient is positive, but small, for the Board of Trade rate.

TABLE 5.12

Grain Shipments into the Eastern Seaboard, 1894 to 1896

The trend of the percentage share of grain tonnage of each railroad has been defined as

$$s = \alpha + \beta T + \gamma X_1 + \delta X_2 + \nu$$

with s = a railroad's percentage of total tonnage of shipments into Montreal, Boston, New York, Philadelphia, and Baltimore; T = month of shipments, numbered from 1 to 156 for the period 1887 to 1899; X_1 = "1" for the winter months in which the lakes were closed to traffic, "0" for the summer months, $X_2 = 0$ for the months before the entry of the Lehigh Valley Railroad, "1" thereafter. The estimates of trend, as calculated by the method of least squares, are as given in the Statistical Appendix to Chapter 5, Table A.2.

The "trend" share for any month is determined by inserting the relevant values of T and X in the computed equation. The actual shares are shown in the table, along with the difference between actual and trend shares for that season.

	The New York Central Railroad		*The Pennsylvania Railroad*		*The Erie Railroad*		*The Lehigh Valley Railroad*		*The Lackawanna Railroad*	
Season and Year	*Actual Share*	*Actual Share Minus "Trend" Share*	*Actual Share*	*Actual Share Minus "Trend" Share*	*Actual Share*	*Actual Share Minus "Trend" Share*	*Actual Share*	*Actual Share Minus "Trend" Share*	*Actual Share*	*Actual Share Minus "Trend" Share*
Winter 1893–1894	20.5	+1	22.0	−1	6.7	−2	4.3	−3*	2.1	−1
Summer 1894	15.6		17.0	+2	4.9	−3*	3.8	−2	1.5	
Winter 1894–1895	19.2		22.8		7.5	−1	5.6	−2	1.8	−1
Summer 1895	23.4	+7*	13.1	−2	15.0	+6*	4.0	−3*	0.6	−1
Winter 1895–1896	20.1		18.5	−6*	8.6		6.0	−2	0.7	−2*
Summer 1896	18.5	−3	12.3	−3	6.5	−1	7.1		4.3	−1

* Greater than the relevant standard error of estimate.
Source: As noted in the Statistical Appendix to Chapter 5, Table A.2.

TABLE 5.13

Profits from Independent Rate-Setting, the Summer of 1895

Total profits are computed from $(R_I - K_2)q$, with "R_I" the independent lake-rail rate of 10.0 cents per hundred pounds, "K_2" the average variable cost of transporting 100 pounds from Buffalo to New York City, "q" the hundreds of pounds of grain received in New York City by the railroad in question.

Cartel profits are the returns from the trend share of the volume shipped at the official rate R_0. These are computed from $(R_0 - K_2) \cdot S \cdot Q$ with "R_0" the official rate, "S" equal to the trend share of shipments from $S = \alpha + \beta T + \gamma X_1 + \delta X_2$ (as described in the Statistical Appendix to Chapter 5, Table A.2), "Q" the total shipments of wheat, corn, oats, and flour into the eastern seaboard cities.

Month	*Year*	*Railroad*	*Total Profits at the Discounted Rate*	*This Railroad's Share of Cartel Profits at the Official Rate*
May	1895	New York Central	$ 94,695	$ 87,576
June	1895		41,654	49,092
July	1895		53,261	53,261
August	1895		91,465	71,829
September	1895		104,528	79,146
	Total		$385,603	$341,084
May	1895	Erie Railroad	$ 42,551	$ 36,326
June	1895		39,955	21,184
July	1895	(24,465)*	38,210	24,465
August	1895		39,527	28,902
September	1895		46,083	31,471
	Total	(192,581)*	$206,326	$142,384

* The trend share, at $R_I = R_0 = 11$ cents per hundred pounds.

Source: Rates are from the text and sources noted in the textual footnotes; realized market shares and total shipments from the *Annual Reports* of the various eastern coast produce exchanges; average variable costs for each road are the sum of fuel expenses, oil and waste expenses, locomotive repair and service expenditures, and train service expenses for that road multiplied by the ratio of total freight train miles over the sum of total passenger train miles and freight miles, reduced to cents per 100 pounds Buffalo to New York (the figures as listed in the *Annual Reports* of the Railroad Commission of the State of Ohio).

shown in Table 5.13, the New York Central received larger (calculated) profits on grain traffic during the first month of the unannounced 10-cent rate, and smaller profits in the succeeding month, relative to profits from the trend shares of east coast tonnage at the official lake-rail rate of 11 cents. The gain in profits during May is estimated as $7,000, while the loss in profits in June is larger; the gains in profits in August and September, however, are estimated as having been from $20,000 to $30,000 per month, so that the total estimated profits for the five-month period equal $385,000 from larger tonnage shares at cut rates as compared to $340,000 from trend tonnage shares at official rates. The Erie Railroad experienced a gain in (calculated) profits in May, and a larger gain in June. Profits from the larger tonnage share at the 10-cent rate were $11,000 greater in August, and $15,000 greater in September, than from the trend of shares at the 11-cent announced rate. The five-month profit at the lower rate was approximately $192,000, as compared to $142,000 from official rates.[198] There would seem to have been an inducement for each railroad to announce the official lake-rail rate of 11 cents, and a stronger inducement to discount this rate as soon as possible.

The state of affairs by the fall of 1895, then, was that the practice of setting "intrastate" rates from Chicago to New York had rendered ineffective the cartel's control of rates. Rate discounts had resulted in a margin between Chicago and New York spot grain prices of less than 14 cents, as compared to margins close to 17 cents in the summer seasons and 23 cents in the winter seasons since 1887. There were discounts in rail rates on provisions that led to declines in margins between spot lard prices as well. Most important, the large gains in tonnage for the New York Central and the Erie during the last rate disruption made it obvious that future agreements on rates and tonnage shares of through traffic could not be maintained.

Regulation and Cartelization: The Strengthening of Cartelization and Effective Control of Rates in 1896

The performance of the trunk-line cartel in 1894 and the summer of 1895 demonstrated the need for reorganization. There was an obvious need to place responsibility for maintenance of rates with railroad employees who set charges on the individual lines; the trunk-line presidents sought to do so: "a governing board will be set up to make differential rates which will be enforced by the respective trunkline presidents. The differential rates will be changed for good reason when any road is dissatisfied."[199] More important, means for penalizing those that set lower

[198] The Erie's profits from the lower rate are shown both on the basis of actual tonnage and, for the month of July when the official rate was maintained, the trend share of tonnage. The first calculation includes profits from having received more than the usual tonnage at official rates as part of "cheaters' " profits, while the second assumes that the increase in tonnage share should not be included in the results from "cheating." The difference in profits between the two calculations is $15,000, but the lower calculation still indicates at least a 20 per cent gain in net receipts from the lower rate.

[199] "The Proposed Strengthening of the Trunkline Agreement," the *Railroad Gazette* (August 2, 1895), 516.

rates had to be found. The system of differential rates was proposed once again; the counterproposal was to set up a "money pool," as in the early 1880's, and to pay fines from the pool to those with deficits in tonnage shares. The working draft of the compromise agreement, circulated in October of 1895, required differential rates to be set by a board of employees from nine trunk-line systems, required fines for violations of the agreement, but also required tonnage or money payments to deficit roads from roads in excess of assigned shares.[200] This new agreement was adopted by the trunk-line roads at a meeting on November 19, 1895.

The Joint Traffic Association which resulted was, as a consequence, more than simply a reorganization of the Trunkline Executive Committee and its commissioner's bureaus. The Board of Managers, consisting of the employees in charge of rate matters on each line, acted upon "all applications for differentials and for changes in rates, fares, charges, and rules. . . ." [201] The managers were to "recommend such changes in rates, fares, charges, and rules as may be reasonable and just and necessary for governing the traffic covered by the agreement . . ." [202] and they were "charged with the duty of securing to each party equitable proportions of the competitive traffic covered by the agreement so far as can be legally done. . . ." [203] An individual railroad was allowed to appeal any controversy before the Board of Managers to "the Board of Control," consisting of the presidents of the trunk-line railroads, for final decision. No railroad was to settle a controversy by quoting an independent rate — to the contrary, rate-cutting subjected a railroad to a fine "not exceeding $5,000 . . . [except] where the gross receipts of the transaction in which this agreement is violated shall exceed $5,000 the offending parties shall forfeit a sum not exceeding such gross receipts." [204] The "duty . . . of securing equitable proportions" had led to the requirement of formal tonnage allocation; the levying of fines was formal pooling of revenues.

The Joint Traffic Association began operations January 1, 1896. The Board of Managers made no immediate revisions of the structure of official

[200] "The Establishment of the Joint Traffic Association," the *Railroad Gazette* (October 18, 1895), 686. The differential rates set the shares, while the tonnage or money payments provided penalties for deviations from set shares.

[201] *The Articles of Organization of the Joint Traffic Association* (November 19, 1895); Article 5, "Relating to the Duties and Powers of the Managers," Section 5 (Archives of the Bureau of Railway Economics, The Association of American Railroads, Washington, D.C.).

[202] *Ibid.*, Article 7, "Rates, Fares, Charges, and Rules," Section 2.

[203] *Ibid.*, Article 8, "Proportions of Competitive Traffic."

[204] *Ibid.*, Article 16, "Forfeitures for Violations of Agreement." Penalties assessed were to have been taken from the deposit of $5,000 made by each of the member railroads at the time the agreement went into effect. (*Ibid.*, Article 17, "Deposits, Expense Fund, and Provision for Forfeiture.")

rates, but affirmed the 20-cent official grain rate announced the previous July. This rate, and comparable (higher) official rates on provisions,[205] were put into effect and maintained for most of the year. In February, it was said that "the new traffic association has been running a month and it seems to have met the expectations of its most hopeful friends. The president of one of the principal roads told a New York reporter that he did not remember the time when rates were so well maintained as during the whole month of January. The correctness of this view is corroborated by reports as to the correct attitude of local soliciting agents toward shippers, and the rigid manner in which city agents maintain rates and freeze out applicants for concessions." [206] A similar report followed in March: "The western correspondents and reporters are complaining loudly of the managers of the Joint Traffic Association who are ruling things with an iron hand" [207]; in May there were complaints from shippers: "the grain merchants in New York have made declarations against the Joint Traffic Association. . . . They seem to forget that the purpose of the Association was not for getting all of the grain to New York . . . but rather to make the competition civilized instead of barbarous." [208] Summarizing the year's experience, the *Railroad Gazette* stated that, "[after January 1, 1896] all reports of irregularity ceased and the officers of every interested road expressed satisfaction with the new scheme. The law-abiding traffic officers had for once gotten the upper hand. The nine principal interests in the Association had nine individual representatives on the Board of Managers and each one had virtually complete control of rates on all the roads that he represented. Thus any representative fearing that one of his big shippers was being drawn away by a competitor could get a true answer to his inquiry on the subject if he brought it before the Board the next morning. Reductions in rates made necessary by legitimate conditions were promptly made, but not until after conferences with competitors; and demands for unreasonable reductions could be resisted because a member receiving such a demand was not worried by fear of secret action by a competing road." [209]

The rate quotations on the Chicago Board of Trade indicated loyalty to the official rates. During the first week of February, there was a report

[205] The rate on lard, as shown in the Statistical Appendix to Chapter 5, Table A.4, was reaffirmed at the 30-cent level announced in 1894 and 1895.

[206] "The Joint Traffic Association," the *Railroad Gazette* (February 7, 1896).

[207] "The Joint Traffic Association," the *Railroad Gazette* (March 20, 1896).

[208] "The Joint Traffic Association," the *Railroad Gazette* (May 1, 1896). The correspondent of the *Railroad Gazette* added: "It is said that prior to January 1, New York roads were secretly cutting rates in order to get grain (and to increase the share transported into New York City); but assuming that to be so, what can be done about it?" (*Ibid.*)

[209] "The Joint Traffic Association," the *Railroad Gazette* (November 4, 1898), 796.

that a 16-cent grain rate had been offered by one New York railroad; in the remaining weeks of the winter season, both official rates and Board of Trade rates were 20 cents per 100 pounds. The summer official rates of 20 cents on wheat and 15 cents on corn (from July 15 to October 31) were quoted on the Chicago Board of Trade without exception.[210] There were no reports of deviations from the posted lake-rail rate during the entire summer season. There were no reports, as well, of provision rates or class rates having been cut during either season of this year.

The new loyalty of the trunk-line railroads to the 20-cent rate had substantial effects on Chicago–New York grain price differences.[211] The cartel had not been able to set rates so that grain price differences were greater than 14 cents per 100 pounds during the summer of 1894, the winter of 1894–1895, or the summer of 1895. The average difference between Chicago and New York spot prices increased to approximately 17 cents per 100 pounds during the winter of 1895–1896 (as shown in Table 5.14). The average declined to slightly more than 13 cents again during the summer of 1896, with the lowest price differences occurring when the published lake-rail rate was 10 cents per 100 pounds. Variation in week-to-week grain price differences was quite small: the standard deviation of the weekly price differences was 1.1 cents per 100 pounds during the winter, and 1.3 cents per 100 pounds during the summer. Approximately 30 per cent of the variance in the weekly differences was explained by changes in the previous week's price difference, in official rates, and in lake-rail rates (as shown by the computed value of $\hat{\rho}^2$ for the lagged regression equation in Table 5.14). The increase in the average price margin, and the relative stability of the weekly margins, were consistent with control of rates.[212]

[210] The 20-cent rate on grain was also maintained during the month of December.

[211] There was some effect as well on shares of grain shipments from Chicago. The summers of 1894 and 1895 were noted for the sharp decline of shares of Chicago shipments of the New York Central, the Pennsylvania, and the Erie railroads (in accord with rate discounts on lake shipments into Erie and Buffalo). During the summer of 1896, there were months in which the New York Central was shipping far less than its usual share from Chicago, the Pennsylvania was shipping far less and then far more than usual, and the Erie was shipping more than usual (as shown in the Statistical Appendix to Chapter 5, Table A.1). The number of such months of deviations from the given trends of shares was less than in the previous year; and deviations were not followed by increases of tonnage shares into the eastern seaboard. The New York Central received somewhat less than its usual share in New York City in January, February, June, July, August, and September (as shown in the Statistical Appendix to Chapter 5, Table A.2). The Pennsylvania received less than its usual share from April to August of that year. The deficit tonnage went to the lake and canal carriers, or perhaps to the Baltimore and Ohio Railroad (for which there is no east coast tonnage information).

[212] The increased control of rates was achieved with the immediate tolerance, and with

TABLE 5.14

Official Rail Rates, Board of Trade Rail Rates, and Differences Between Chicago and New York Grain Prices, 1896 to 1899

$$R_T = \alpha + \beta R_0 + \nu \quad (1)$$

R_T, the rail rate, in cents per hundred pounds Chicago to New York, from the Board of Trade *Daily Commercial Bulletin;* R_0, the rail rate from compilations of posted tariffs and from announcements of the various traffic associations.

$$P_g^* = \alpha + \beta R_0 + \gamma R_L + \nu; \qquad P_g^* = \alpha + \beta R_T + \gamma R_L + \nu \quad (2)$$

P_g^*, the weekly average of the daily differences between Chicago and New York spot prices, in cents per hundred pounds, for the number 2 "western" grade of corn and oats; the official rate R_0 or the Board of Trade rate R_T; the lake-rail rate R_L from Chicago to New York via steamship to Buffalo and rail from Buffalo to New York.

$$P_{g,t} = P_{g,t-1} + \epsilon(P_g^* - P_{g,t-1}) \quad (3)$$

$P_{g,t}$, the average of the daily Chicago–New York grain price differences during week t and $P_{g,t-1}$, the average for the previous week; $P_g^* = \alpha + \beta R + \gamma R_L$ for the summer season and $P_g^* = \alpha + \beta R$ for the winter season from an equilibrium adjustment to changes in transport rates.

The standard error of each coefficient is shown in parentheses below the coefficient.

SUMMER SEASON, 1896:

(36 weeks); $\bar{R}_0 = \bar{R}_T = 18.833$ and $R_0 = R_T$ each week; $\bar{R}_L = 11.194$

$\bar{P}_g = 13.165$; $SP_g = 1.284$

(35 weeks); $P_g = 4.833 + \underset{(0.134)}{0.422}R + \underset{(0.349)}{0.052}R_L$; $\hat{\rho}^2 = .243$

(35 weeks); $P_{g,t} = \underset{(0.148)}{0.616}P_{g,t-1} + 0.384(4.405 + 0.382R + 0.142R_L)$; $\hat{\rho}^2 = .298$

WINTER SEASON, 1896–1897: Twelve weeks for which there are complete data; the statistics are as follows:

Official Rate	*Board of Trade Rate*	*Average Corn-Oats Price Difference, Chicago–New York*	*Date, Last Day of Week*
20.0	20.0	15.5	January 16
20.0	20.0	15.0	January 23
20.0	20.0	14.6	January 30
20.0	20.0	16.5	February 6
20.0	12.0	14.1	February 13
20.0	12.0	12.2	February 20
20.0	12.0	13.6	February 27
20.0	20.0	14.4	March 6
20.0	20.0	13.2	March 13
20.0	20.0	13.4	March 20
20.0	11.0	11.2	March 27
20.0	11.0	10.8	April 3

$\bar{R}_0 = 19.2$; $\bar{R}_T = 16.6$; $\bar{P}_g = 13.7$

SUMMER SEASON, 1897:

(32 weeks); $R_T = 14.186 - \underset{(0.101)}{0.004}R_0$; $\hat{\rho}^2 = .001$

$\bar{R}_0 = 17.250$; $\bar{R}_T = 14.125$; $\bar{R}_L = 12.625$; $\bar{P}_g = 11.586$; $SP_g = 0.996$

(32 weeks); $P_g = 8.676 + \underset{(0.061)}{0.018}R_0 + \underset{(0.156)}{0.206}R_L$; $\hat{\rho}^2 = .066$

(32 weeks); $P_g = 8.513 + \underset{(0.113)}{0.038}R_T + \underset{(0.159)}{0.201}R_L$; $\hat{\rho}^2 = .065$

(32 weeks); $P_{g,t} = 0.943P_{g,t-1} + \underset{(0.115)}{0.058}(43.340 + 0.507R_0 - 3.204R_L)$; $\hat{\rho}^2 = .184$

(32 weeks); $P_{g,t} = 0.964P_{g,t-1} + \underset{(0.112)}{0.036}(47.223 + 2.989R_T - 6.159R_L)$; $\hat{\rho}^2 = .246$

WINTER SEASON, 1897–1898:

(17 weeks); $R_T = 22.916 - \underset{(0.796)}{0.242}R_0$; $\hat{\rho}^2 = .005$

$\bar{R}_0 = 20.588$; $\bar{R}_T = 17.941$; $\bar{P}_g = 13.766$; $SP_g = 0.650$

(17 weeks); $P_g = 20.822 - \underset{(0.163)}{0.343}R_0$; $\hat{\rho}^2 = .218$

(17 weeks); $P_g = 13.149 + \underset{(0.059)}{0.034}R_T$; $\hat{\rho}^2 = .213$

(17 weeks); $P_{g,t} = 0.609P_{g,t-1} + \underset{(0.217)}{0.391}(12.591 + 0.052R_0)$; $\hat{\rho}^2 = .331$

(17 weeks); $P_{g,t} = 0.586P_{g,t-1} + \underset{(0.154)}{0.414}(12.687 + 0.055R_T)$; $\hat{\rho}^2 = .343$

SUMMER SEASON, 1898:

(34 weeks); $R_T = 44.026 - \underset{(0.469)}{1.514}R_0$; $\hat{\rho}^2 = .270$

$\bar{R}_0 = 18.882$; $\bar{R}_T = 15.471$; $\bar{R}_L = 8.118$; $\bar{P}_g = 11.418$; $SP_g = 1.454$

(34 weeks); $P_g = 14.144 - \underset{(0.286)}{0.106}R_0 - \underset{(0.385)}{0.092}R_L$; $\hat{\rho}^2 = 0.003$

(34 weeks); $P_g = 13.445 - \underset{(0.105)}{0.044}R_T - \underset{(0.409)}{0.168}R_L$; $\hat{\rho}^2 = 0.008$

(30 weeks); $P_{g,t} = 0.637P_{g,t-1} + \underset{(0.137)}{0.363}(12.254 - 0.486R_0 - 0.213R_L)$; $\hat{\rho}^2 = .210$

(30 weeks); $P_{g,t} = 0.701P_{g,t-1} + \underset{(0.138)}{0.300}(18.045 - 0.175R_T - 0.503R_L)$; $\hat{\rho}^2 = .209$

WINTER SEASON, 1898–1899:

(14 weeks); $R_T = 61.746 - \underset{(1.331)}{2.249}R_0$; $\hat{\rho}^2 = .192$

$\bar{R}_0 = 19.286$; $\bar{R}_T = 18.357$; $\bar{P}_g = 16.779$; $SP_g = 2.414$

(14 weeks); $P_g = 97.552 - \underset{(0.957)}{4.188}R_0$; $\hat{\rho}^2 = .615$

WINTER SEASON, 1898–1899: (Continued)

(14 weeks); $P_{g.} = 5.556 + 0.612R_T$; $\hat{\rho}^2 = .344$
(0.243)

(14 weeks); $P_{g,t} = 0.734P_{g,t-1} + 0.267(57.848 - 2.063R_0)$; $\hat{\rho}^2 = .286$
(0.162)

(14 weeks); $P_{g,t} = 0.736P_{g,t-1} + 0.264(6.642 + 0.623R_T)$; $\hat{\rho}^2 = .365$
(0.105)

The values of R_0 in this period are not official rates, but rather the rates filed by the individual railroads with the Interstate Commerce Commission.

SUMMER SEASON, 1899: No week for which there is information on New York corn prices; the information for oats price differences is as follows:

Average Rate Filed at the Interstate Commerce Commission	*Board of Trade Rate*	*Board of Trade Lake-Rail Rate*	*Average Chicago–New York Oats Price Difference*	*Date: Month and Last Day of the Week*
17	17	10	18.0	April 24
17	17	9	16.1	May 1
17	17	10	16.5	May 8
17	17	10	14.9	May 15
17	17	10	17.6	May 22
17	17	10	—	May 29
17	17	10	18.7	June 5
17	17	10	17.7	June 12
17	$12\frac{1}{2}$–15	9	15.6	June 26
17	$12\frac{1}{2}$–15	9	17.2	July 3
17	$12\frac{1}{2}$–15	9	17.5	July 10
17	$12\frac{1}{2}$–15	9	15.0	July 17
17	$12\frac{1}{2}$–15	9	14.5	July 24
17	$12\frac{1}{2}$–15	9	21.3	July 31
16	16	10	19.8	August 7
16	16	10	17.5	August 14
16	16	11	15.0	August 21
16	16	10	15.9	August 28
16	16	11	13.5	September 4
16	16	11	14.2	September 11
19	19	12	15.0	September 18
19	19	12	17.4	September 25
19	19	12	19.6	October 2
19	19	15	19.3	October 9
19	19	14	19.6	October 16
19	19	14	18.4	October 23
19	19	14	17.9	October 30
22	20–25	14	19.5	November 6
22	20–25	14	19.1	November 13
22	20–25	14	18.6	November 20
22	20–25	14	21.7	November 27

The Destruction of Regulation and Cartelization, 1897 to 1899

The Joint Traffic Association had restored 20-cent official rates, by means of a strong money pool, within the framework of regulation. The Association lost control in the following two years, because of court rulings against regulation and against pooling. A series of United States Supreme Court decisions drastically reduced the scope of regulation in 1896, to the detriment of national regulation of discrimination.

An adverse, but preliminary, decision was handed down March 30, 1896, in the long-standing *Social Circle* case.[213] The Supreme Court saw the problem at issue to be as follows: "The James N. Mayer Buggy Company [in October, 1889] filed a complaint before the Interstate Commerce Commission against [a number of Cincinnati to Georgia railroad companies] alleging that [they] charge the same rate for transporting vehicles shipped from Cincinnati to Atlanta . . . or to Augusta . . . and charge 30 cents per hundred pounds more on such vehicles shipped to Social Circle, Georgia [between Atlanta and Augusta]. . . . They also contended that the charge to Social Circle was excessive and undue. . . ." [214] A partial defense was that the Social Circle rate was not comparable to the others, since it consisted of an interstate rate to Atlanta and a separate intrastate rate from Atlanta to Social Circle not subject to regulation. The court disposed of this argument by stating that "having elected to enter into the carriage of interstate freights and thus subject itself to the control of the

---

the assistance, of prevailing regulation. Pooling was seemingly allowed during these months. The tolerance to pooling did not last — in fact, it existed only as a result of the delay of the courts (as is indicated below). The assistance followed from strengthening one regulatory rule. It became more difficult to conceal departures from the existing structure of rates after the decision in *Brown* vs. *Walker* affirmed the Interstate Commerce Commission's right to compel disclosure of rates in investigations for violations of Sections 1–3 of the *Act to Regulate Commerce* (even when the rates under consideration as "just and reasonable" were secret rates). The Interstate Commerce Commission's view was that when "the *Brown* decision was announced in April 1896 . . . railroad managers immediately became loud in their protests that whatever might have occurred in the past upon the strength of the notion that it could not be discovered, should no longer occur and that rates from then on would be scrupulously maintained." [*The Eleventh Annual Report of the Interstate Commerce Commission, December 6, 1897* (Washington: Government Printing Office, 1897), p. 47.] This suggests that regulation was "strengthened" during the first year of the Joint Traffic Association, and that the "strengthening" made it less desirable to secretly cut the new association's announced rates.

[213] Cf. *Cincinnati, New Orleans and Texas Pacific Railway Company* vs. *Interstate Commerce Commission et al.*, 162 U.S. 184 (1896).

[214] *Ibid.*, at pp. 185–186.

Commission, it would not be competent for the [Georgia Railroad and Banking Company, from Atlanta to Social Circle] to limit that control in respect to foreign traffic to certain points on its road and exclude other points." [215] There was the further question,[216] however, as to whether the Interstate Commerce Commission could compare rates from Cincinnati to Birmingham with those from Cincinnati to Atlanta, and require that since "there is apparently nothing in the nature and character of the service to justify [a 30-cent] difference . . ." the Cincinnati to Atlanta rate had to be decreased.[217] Commission proceedings which set a maximum "reasonable" rate, in the process of reviewing present rates, were not allowed: the court stated that "we do not find any provision of the *Act* that expressly or by necessary implication confers such a power. . . . It is argued on behalf of the Commission that the power to pass upon the reasonableness of existing rates implies a right to prescribe rates. This is not necessarily so. The reasonableness of the rate in a given case depends on the facts, and the function of the Commission is to consider these facts and give them their proper weight." [218] It was not evident, however, that the court was denying the Commission the right to set the rate in special hearings with the purpose of "consider[ing] these facts and [giving] them their proper weight."

In May of 1897 in the *Maximum Freight Rate* case,[219] the Supreme Court paraphrased the argument that the *Social Circle* decision did not preclude the setting of maximum rates: "It is thought that this court meant that while the Commission was not in the first instance authorized to fix a rate, yet that it could whenever complaint of an existing rate was made give notice and direct a hearing, and upon such hearing determine . . . what would be a reasonable rate if the one prescribed was found not to

[215] *Ibid.*, p. 192.

[216] This point was appealed from the lower court by the Interstate Commerce Commission.

[217] *Cincinnati, New Orleans and Texas Pacific Railway Company* vs. *Interstate Commerce Commission et al.*, *op. cit.*, p. 195, as quoted from the Interstate Commerce Commission order of June 29, 1891, requiring the defendant railroads "to cease and desist from making any greater charge in the aggregate . . . from Cincinnati to Social Circle than charged on such freights from Cincinnati to Augusta and to cease and desist from making any charge for the transportation of such freight from Cincinnati to Atlanta in excess of $1.00 per hundred pounds."

[218] *Ibid.*, at p. 197.

[219] *Interstate Commerce Commission* vs. *Cincinnati, New Orleans, and Texas Pacific Railway Company*, 167 U.S. 479 (1897). The issue was an order of the Commission requiring the members of the Southern Railway and Steamship Association to "abstain from charging, demanding, collecting, or receiving any greater aggregate rate" than shown in a schedule set up by the Commission for transportation from Cincinnati to various Tennessee and Georgia cities. Cf. *ibid.*, pp. 480 to 482.

be. . . ." [220] The argument was incorrect: "The vice of this argument is that it is building up indirectly and by implication a power which is not in terms granted." [221] The Court found nothing in the *Act to Regulate Commerce* directing the Commission to set rates: "That Congress has transferred such a power to any administrative body is not to be presumed or implied from any doubtful or uncertain language . . . the words and phrases efficacious to [making] such a delegation of power are well understood and have been frequently used, and if Congress had intended to grant such a power to the Interstate Commerce Commission it cannot be doubted that it would have used language open to no misconstruction but clear and direct." [222] The Supreme Court concluded that "incorporating into a statute the common law obligation resting upon the carrier to make all its charges reasonable and just, and directing the Commission to execute and enforce the provisions of the *Act*, does not by implication carry to the Commission or invest it with the power to exercise the legislative function of prescribing rates which shall control in the future." [223] This was clear and explicit notice to the Commission that it was not to set the rates charged by the railroads.

Later in 1897 the same court decided another case which not only reaffirmed the *Maximum Freight Rate* decision, but also struck down the existing standards for a violation of the long haul–short haul clause of the *Act to Regulate Commerce*. The decision in the *Troy* case arose from a complaint of the Board of Trade of Troy, Alabama, that the Alabama Midland Railway and other railroads "forming lines with it from Baltimore . . . to Troy and Montgomery [Alabama] charged and collected a higher rate on shipments of class goods from those cities to Troy than on such shipments through Troy to Montgomery, the latter being a longer distance point by 52 miles." [224] The Interstate Commerce Commission concluded that the rate to Troy violated the provision in Section 4 of the *Interstate Commerce Act* against a higher rate for a shorter distance, since "circumstances were similar" and the transport was over the same railroads in the

[220] *Ibid.*, p. 508.

[221] *Ibid.*

[222] *Ibid.*, p. 505. The Congress, in fact, made it specific that the Commission lacked such power in the wording of the *Act to Regulate Commerce*. The court found, referring to Section 6, that the framers of the *Act* "contemplate the fixing of rates and recognize the authority in which the power exists" by reference to "every common carrier [printing and keeping open] to public inspection schedules showing the rates . . . which any such common carrier has established." (*Ibid.*, p. 502; as quoted from Section 6 of the *Act to Regulate Commerce*, *op. cit.*)

[223] *Ibid.*, at p. 506.

[224] Cf. *Interstate Commerce Commission* vs. *Alabama Midland Railway Company*, 168 U.S. 144 (1897), at p. 146.

same direction. The Circuit Court of Appeals noted that "only two railroads, the Alabama Midland and the Georgia Central reached Troy. . . . There are many more railway lines running to and through Montgomery, connecting with all the distant markets. . . . The volume of trade to be competed for, the number of carriers actively competing for it, and the constantly open river . . . seem to us, as they did to the Circuit Court, to constitute circumstances and conditions at Montgomery substantially dissimilar from those existing at Troy and to relieve the carriers from the charges preferred against them by the Board of Trade of Troy." [225] The Supreme Court affirmed these findings: "It seems undeniable as the effect of evidence on both sides that an actual dissimilarity of circumstances and conditions exists between the two cities concerned, both as respects the volume of their respective trade and the competition, affecting rates, occasioned by rival routes by land and water." [226] Circumstances were not only "dissimilar" as a result of competition from water transport at Montgomery, but also "competition between rival routes is one of the matters which may lawfully be considered in making rates, and that substantial dissimilarity of circumstances and conditions may justify common carriers in charging greater compensation for the transportation of like kinds of property for a shorter than for a longer distance over the same line. . . ." [227] The Interstate Commerce Commission, then, could no longer require lower short-haul rates when there were differences between short-haul and long-haul transport in the extent of railroad competition.

The *Troy* decision also reinterpreted the meaning of "just and reasonable" rates. The court's decision implied that standards for "dissimilar conditions" in the fourth section applied for the first and third sections of the *Act* as well: "In the other sections [excepting Section 2] a meaning must be given to the phrase ["under substantially similar circumstances and conditions"] wide enough to include all the facts that have a legitimate bearing on the situation — among which we find the fact of competition when it affects rates." [228] The Commission could not require comparable rates for similar distances unless railroad competition was the same at all of the compared locations. From the Commission's view, the result was to make control of the structure of rates impossible: "competition between carriers creates the necessary dissimilarity of circumstances and conditions, and this necessarily overturns the procedure of the Commission up to the present time and virtually nullifies the sections." [229]

[225] *Ibid.*, at pp. 171–172.
[226] *Ibid.*, at p. 175.
[227] *Ibid.*, at p. 170.
[228] *Ibid.*, at p. 167.
[229] *The Eleventh Annual Report of the Interstate Commerce Commission, December 6, 1897* (Washington: Government Printing Office, 1897), p. 42.

With no basis for comparing rates, and no power to prescribe future rates, the Interstate Commerce Commission could not regulate the structure of trunk-line official rates, or the conduct of those railroads charging official rates. Regulation, as experienced from 1887 to 1896, had been destroyed. Cartelization was eliminated at the same time.

The Joint Traffic Association's formal devices for dividing traffic and levying penalties had contained, from the beginning, all of the trappings of a pooling agreement contrary to the *Act to Regulate Commerce*. The Interstate Commerce Commission argued that there was such a violation: upon receiving a copy of the new contract agreement, the Commission informed the Justice Department that "this contract, agreement, or arrangement is, we believe, in conflict with the *Act to Regulate Commerce* and cannot be carried into effect without violating the provisions of said *Act*." [230] The Justice Department filed suit for an injunction in the United States Circuit Court "enjoining the further execution" of the Joint Traffic Association agreement to obtain relief from violation of the *Interstate Commerce Act*, and also because the agreement was "a combination and conspiracy entered into in order to eliminate all competition among the parties to it for freight and passenger traffic" in violation of the *Sherman Anti-trust Law of 1890*.[231]

The court proceedings extended over a considerable period of time. The Circuit Court dismissed the suit early in 1897, and the Court of Appeals for the Second Circuit affirmed the dismissal on March 20, 1897.[232] Three days later, in proceedings in a suit to dissolve the Trans-Missouri Freight Rate Association, the Supreme Court found that railroad "freight association" tonnage allocation and rate-setting were in restraint of trade and contrary to Section 1 of the Sherman Act. The court stated that "the question is one of law in regard to the meaning and effect of the [Trans-Missouri] agreement itself, namely: does the agreement restrain trade or

[230] Communication from the Interstate Commerce Commission to the Attorney General of the United States, December 26, 1895, as reproduced in *The Tenth Annual Report of the Interstate Commerce Commission, December 1, 1896* (Washington: Government Printing Office, 1896), pp. 87–88. The reason given was that "these agreements are skillfully drawn up to enable the parties to them to apportion and divide competitive traffic without incurring the penalties of the fifth section [the antipooling clause of the *Act to Regulate Commerce*]." (*The Tenth Annual Report of the Interstate Commerce Commission, December 1, 1896, op. cit.*, p. 90.)

[231] According to the statement of the facts by Mr. Justice Peckham in *United States* vs. *The Joint Traffic Association*, 19 Sup. Ct. 506 (1898) at p. 509.

[232] The opinion of the Appeals Court was that the suit involved questions only in the domain of Interstate Commerce Commission regulation: "If there has been a violation of the pooling section of that Act because of the existence of contracts, the United States has no right under that Act by injunction. Such an action can be maintained only by the Interstate Commerce Commission. . . ." (As reproduced in *The New York Times*, March 20, 1897, p. 12.)

commerce in any way so as to be a violation of the [Sherman] Act? We have no doubt that it does." [233] Since the Trans-Missouri and Joint Traffic agreements were quite similar, the trunk-line railroads' case on appeal to the Supreme Court was not promising. On October 24, 1898, this last court of appeals declared null and void the Joint Traffic Association agreement. It was stated that the Trans-Missouri agreement "did in fact, and by its terms, restrain trade" and that "we do not see any substantial difference between this agreement [of the Joint Traffic Association] and the one set forth in the Trans-Missouri case." [234] The formal activities of the Association were discontinued.

The destruction of the actual rate-setting and rate-maintaining activities of the Joint Traffic Association extended over much of this period from March of 1897 to October of 1898. The announcement of the decision in the Trans-Missouri case brought the activities of the Association to the point of "general inertia"; it was believed that "practically all the associations are in a state of suspended animation looking over the ground and seeing how they stand." [235] The Board of Trade recorded the results from Association inactivity: the listed rate of 20 cents per 100 pounds was "cut as much as $8\frac{1}{2}$ cents and . . . contracts have been made at this reduction for the balance of the present month . . . eastbound freight rates from Chicago to the seaboard are in a demoralized condition by reason of the disintegration of the freight associations." [236] Rate demoralization continued during the summer season. The Board of Managers conducted the formalities of rate-setting, but official rates were not in effect most of the time: the average official rate for the season was 17.3 cents per 100 pounds, while the average rate quoted on the Board of Trade was 14.1 cents per 100 pounds; changes in official rates explained no more than 0.1 per cent of the variance in Board of Trade rates (as shown, for the summer of 1897, from $\hat{\rho}^2$ in the equation $R_T = \hat{\alpha} + \hat{\beta} R_0$ in Table 5.14). Announced lake-rail rates also were cut; posted rates were close to 14 cents per 100 pounds when the season opened, and approximately 12 to 13 cents per 100 pounds from July through November, but the Board of Trade listed "suspected" rates 2 cents less during June and July and again in November. [237] The low

[233] As quoted from the decision in "Anti-trust Law Decision," *The New York Times*, March 23, 1896, p. 8.

[234] *United States* vs. *Joint Traffic Association*, *op. cit.*, at p. 562. The economic results of the agreement were evident to the court: "viz. to maintain rates at a higher level than would otherwise prevail." (*Ibid.*, at p. 565.)

[235] "News of the Railroads," *The New York Times*, March 25, 1897, p. 5.

[236] "Weekly Trade Review," the *Chicago Daily Commercial Bulletin*, March 26, 1897.

[237] Cf. the "Weekly Trade Review," the *Chicago Daily Commercial Bulletin* during the months of June to November, 1897. The "suspected" cutting was on New England rates. In previous years the New England rate had been from 2 to 5 cents per 100 pounds greater than the New York City rate; given this customary difference, it would appear

point was reached in July: "Eastbound rates are demoralized. . . . Grain is said to be taken from Chicago to New York at 10 cents per hundred pounds or one-half the regular tariff. . . . The rate on provisions is also being cut . . . and through rates, lake and rail, from Chicago to New England are at 10 cents per bushel on wheat, 8¼ cents on corn, and 5½ cents on oats." [238] There was some improvement upon this experience in August and September, but Board of Trade rates did not exceed 17 cents for the rest of the season.

The effects of the rate breakdown were similar to those observed in the year before the organization of the Joint Traffic Association. When there was cutting of the rates on lake-rail transport via Buffalo, the Erie Railroad and the New York Central Railroad experienced sharp declines in grain shipments from Chicago, but increased shares of grain receipts on the eastern seaboard. There were gains from setting intrastate rates from Chicago of 3 per cent to 7 per cent of the total east coast tonnage.[239] The breakdown in rates reduced the margin between spot grain prices in Chicago and New York to 11.6 cents per 100 pounds, the lowest level yet experienced, and reduced the variance in week-to-week grain price differences "explained" by official rail rates and posted lake-rail rates.[240]

The winter season of 1897–1898, and the summer of 1898, were no different. Official rates were established at 22 cents in December of 1897, and at 20 cents for the period from January through June of 1898; Board of Trade rates averaged 17.9 cents per 100 pounds for the entire winter season, and were never higher than 15 cents per 100 pounds from the opening of navigation to the end of June. Official rates averaged 18.9 cents per 100 pounds for the remainder of the year, while Board of Trade rates averaged 15.5 cents. The announced lake-rail rates during the summer were

---

that steamer lines and railroads on the Chicago–Buffalo–Boston route were cutting posted rates by offering the same rate as to New York City.

[238] "Weekly Trade Review," the *Chicago Daily Commercial Bulletin*, July 30, 1897.

[239] As can be seen from the excess of 1897 tonnage shares over trend shares of east coast receipts of these two railroads, in the Statistical Appendix to Chapter 5, Table A.2, *Grain Shipments into the Eastern Seaboard, 1887 to 1899*.

[240] The equations for simultaneous adjustment and for a lagged price difference both indicate a *slight amount* of additional variance explained by Board of Trade rates, however. As shown in Table 5.14, the adjustment period to a change in a rail or a lake rate was up to 25 weeks, and grain price differences were inversely related to changes in announced lake-rail rates. It is also shown that grain price differences were directly related to changes in Board of Trade rates, and that R_T explained a greater proportion of the variance in week-to-week grain price differences in the lagged adjustment equation than did R_0.

There were no tests made of the "explainability" of changes in the "cut" lake-rail rates to New England, because of the lack of day-by-day quotations in this series. It would seem likely that variations in these rates explain much of the variance in price differences.

never more than 10 cents per 100 pounds; but there were 2-cent discounts on transport via Buffalo until October (and, in November, a 1-cent discount was quoted for shipment via Philadelphia, Baltimore, and Newport News, while New York rates were "suspected" to have been lower).[241]

The systematic rate-cutting in these seasons seems to have worked mostly to the advantage of the New York Central Railroad. During the months of March and April, when "eastbound rates have gone to pieces and are down to the lowest on record at 12 cents per hundred pounds on grain from Chicago to New York," the Central's share of the tonnage out of Chicago was appreciably greater than its long-term trend of such shares (as shown in the Statistical Appendix to Chapter 5, Table A.1). The New York Central's shares of tonnage into the eastern seaboard in April and May were also from 6 to 10 per cent greater than its trend of these shares (as shown in the Statistical Appendix, Table A.2). Later in the summer, when there were discounts of the lake-rail rate, the New York Central's shares of Chicago tonnage were below the trend shares, while the road's shares of east coast receipts were greater than the long-term trend of such shares. The shifts of tonnage to this line took place only in the periods of rate demoralization.[242]

The margins between Chicago and New York spot grain prices continued at very low levels, and the variance in such margins was not explained by listed rates to as great an extent as in the early 1880's. The average of the weekly price differences during the winter was less than 14 cents per 100 pounds; the average price difference for the summer season was 11.4 cents per 100 pounds, which was less than any previous average difference. The variance in the winter's grain price differences was less than 0.36 cent and 20 to 30 per cent of this amount was explained by either official or Board of Trade rates. During the summer, the variance was greater and the explained portion was smaller; moreover, grain price differences decreased when rail rates and the announced lake-rail rates increased (as shown by the negative $\hat{\beta}$ in the summer regression equations in Table 5.14). Since the winter price decreases followed a period of discounts in the lake-rail rates, and the summer price decreases occurred during more such discounts, this behavior may have resulted entirely from the "intrastate" rates via Buffalo.

The formal demise of the Joint Traffic Association, following the Su-

[241] Cf., for example, "Weekly Trade Review," the *Chicago Daily Commercial Bulletin*, October 14, 1898, where it is stated that "it is possible for a few grain shippers to ship to Philadelphia, Baltimore and Newport News at 10 cents per hundred pounds . . ." when the *announced* lake-rail rate from Chicago to New York was 10 cents per 100 pounds.

[242] At this point, however, neither the railroad news journals nor the Joint Traffic Association were concerned with identifying the cheater, so that there was no independent accusation that the New York Central was cheating.

preme Court decision of October 28, 1898, was the occasion for general acknowledgment of the failure of the agreement. The means for recognizing the end of the agreement was rate-cutting: as noted in the *Railroad Gazette*, "the very day of this decision, the newspapers report that 8 million bushels of grain are being taken from Chicago to Baltimore on secret rates said to be only 10 cents per hundred pounds . . ." [243] while the Board of Trade reported that "the roads are making rates to suit the occasion . . . grain is being taken from Chicago to New York at 13 cents per hundred pounds, wheat to Baltimore at 9 cents and corn at 8 cents per hundred pounds with 200,000 bushels of corn sold for all-rail shipment on Thursday." [244] There was no suggestion, however, that the dissolution of the Association made an immediate difference; to the contrary, "this cannot be a matter of much consequence for the Association has had little or no influence on rates since the Trans-Missouri decision. . . . Every road knew that the Association was repugnant to the antitrust laws and therefore placed no dependence upon its work. . . . It is gravely remarked by one reviewer that the effects of this decision can only be learned by experience; but, as for a year and a half each road has acted independently, making secret rates whenever it saw fit, we may safely say that the lesson of experience is already before us." [245] There was a dispersal of the bureaucracy of the Joint Traffic Association in November to various committees in charge of traffic transfers, ticket exchange, etc., at separate business offices in the trunk-line territory, and official rates were discontinued.[246]

[243] "The Joint Traffic Association," the *Railroad Gazette* (October 28, 1898), 778.

[244] The *Chicago Daily Commercial Bulletin*, October 28, 1898, p. 2.

[245] "The Joint Traffic Association," the *Railroad Gazette* (October 28, 1898), 778. The *Railroad Gazette's* view was that "rates on the principal competitive commodities cannot be much lower than they have been during the recent wheat movements when roads which had difficulty in getting cars still took the freight at about cost." (*Ibid.*)

[246] The trunk-line railroads were left without formal organization, while they had perhaps been without a substantive agreement since March of 1897. Each road henceforth set individual rates which it submitted to the Interstate Commerce Commission in accordance with regulatory procedures or which it quoted to individual shippers secretly contrary to law. The Board of Trade listed the rates which it estimated were in effect; at times these equaled the posted rates of the individual railroads but in most instances they did not. Posted rates, as shown in Table 5.14, went as low as $9\frac{1}{2}$ cents per hundred pounds on grain for export, as high as 22 cents at the close of navigation, and averaged 19.3 cents per hundred pounds. Average Board of Trade rates were somewhat lower during the winter of 1898–1899 (at 18.4 cents per 100 pounds); the average Chicago–New York grain price difference, and variance in week-to-week grain price differences, were explained to some extent by Board of Trade rates (as shown by $\hat{\beta}$ and $\hat{\rho}^2$ for the R_T regression equations in Table 5.14). During the summer, Board of Trade rates varied from line to line; the lowest quotation was $12\frac{1}{2}$ cents per 100 pounds, while the highest may have been 25 cents per 100 pounds at the close of navigation. The higher rates were accompanied by higher grain price differences (for oats alone) than in the previous five years, so that it might appear that rates were subject to more control after

the destruction of the Joint Traffic Association. This was not the impression of the Board of Trade, however: the grain tonnage through Chicago during 1899 had been the largest in the century but "inducements in the way of cut rates and free storage attracted large quantities of grains to Gulf and south Atlantic ports and diverted fair supplies from the center. . . . More equitable freight rates and a more stringent Interstate Commerce Law will place all markets on an equal footing and under such conditions Chicago commission merchants are willing to accept their chances in maintaining their position in the general trade." ("Produce Trade of Chicago, 1899," the *Chicago Daily Commercial Bulletin*, December 30, 1899, p. 1.)

6

The Effects of Regulation and Cartelization, 1887 to 1899

The destruction in 1898 of the Joint Traffic Association had little effect on long-distance rates or on shippers' cost margins. The rates available on the Chicago Board of Trade, for shipments to the eastern seaboard, had been 3 cents less than the official rates of the Traffic Association during most of the seasons since 1896. At the same time, spreaders' margins between prices for spot grain in Chicago and New York had declined from 15 to 20 cents in 1896 to 13 cents or less. Railroad control of through rates on grain was not weakened by the court remedy of October, 1898, because there was little or no control.

Grain shippers reviewing the destruction of the cartel and the earlier weakening of regulation must have approved. As shown in Table 6.1, the published rates did not increase in the period of strong regulation and cartelization from 1887 to 1893, but both Board of Trade rates and Chicago–New York grain price differences were 2 to 7 cents greater per 100 pounds than during the rate disruptions of 1885–1886 and of 1893 to 1895. Higher actual (or Board of Trade) rates meant increased costs for the spreaders seeking profits from transporting between cash markets. Higher grain price differences reflect these increased costs of transport. As shown by grain price differences 3 to 10 cents higher during 1887 to 1893 than in subsequent years of cartel breakdown, the long-distance shippers benefited from the removal of cartel-regulatory control of rates.

Shippers between local terminals on any one of the trunk-line railroads did not clearly benefit from the destruction of regulation and cartelization. From 1893 to 1898 these shippers had been quoted local rates of 15 cents for 450 miles, according to the scale of uniform mileage rates put into effect in 1893.[1] There were no disruptions of this scale following from the dissolution, nor had there been during 1894 to 1898, since local rates had been mostly within the influence of one or the other of the individual railroads. Moreover, short-distance local rates in the scale were lower than in previous schedules, as were many of the longer-distance local rates (as

[1] As shown in Table 5.7, *Central Traffic Association Scale of Uniform Class and Mileage Rates for All Railroads, Circa 1893.*

TABLE 6.1

Grain Rates and the Average Difference Between Spot Prices in Chicago and New York Grain Markets, 1880 to 1899

	Average Official Rate $\overline{R}_0$ (Cents per 100 Pounds)	*Average Board of Trade Rate $\overline{R}_T$ (Cents per 100 Pounds)*	*Average Grain Price Difference $\overline{P}_g$ (Cents per 100 Pounds)*	*Average Announced Lake-Rail Rate $\overline{R}_L$ (Cents per 100 Pounds)*
S80	30.00	30.00	31.67	26.81
W80–81	33.80	33.62	37.28	
* S81	19.10	17.14	17.48	17.07
*W81–82	19.63	19.16	19.96	
S82	25.00	25.00	23.22	18.34
W82–83	30.00	30.00	28.06	
S83	25.30	25.30	20.60	19.64
*W83–84	24.86	24.14	19.30	
* S84	21.89	21.25	18.68	16.82
*W84–85	23.61	22.16	25.93	
* S85	20.52	17.14	15.51	15.24
*W85–86	25.00	24.23	22.80	
* S86	25.00	24.82	17.79	20.00
*W86–87	28.33	24.21	27.45	
S87	25.00	25.00	19.87	20.93
W87–88	26.14	26.14	25.71	
S88	23.64	22.94	17.73	17.89
W88–89	25.00	25.00	22.22	
S89	23.00	22.87	18.11	14.07
W89–90	22.00	22.00	21.28	
S90	21.38	21.32	16.35	15.07
W90–91	24.53	24.53	21.85	
S91	25.00	24.43	18.23	11.00
W91–92	25.00	25.00	29.55	
S92	23.41	23.41	15.20	12.81
W92–93	25.00	25.00	23.53	
S93	24.85	24.44	19.00	13.94
*W93–94	22.06	18.75	16.81	
* S94	20.43	20.00	13.92	12.00
*W94–95	22.35	19.76	13.96	
* S95	19.10	18.72	13.22	11.80
W95–96	20.00	19.73	16.96	
S96	18.83	18.83	13.17	11.19
*W96–97	19.17	16.67	13.72	
* S97	17.25	14.13	11.58	12.68
*W97–98	20.59	17.94	13.76	
* S98	18.88	15.47	11.42	8.12
†W98–99		18.35	16.78	
† S99		17.26	17.42‡	11.09

* Seasons in which there were suspected breakdowns of official rates.
† Seasons in which there were no official rates.
‡ Oats average only.
Source: The preceding tables in Chapters 4 and 5.

can be seen from comparing local rates in 1886, in 1887, and in 1898). The rationale for the decreases had been to obtain less discriminatory rates in accord with Sections 1, 3, and 4 of the *Act to Regulate Commerce* — which was possible only by maintaining through rates and reducing local rates. The disruptions in through rates after the Trans-Missouri decision had reduced Board of Trade rates to less than 15 cents per 100 pounds in the summer of 1897, which was less for transport 1,000 miles from Chicago to New York than for 500 miles of local transport.[2] From the long view, the local shipper might have expected some "good" from regulation and cartelization that, in the tradition of 1887 or 1893, brought local rates down to parity with through rates.

From the view of the railroads, regulation and cartelization had been of some benefit. The trunk lines had been able to make the agreements on through rates work under "strong" regulation — seven years, from 1887 to 1893, of stable rates and high grain price differences provided testimony to this success — and there had been a small measure of similar success under "weak" regulation, but strong cartelization, with the Joint Traffic Association agreement in 1896.[3] The working agreements in the "seven good years" had resulted in an increase in profits for most of the railroads. The Pennsylvania Railroad experienced fluctuations in profits for reasons

[2] This may not have been very important for most shippers. But if the shipper was a grain miller or oats farmer in Ohio producing for the New York market, then it was relevant that other sources of supply at much farther locations had lower costs of transport. This local shipper should have objected to such increased geographical rate discrimination.

[3] An alternative hypothesis, mentioned in Chapter 3, is that the recorded breakdowns in rates occurred each time costs of transport decreased or demand for transport decreased so that there was "excess capacity." That is, the agreements never "worked," and the only reason for the lack of continuous breakdowns was that there were periods in which rates were maintained because each railroad could not transport more tonnage from cheating in those periods. This hypothesis would not seem to explain the timing, nor the extent, of the breakdowns. Changes in costs of transport were continuous and small after 1881, as a result of decreased wages and prices of materials, but there were no sharp and pronounced decreases in 1883 to 1886, or 1893 to 1898. Changes in demand cannot explain the occurrence of breakdowns. Summer demand for transport was always less than winter demand; all of the extended wars in rates from 1874 to 1898, with the exception of that in 1881, began in the winter season. Demand for the transport of grain for export was an important part of total demand. The Statistical Appendix to Chapter 6, p. 265, indicates that exports were smaller, and the declared value of exports smaller, in 1881–1882 than in the previous year — so that demand for export-grain transport had fallen in this year of cartel breakdown. This is true of 1883–1884 as well, but not of 1884–1885; and grain exports in 1886–1887, another disastrous year for the cartel, were the largest of the decade. Demand for the transport of export-grain appears to have greatly declined in 1887–1888, but there was no disloyalty to the agreement this first year of regulation. In the 1890's, demand declined in the 1893 to 1895 period, as did the success of the cartel; demand increased from 1896 to 1898, while the cartel continued to be ineffective.

of fluctuations in sales of finished products, of fluctuations in the sizes of crops, and in livestock population; taking account of these reasons, the railroad also received more profits when the agreement on through traffic was operating. In the first of the rate disruptions of the 1880's, from the winter of 1883–1884 to the summer of 1886, the Pennsylvania carried more tonnage than in preceding seasons but for less gross receipts and less profits[4]; in the seven years of strong regulation, the railroad carried more tonnage each month and received much greater profits (as shown in Table 6.2).[5] The Erie Railroad reported sharply decreased profits and increased tonnage during 1885 and 1886, as compared with the immediately preceding and following years. The New York Central Railroad had higher profits during the rate demoralizations of 1894–1895, and increased profits after 1896 (when its tonnage out of Buffalo increased with the introduction of Chicago–New York "intrastate" rates). The Baltimore and Ohio participated in the general decline of tonnage and profits in 1894, and when its tonnage increased in 1895–1896 its profits continued to decline.[6] Three of the four large trunk lines had higher returns during strong regulation and cartelization.

Profits from regulation and cartelization seem to have been realized by the stockholders of the trunk-line railroads. If the present value of expected future profits per share of stock of any of the companies equaled the sales price of the stock, then the owners of most of the railroads held expectations of greater profit during periods of strong regulation-cartelization. The Pennsylvania Railroad's common stock price declined in 1884 and remained below the long-term trend of such prices until the summer of 1887; the next large decrease was in the summer of 1893. The common

[4] Lower profits, a lower "profit margin" (as shown in the last column of Table 6.2), and larger tonnage could only have followed from the railroad having been required to set rates below maximum profit levels. Lower profits were also experienced during the breakdown of through rates in 1894–1895, but tonnage transported declined as well (perhaps as a result of the general downturn of industrial production of 1893), so that there is not an unambiguous indication of less profits as a result of cartel breakdown.

[5] The profits are accounting profits as given monthly by this railroad. They are subject to all of the errors mentioned in Chapter 3 — those inherent in arbitrary division of costs between "fixed" and "variable," and in the reporting of revenues long after transport costs have been incurred. The errors have been kept as small as possible, primarily by including only operating costs in the calculations, and by averaging monthly profits over the season.

[6] This was not the tendency of profits from local traffic alone; the recorded tonnage and profits for the Baltimore and Ohio's "Straitsville Division" were higher in 1895 and 1896 (after declines in tonnage and profits in 1891, 1892, and 1893), as shown in Table 6.2. The Pennsylvania and the Erie experienced reduced profits and reduced tonnage as well after 1894. For these railroads it is not possible to separate the effects of cartel breakdown from those following a general decrease in the demand for transport; for the Baltimore railroad, it is clear that rate reductions, not tonnage reductions, reduced profits.

TABLE 6.2

Accounting Profits from Freight Transportation, 1880 to 1898

Season	Tons Average per Month (Thousands)	Gross Receipts, Average per Month (Thousands of Dollars)	Net Receipts, Average per Month (Thousands of Dollars)
	*The New York Central Railroad**		
Winter 1887	1187	2786.60	924.60
Summer 1887	1219	3153.28	1192.42
Winter 1888	1246	2825.60	801.80
Summer 1888	1253	3054.00	1006.57
Winter 1889	1256	2754.80	866.80
Summer 1889	1257	3185.00	1111.57
Winter 1890	1332	2915.00	931.80
Summer 1890	1351	3082.00	979.42
Winter 1891	1378	3082.00	1068.80
Summer 1891	1385	3522.50	1238.00
Winter 1892			
Summer 1892	1727	2916.85	892.85
Winter 1893	1760	3663.20	1173.40
Summer 1893	1768	4093.57	1328.00
Winter 1894†	1602	3314.60	1090.20
Summer 1894†	1561	3593.42	1212.42
Winter 1895†	1628	3385.60	1054.00
Summer 1895†	1645	3861.28	1229.28
Winter 1896	1804	3542.80	1105.00
Summer 1896	1844	3819.57	1271.85
Winter 1897†	1746	3454.80	1106.60
Summer 1897†	1721	3851.00	1409.50
	The Pennsylvania Railroad		
Summer 1880	2170.91	2479.16	1021.66
Winter 1881	2493.84	2571.03	1027.99
Summer 1881†	2574.58	2594.00	1029.58
Winter 1882†	2793.64	2799.13	1040.58
Summer 1882	2848.41	2850.41	1043.33
Winter 1883	2948.61	2954.34	1072.46
Summer 1883	2973.66	2980.33	1079.75
Winter 1884†	3036.86	2812.19	1005.48
Summer 1884†	3052.66	2770.16	986.91
Winter 1885†	3242.59	2613.69	894.71
Summer 1885†	3290.08	2574.58	871.66
Winter 1886†	3513.54	2823.11	978.13
Summer 1886†	3569.41	2885.25	1004.75
Winter 1887†	4049.41	3115.71	1012.15
Summer 1887	4169.41	3173.33	1014.00
Winter 1888	4547.74	3312.99	1018.20
Summer 1888	4642.33	3347.91	1019.25
Winter 1889	4819.99	3489.71	1078.51
Summer 1889	4864.41	3525.16	1093.33

TABLE 6.2 (Continued)

Season	*Tons Average per Month (Thousands)*	*Gross Receipts, Average per Month (Thousands of Dollars)*	*Net Receipts, Average per Month (Thousands of Dollars)*
	The Pennsylvania Railroad (Continued)		
Winter 1890	5416.08	3757.23	1110.59
Summer 1890	5554.00	3815.25	1114.91
Winter 1891	5544.13	3873.05	1175.44
Summer 1891	5541.66	3887.50	1190.58
Winter 1892	5849.66	3941.50	1103.51
Summer 1892	5926.66	3955.00	1081.75
Winter 1893	5793.93	3831.40	1042.68
Summer 1893	5760.75	3800.50	1032.91
Winter 1894†	5416.95	3454.23	992.04
Summer 1894†	5331.00	3367.66	981.83
Winter 1895†	6239.40	3734.99	1101.96
Summer 1895†	6466.50	3826.83	1132.00
	The Erie Railroad		
Winter 1883	435.80	444.80	92.40
Summer 1883	498.42	511.28	176.14
Winter 1884†	411.60	473.80	139.60
Summer 1884†	421.71	637.14	217.57
Winter 1885†	372.40	460.00	146.80
Summer 1885†	414.57	484.71	154.57
Winter 1886†	452.00	403.00	128.40
Summer 1886†	527.85	437.00	139.57
Winter 1887†	474.80	470.00	150.00
Summer 1887	520.14	557.28	182.28
Winter 1888	446.20	502.40	160.60
Summer 1888	558.00	561.71	186.42
Winter 1889	474.80	485.80	155.20
Summer 1889	622.57	544.57	182.00
Winter 1890	620.60	482.80	154.20
Summer 1890	664.85	596.71	197.71
Winter 1891	492.40	580.00	185.60
Summer 1891	658.42	621.42	187.71
Winter 1892	622.20	532.60	170.00
Summer 1892	680.14	630.14	191.85
Winter 1893	653.00	578.80	185.00
Summer 1893	573.80	627.00	181.80
	The Baltimore and Ohio Mainline		
Winter 1887	546.80	601.46	86.06
Summer 1887	548.88	610.30	103.52
Winter 1888	554.10	632.39	147.15
Summer 1888	570.33	629.04	138.91
Winter 1889	610.90	620.67	118.31
Summer 1889	649.60	651.06	122.83
Winter 1890	746.34	727.02	134.12
Summer 1890	764.60	730.57	146.45

TABLE 6.2 (Continued)

Season	Tons Average per Month (Thousands)	Gross Receipts, Average per Month (Thousands of Dollars)	Net Receipts, Average per Month (Thousands of Dollars)
	The Baltimore and Ohio Mainline (Continued)		
Winter 1891	810.25	739.45	177.27
Summer 1891	801.80	738.75	163.32
Winter 1892	780.65	737.02	128.45
Summer 1892	764.37	724.83	118.66
Winter 1893	757.85	719.96	114.75
Summer 1893	671.07	649.15	123.05
Winter 1894†	636.35	620.82	126.37
Summer 1894†	692.29	639.45	137.76
Winter 1895†	714.66	646.90	142.32
Summer 1895†	759.35	683.65	130.53
Winter 1896	777.23	698.35	125.81
	The Baltimore and Ohio Railroad — Straitsville Division		
Winter 1887	46.95	13.21	−.53
Summer 1887	51.40	14.62	−.38
Winter 1888	62.55	18.15	−.01
Summer 1888	61.73	17.78	.51
Winter 1889	59.70	16.86	1.85
Summer 1889	55.83	15.71	1.33
Winter 1890	46.15	12.82	.02
Summer 1890	42.91	12.20	−.19
Winter 1891	34.81	10.65	−.74
Summer 1891	35.14	10.63	−1.01
Winter 1892	35.96	10.61	−1.67
Summer 1892	38.90	10.88	−1.60
Winter 1893	40.07	10.99	−1.57
Summer 1893	28.88	8.22	−1.73
Winter 1894	24.40	7.11	−1.79
Summer 1894	29.79	8.75	−.99
Winter 1895	31.95	9.40	−.68
Summer 1895	37.95	10.37	−.17
Winter 1896	40.35	10.75	.03

* Monthly averages of annual statistics for New York state traffic only; the receipts are those for transport throughout the system.

† Seasons of suspected breakdowns in rates.

Source: Accounting statistics on Tonnage and Gross Receipts are presented in the *Annual Reports* of the Pennsylvania Railroad and the New York Central Railroad. The figures for the Erie Railroad are those for the "New York, Pennsylvania and Ohio" Railroad only, in the archives of the Erie-Lackawanna Railroad, Cleveland, Ohio. Each of these sources contains monthly statistics; the Baltimore and Ohio Railroad statistics (as shown in the *Annual Reports*) are yearly statistics from which seasonal averages were forced by weighting annual averages by the number of months in a season.

Accounting statistics on "transport costs" for each railroad are from the same sources. The total operating costs were multiplied by the ratio of freight train mileage to total train mileage to obtain an estimate of "freight transport costs." This estimate, in assuming that mileage costs for operating a freight and passenger train were the same, may not result in an accurate estimate of costs each month. It is difficult to conceive of means by which it would bias month-to-month comparisons, however.

Gross Receipts minus "freight transport costs" provide the estimate of Net Receipts.

stock of the Erie Railroad, and the Lehigh Valley Railroad, experienced price changes quite similar to those in the Pennsylvania stock: large decreases in the 1884 to 1886 period of rate breakdowns, small increases upon the advent of regulation, rising prices until 1893, and sharp declines during the rate breakdowns thereafter.[7] Prices of Baltimore and Ohio stock fluctuated most widely. The Baltimore road's average seasonal price increased, and was above the long-term trend, from 1881 to 1884, then decreased 12 points in 1884. The price increased sharply in 1887, but decreased even more sharply in 1888; steady gains were made thereafter until large decreases were experienced during the rate breakdowns in 1893 and in 1896. All of these changes (except the decline in 1888) followed closely the profit losses or gains from cartel breakdown or cartel revival.

The stockholders in the New York Central system (the Lake Shore and Michigan Southern affiliate) had different profit expectations. Stock prices were above the long-term trend from 1881 to the summer of 1884; then the price was from 13 to 30 points below trend until the beginning of regulation. Under regulation, the price remained below trend but increased by a few dollars each season until the summer of 1893 when there was a large decrease. This price decline was reversed by increases each season of five dollars during the cartel breakdowns. At the time of the advent of the Joint Traffic Association there was some gain in price; but large gains were not realized until the breakdowns in through rates in 1897–1898.[8]

[7] The interesting exception to this general pattern was the 30-point increase in the Erie's price during the winter of 1886–1887. This occurred at the time this road was accused of cheating on the cartel agreement on through rates, and at the time statistics on through tonnage indicated that the railroad was profiting from cheating.

[8] The trend in part reflects the decline in wholesale prices: the "Index of Wholesale Prices" declined from 110.3 in 1881 to 65.4 in 1896, then increased to 70.5 in 1898 [cf. C. Snyder, *Business Cycles and Business Measurements* (New York: 1929), p. 288]. The deviations from trend might be explained by declines in the demand for transport: tonnage declined sharply in 1893, as noted in the discussion of Table 6.2, concurrent with declines in stock prices and with the breakdown in official rates.

A test of the effects of general price changes, and of changes in tonnage transport, on profit expectations can be made if it is assumed that an index of all railroad stock prices was affected by these factors in the same manner as was the price of Lake Shore and Michigan Southern. Then calculation of the equation $(P_t - P_{t-1}) = \alpha + \beta(I_t - I_{t-1})$ provides an estimate of the effects of "industry-wide" profitability; the deviations from the calculated equation can be assumed to be the result of "special conditions" of profitability for this railroad. Using the Cowles Index of railroad stock for I_t, I_{t-1} [cf. Alfred Cowles, *Common Stock Indexes, 1871–1937* (Bloomington, Ind.: Principia Press, 1938)] for each month "t," and the monthly average of the Wednesday prices of Lake Shore stock for 1871 to 1898, the calculated equation is $(P_t - P_{t-1}) = 0.259 + 1.786(I_t - I_{t-1})$ with $\hat{\rho}^2 = .157$. For the seasons in 1894–1895, beginning with the winter of 1894, the average "residuals" $[(P_t - P_{t-1}) - 0.259 - 1.786(I_t - I_{t-1})]$ were as follows: +.819, +.699, +.785, +.950. For the seasons in 1897–1898, the average residuals were +2.261, −0.498, +2.398, +0.455. The first series suggests that, while general

The owners' expectations for New York Central profits were greater at the time of cartel failure (at the time the company was gaining tonnage from the secret New York–Chicago "intrastate" rates). Comparing this experience with that of the other trunk-line railroads, it would appear that there were abrupt declines in stock prices during the years of the wars in through rates for all the railroads except the alleged perpetrator.

Thus the 25 years of formal rate-setting organizations, and the 13 years of regulation, had provided benefits for some and losses for others.

Through shippers were deprived of any observable benefits. When rates were maintained, the grain spreaders' costs were increased and Chicago–New York grain price differences were greater. During the seven years of strong regulation, the price difference averaged 6 cents per 100 pounds more than in the succeeding six years of weakened (or nonexistent) regulation (as can be seen from Table 6.1). But local shippers benefited from the prohibitions against discrimination. Reductions in short-distance rates in the Official Classification of 1887, and in the Uniform Mileage Scale of 1893, were not dictated by market conditions but rather followed from the imposition of regulatory rules. They provided the smallest of the shippers, seeking the most limited service, with somewhat lower costs of transport.

The trunk-line railroads benefited from regulation — the rules of the Interstate Commerce Commission against personal and locational discrimination, and against setting secret rates, had made it possible for the cartel to maintain basic long-distance grain rates of 23 to 25 cents for seven years. It had not been possible to maintain 21–22 cent rates previously, nor was it possible to maintain 17–20 cent rates in the years following the weakening of regulation. There had been disadvantages for the trunk-line railroads from regulation: for one, reductions in short-distance local rates on each road had been required. But the benefits seem to have prevailed, given that the railroads were generally more prosperous in the seven years of strong regulation–cartelization (as shown in Tables 6.2 and 6.3).

It seems impossible, because of these contrasting results, to indicate "net" benefits or disadvantages from regulation and cartelization. To be sure, less discrimination between local and through transport would have been desirable in a world of perfect competition in all other markets[9]; but

---

prices and transport demand declined, New York Central profits did not decline as much; the second series suggests that this railroad's change in expected profits was greater than justified by transport demand increases. It is possible that the New York railroad profited from the breakdowns.

[9] Since this is no more than a requirement for Pareto optimum conditions of production and exchange; cf. I. M. D. Little, *A Critique of Welfare Economics* (Oxford Paperbacks), Chapter IX, p. 152 *et seq.*

TABLE 6.3

Stockholder's Profits from Freight Transport, 1880 to 1898

(1) The trend of Wednesday closing prices of common stock $P = \alpha + \beta T$ for each week "T" numbered consecutively from 1880 to 1898, for the following railroads:

The New York Central Railroad:	$P = 81.80 + 0.073T$	$\hat{\rho}^2 = .501;$	$S_u = 20.96$
The Pennsylvania Railroad:	$P = 57.755 - 0.006T$	$\hat{\rho}^2 = .195;$	$S_u = 3.588$
The Erie Railroad:	$P = 28.56 - 0.016T$	$\hat{\rho}^2 = .202;$	$S_u = 9.49$
The Baltimore and Ohio Railroad:	$P = 208.80 - 0.194T$	$\hat{\rho}^2 = .869;$	$S_u = 21.417$
The Lehigh Valley Railroad:	$P = 73.24 - 0.045T$	$\hat{\rho}^2 = .790;$	$S_u = 6.277$

(2) Deviations from Trend: Average difference between actual price and "P" expected from $(\alpha + \beta T)$ as shown above.

	The Erie Railroad		*The New York Central Railroad*		*The Lehigh Valley Railroad*		*The Pennsylvania Railroad*		*The Baltimore and Ohio Railroad*	
	Average Price	*Average Deviation from Trend*	*Average Price*	*Average Deviation from Trend*	*Average Price*	*Average Deviation from Trend*	*Average Price*	*Average Deviation from Trend*	*Average Price*	*Average Deviation from Trend*
Winter 1880	32.29	2.87	105.47	23.00	116.00	43.16	51.53	−6.15	146.26	−60.70
Summer 1880	30.12	1.11	107.96	23.74	116.00	43.57	56.14	−1.40	162.07	−40.22
Winter 1881	47.80	19.21	128.90	42.74	59.53	−10.90	65.18	7.79	191.25	−6.05
Summer 1881*	54.00	25.86	125.06	36.99	61.03	−8.36	64.50	7.26	197.62	5.40
Winter 1882*	33.42	5.72	114.00	24.03	60.94	−7.25	61.22	4.14	194.21	7.04
Summer 1882	35.52	8.25	110.90	19.02	61.95	−5.12	60.44	3.52	193.54	11.25
Winter 1883	29.90	3.03	111.95	18.17	64.05	−1.83	60.21	3.44	199.05	21.86
Summer 1883	24.35	−2.07	104.77	9.05	68.47	3.76	57.92	1.30	195.70	23.45
Winter 1884*	17.95	−8.02	99.68	2.10	69.84	6.37	58.77	2.33	196.00	29.09
Summer 1884*	11.89	−13.66	79.19	−20.31	63.03	.74	53.66	−2.60	175.79	13.93
Winter 1885*	13.05	−12.04	63.04	−38.36	58.15	−3.00	51.75	−4.36	171.14	14.43
Summer 1885*	7.75	−16.90	68.12	−35.19	56.60	−3.25	51.16	−4.78	170.80	19.46

Winter 1886*	13.80	−10.43	85.57	−19.65	56.72	−2.03	53.50	−2.29	175.66	28.88
Summer 1886*	12.26	−11.56	87.51	−19.60	56.48	−1.10	55.20	−.43	153.96	12.47
Winter 1887*	44.72	21.37	95.00	−14.06	55.76	−.66	56.68	1.19	164.50	27.96
Summer 1887	18.50	−4.43	95.03	−15.93	55.70	.51	55.19	−.12	139.56	8.39
Winter 1888	14.57	−7.93	92.00	−20.86	54.19	.12	53.85	−1.30	98.57	−27.76
Summer 1888	16.29	−5.76	96.57	−18.24	53.16	.29	53.36	−1.63	90.86	−30.35
Winter 1889	16.90	−4.73	102.27	−14.40	53.66	1.94	53.85	−.99	89.55	−26.68
Summer 1889	18.43	−2.76	104.76	−13.88	52.93	2.43	52.25	−2.42	85.30	−25.65
Winter 1890	17.42	−3.34	106.00	−14.57	51.84	2.54	53.47	−1.04	98.90	−7.13
Summer 1890	16.87	−3.47	108.67	−13.74	51.60	3.46	52.60	−1.76	102.44	1.68
Winter 1891	13.85	−6.05	109.04	−15.28	49.00	2.01	50.71	−3.49	90.65	−5.26
Summer 1891	16.00	−3.48	115.00	−11.23	48.48	2.67	51.83	−2.21	93.96	3.44
Winter 1892	23.36	4.31	127.68	−.49	54.27	9.66	55.45	1.55	93.68	8.02
Summer 1892	24.17	5.54	133.03	2.97	59.03	15.60	54.46	.73	96.53	15.92
Winter 1893	22.66	4.47	129.19	−2.78	53.80	11.54	53.76	.18	93.11	17.42
Summer 1893	16.36	−1.38	121.10	−12.78	40.00	−1.07	50.13	−3.27	71.96	1.41
Winter 1894*	15.22	−2.08	126.68	−9.10	39.00	−.90	49.36	−3.88	73.38	7.96
Summer 1894*	16.57	−.30	132.89	−4.83	36.85	−1.84	50.00	−3.08	72.57	12.17
Winter 1895*	16.68	.22	137.09	−2.50	33.09	−4.45	50.45	−2.47	61.18	5.85
Summer 1895*	24.09	8.07	149.19	7.65	38.74	2.38	53.80	1.02	62.48	12.40
Winter 1896	20.35	4.78	146.47	2.93	36.11	.99	53.23	.62	29.33	−15.43
Summer 1896	16.87	1.71	146.90	1.56	31.22	−2.77	51.58	−.87	15.78	−24.28
Winter 1897*	17.14	2.41	158.90	11.65	27.19	−5.63	52.04	−.25	14.50	−20.39
Summer 1897*	16.51	2.22	171.29	22.13	28.00	−3.64	54.38	2.24	13.20	−16.73
Winter 1898*	15.80	1.96	183.00	31.91	23.71	−6.72	57.09	5.11	15.31	−9.25
Summer 1898*	15.89	2.48	192.21	39.18	20.83	−8.44	57.90	6.07	27.11	7.39

* Seasons of suspected breakdowns in rates.
Source: Wednesday closing prices for common stock of these railroads in New York as listed in the *Financial Chronicle* (Chicago, Illinois), 1880 to 1898. The statistics for the New York Central Railroad are for the Lake Shore and Michigan Southern Railroad only.

raising through transport rates, and lowering some local rates, may or may not have increased economic welfare in the framework of these transport markets. Some long-distance transporters, such as the Chicago meat packers, were operating producers' cartels; adding to producers' (cartel) prices the (cartel) rates on lard and provision transport after 1887 may or may not have increased social benefits.[10] The losses of grain spreaders cannot be compared with the gains of local shippers with the lower rates, because there was not compensation of the losers by those who benefited. Both the effects on the rate of use of resources, and upon the distribution of income, are diverse, then. Only with quantitative estimates of each of these effects can an assessment be made: "there is no *a priori* way to judge as between various situations in which some of the . . . optimum conditions are fulfilled while others are not." [11]

It is possible, however, for a society to choose regulation and cartelization or their absence on grounds of "justice" or "welfare," politically defined. The experience of the trunk-line railroads with rates, market shares, and profits indicates that an antidiscrimination law rigidly imposed on a cartelized market provides the means for effective cartel control. The Congress and the courts can decide whether or not to make these means available.

In the present case, by the middle of the 1890's, the United States courts seem to have decided that the Interstate Commerce Commission did not have jurisdiction to police the structure and level of rates, at least not in a manner having the effect of maintaining official rates. The courts also ruled that the trunk-line railroad cartel could not operate with its own rules for enforcing official rates, for reasons which included the adverse effects on long-distance rates: "the natural and direct effect of the [Trans-Missouri and Joint Traffic] agreements is the same, namely, to maintain rates at a higher level than would otherwise prevail. . . ." [12] For reasons of "justice" under the *Interstate Commerce Act*, rates were to be determined by the railroads; for reasons of "welfare" under the common law and the *Sherman Act*, competition was to be the means.

[10] The problem of "multiple-layer" optimum, when there are imperfections in supply markets, can be solved by introducing the presence of particular noncompetitive profit margins at the transport level (for example). But the new imperfection has to be "offsetting" in the sense of providing a counteracting power to control price — and there is little indication of this effect here, other than that contained in the instability of the agreement on dressed beef in 1888 and 1890. Cf. A. Fishlow and P. A. David, "Optimum Resource Allocation in an Imperfect Market Setting," *Journal of Political Economy*, *LXIX* (1961), 529 to 546.

[11] R. G. Lipsey and Kelvin Lancaster, "The General Theory of Second Best," *Review of Economic Studies*, *XXIV* (1), No. 63 (December 1956), 11.

[12] *U.S.* v. *Joint Traffic Association*, 19 Sup. Ct. 26, at 35.

Appendix to Chapter 2

Price Strategies to Promote Cartel Stability*

DANIEL ORR AND PAUL W. MACAVOY

1. *Introduction*

A cartel cannot control the price of a product unless all member firms adhere to the agreed price. If collusive contracts are enforceable, price maintenance is straightforward; if contracts "in restraint of trade" are unenforceable, or illegal, this effective control of cheating is eliminated. Yet even without recourse to court-enforced sanctions, cartels have occasionally prospered.[1]

A cartel can survive by policing the members' behavior and threatening believable and punitive retaliation against cheating. In this paper, the requisite thoroughness of policing and the deterrent powers of certain pricing strategies are explored. In particular, it is shown that, in an idealized homogeneous market, the cartel can make cheating unprofitable simply by threatening to establish a price that maximizes the profits of loyal member firms in the presence of cheaters. This pricing strategy is certainly believable; that it is also punitive to cheaters is shown in Sections 3 to 5.

Another interesting pricing strategy is the matching of price cuts. This seems a natural course for a cartel confronted by a cheater: it was advocated by the Western Traffic Association in the United States when that cartel faced a breakdown of its rate structure,[2] and is suggested in recent theoretical literature as a viable means of enforcement.[3] This pricing strategy is examined in Section 2.

2. *A Cartel Response That Invites Breakdown*

Assume an industry in which the suppliers of a homogeneous good are differentiated by location. Price information is not transmitted instantaneously through the market. The market is spatially homogeneous so

* This Appendix, with minor changes, originally appeared in *Economica* (May, 1965), pp. 186–197.

that equal division among member firms at a single price is a practicable cartel objective.

If there are $n + 1$ firms, the market demand (in units per unit of time) is given by

$$Q = (n + 1)\alpha - (n + 1)\beta p \qquad \alpha, \beta > 0$$

when each firm charges the same price p. If variable costs are zero and the cartel maximizes joint profits, the optimal agreed price is

$$p_0 = \alpha/2\beta$$

With the market apportioned equally, profit per unit time to each member is[4]

$$\alpha^2/4\beta$$

so the present value of future profits to each firm is

$$(\alpha^2/4\beta) \int_0^\infty e^{-\lambda t}\, dt = \alpha^2/4\beta\lambda \tag{1}$$

(where λ is the appropriate rate of discount) when all firms remain loyal to the agreement.

The cartel reacts to cheating by meeting the price cut (from p_0 to p^*) when it is discovered. A time interval of length τ elapses between the price cut and its discovery by the cartel; the existence and length of this reaction lag is known to at least one firm.[5]

The cartel's price p_0 and the cheater's price p^* both hold during the interval $(0, \tau)$ beginning with the cut. During this time the cheater's demand is $Q^* = \alpha - \beta p_0 + \gamma(p_0 - p^*)$ — an increment proportional to the cut in price is added to the firm's quantity — and his revenue rate is

$$p^*Q^* = p^*[\alpha - \beta p_0 + \gamma(p_0 - p^*)] \tag{2}$$

where $\gamma > \beta$.[6] When the cartel meets the price cut, the cheater's revenue rate is

$$p^*Q^* = p^*(\alpha - \beta p^*) \tag{3}$$

Under these circumstances, matching a cut in price is never a punitive cartel strategy: a cheater can always choose some $p^* < p_0$ that will yield him a profit. This is true, even though no other firm would have upset the collusive agreement,[7] and even though no other firm embarks on a subsequent round of cheating.

The joint profit maximum share is $\alpha^2/4\beta\lambda$ from Equation 1. From Equations 2 and 3, the present value of a chiseler's earning stream is

$$\frac{p^*}{\lambda} \{(1 - e^{-\lambda t})[\alpha - \beta p_0 + \gamma(p_0 - p^*)] + e^{-\lambda t}(\alpha - \beta p^*)\} \tag{4}$$

The maximizing p^* ($= p_0^*$) is

$$p_0^* = \frac{\alpha[\beta + \gamma + (\beta - \gamma)e^{-\lambda t}]}{4\beta[\gamma + (\beta - \gamma)e^{-\lambda t}]} \quad (5)$$

and, by substitution in Expression 4, maximum profit is

$$\frac{\alpha^2}{16\beta^2\lambda} \frac{[\beta + \gamma + (\beta - \gamma)e^{-\lambda t}]^2}{[\gamma + (\beta - \gamma)e^{-\lambda t}]} \quad (6)$$

A firm breaks even by defecting if Expression 6 and the right side of Equation 1 are equal, i.e., if

$$\left[\left(1 + \frac{\gamma}{\beta}\right) + \left(1 - \frac{\gamma}{\beta}\right) e^{-\lambda t}\right]^2 = \frac{4\gamma}{\beta} + 4\left(1 - \frac{\gamma}{\beta}\right) e^{-\lambda t}$$

The solution of this equation for τ indicates the minimum lag necessary for the cheater to break even, and thus the minimum condition for cheating to be profitable. Multiplying out the left side, regrouping terms, and dividing by $[1 - (\gamma/\beta)]^2$ yields

$$(e^{-\lambda t} - 1)^2 = 0$$

an equation with a double root at $e^{-\lambda t} = 1$. The only value that satisfies the equation is the uninteresting case $\tau = 0$ (where it follows immediately that $p_0^* = p_0$). Under all other circumstances, the profits from calculated cheating exceed those obtained from an equal share of the cartel profit.

These somewhat surprising results may be weaker in their implications than the foregoing statement conveys, however, because they assume prior knowledge of the cartel's reaction time. (The optimal p^* depends on τ in Equation 5.) To assess the importance of the assumption of known reaction time, suppose one firm cuts price to p^* and that this cut is made in ignorance of the exact time until the cartel matches. Two questions are asked: (1) for given values α, β, γ, λ, and τ, how large a price cut can one firm make without *reducing* its earnings? (2) What estimate of τ would yield a cut of that size, given the optimal p^* calculation? The answers to these questions show by how much a cheater can overestimate τ and still profit from breaking the agreement.

Collusion assures a profit of $\alpha^2/4\beta\lambda$. To find the reduced price p_b^* that yields the same profit, substitute $\alpha/2\beta$ for p_0 in Expression 4, cancel $1/\lambda$ from both sides, and solve the resulting equation:

$$p_b^*\{[\alpha/2 + \gamma(\alpha/2\beta - p_b^*)](1 - e^{-\lambda t}) + (\alpha - \beta p_b^*)e^{-\lambda t}\} = \alpha^2/4\beta$$

This is a quadratic equation in p_b^* with roots[8]

$$p_b^* = \frac{-\alpha[\beta + \Gamma] \pm \{\alpha^2[\beta + \Gamma]^2 + 4\alpha^2\beta\Gamma\}^{1/2}}{-4\beta}$$

where $\Gamma = (\beta - \gamma)e^{-\lambda t} + \gamma$. Table A.1 presents several cases of "large"

TABLE A.1

Case	α	β	γ	τ	p_0^*	p_b^*	$t(p_b^*)$	$t(p_b^*) \div \tau$
1	100	1	5	.11	49.04	48.07	.233	2.1
2	100	1	10	1	39.086	28.175	16.022	16
3	100	1	2	1	48.019	46.038	2.322	2.3
4	100	.1	2	.11	420.168	.278	.278	2.5
5	100	10	12	.11	4.995	4.990	.227	2.0

Note: All cases use the same λ value of .09.

errors that do *not* preclude the success of cheating. In the headings, p_0^* is the optimal cheater's price and p_b^* is the lowest price (given τ) at which he can break even. The column $t(p_b^*)$ contains the mistaken estimates of the reaction lag for which p_b^* would be the optimal price; these are obtained by substituting p_b^* in the left side of Equation 5, and solving for τ. Comparing τ to $t(p_b^*)$ shows that large percentage errors in estimation of the lag can be tolerated without loss to the cheater. In case 1, for instance, a cheater who estimates the cartel reaction lag at .233 period will set a price of 48.07. The true lag is .11. Despite the overestimate of τ by a factor of 2.1, he breaks even. Case 2 indicates that a large true value of τ in conjunction with a high ratio of γ to β is a combination extremely insensitive to error in the estimate of τ. In no tabulated instance does a firm suffer lower profits from cheating than from adhering, despite overestimating τ by a factor of two. These cases suggest that price-cut matching is not effective, even if the cartel's reaction time is not known exactly.

3. *The Optimal Pricing Strategy for the Cartel in the Presence of Cheating*

Rather than follow the disastrous policy of matching cuts, the cartel might continue to price optimally, taking the cheater's presence into account.[9] *The cartel should set its price to yield maximum joint profit for the loyal members, given that a cheater is maximizing his own profits in the face of the cartel policy.* Before such a policy can be used, it must be established that the resulting price pair is feasible and stable. That is, prices of cartel and chiseler must be positive and must yield positive sales volumes. The process is stable if the market will establish and maintain equilibrium prices.

The effect of the price cut on the cartel's demand has not yet been specified. If the cartel reacts to cheating by setting a price P, and the cheater moves from p^* to p^{**}, the loyal firms' demand is

$$Q = n\alpha - n\beta P - \delta(P - p^{**}) \quad 0 < \delta < \gamma \tag{7}$$

The bounds on δ assure that the cohesive firms lose sales to the cheater but that their loss is less than his gain; market demand increases when one firm cuts price.

The specific value of P chosen by the cartel, P_0, is at the intersection of the reaction functions of the cheater and the cartel, and the cheater's final price p_0^{**} is the intersection's other coordinate (Diagram 1). With this revision of cartel strategy, the cheater will not adopt the p_0^* of Equation 5, the optimal price of Section 2, because his initial price deviation does not affect the profit stream after the time τ. (In Section 2, the cartel adopted p^* beginning at time τ.)

The profit functions manipulated by the cartel and the cheater are, respectively,

$$\pi_c = P[n\alpha - n\beta P - \delta(P - p^{**})]$$
$$\pi_r = p^{**}[\alpha - \beta P + \gamma(P - p^{**})]$$

and the reaction functions are given by the conditions

$$\partial\pi_c/\partial P = \partial\pi_r/\partial p^{**} = 0$$

that is,

$$p_0^{**} = \frac{\alpha + (\gamma - \beta)P}{2\gamma} \tag{8a}$$

is the equation of the reaction of the cheater to the cartel price changes, and

$$p^{**} = \frac{-n\alpha + 2(n\beta + \delta)P_0}{\delta} \tag{8b}$$

is the reaction function of the cartel; in Diagram 1 these functions are plotted in the (p^{**}, P) plane.

The cheater's function has positive slope $(\gamma - \beta)/2\gamma$ and intercept $\alpha/2\gamma$, while the cartel's function has positive slope $2(n\beta + \delta)/\delta$ and negative intercept $-n\alpha/\delta$. For the reaction functions to intersect in the first quadrant with $\gamma > \beta$, it is necessary and sufficient that

$$\frac{2(n\beta + \delta)}{\delta} > \frac{\gamma - \beta}{2\gamma}$$

which is the tautology that $2n\beta/\delta > -\frac{3}{2} - \beta/2\gamma$: both prices will be positive at the intersection (as in Diagram 1).

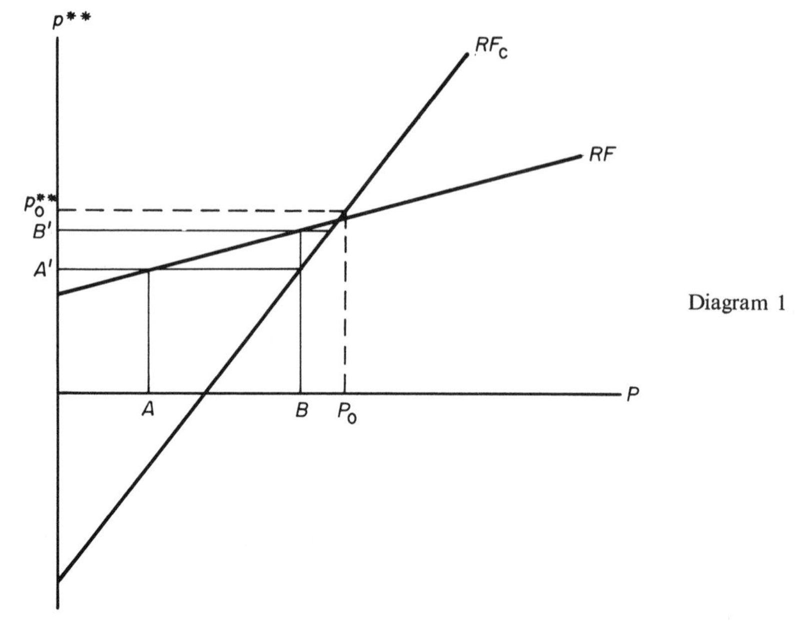

Diagram 1

This solution is feasible if the prices $p_0{}^{**}$ and P_0 are low enough to yield sales. Both should be lower than the collusive price $p_0 = \alpha/2\beta$. This is in fact the case as is seen by solving Equation 8. This yields

$$p_0 = \frac{\alpha(2n\gamma + \delta)}{4n\beta\gamma + \beta\delta + 3\gamma\delta} \qquad p_0{}^{**} = \frac{\alpha[n(\beta + \gamma) + 2\delta]}{4n\beta\gamma + \beta\delta + 3\gamma\delta}$$

We get $P_0 > p_0$ only if

$$4n\alpha\beta\gamma + 2\alpha\beta\delta > 4n\alpha\beta\gamma + \alpha\beta\delta + 3\alpha\gamma\delta$$

i.e., $P_0 > p_0$ if $\beta > 3\gamma$, which is ruled out by the condition $\gamma > \beta$. Similarly, $p_0{}^{**} > p_0$ only if

$$\alpha\beta(2n\beta + 3\delta) > \alpha\gamma(2n\beta + 3\gamma)$$

another condition precluded by $\gamma > \beta$. Both equilibrium prices are lower than the collusive price when $\gamma > \beta$, and yield a positive sales volume.

If $\gamma > \beta$, the equilibrium will be stable, as can be seen from Diagram I. A price time path like $A \to A' \to B \to B'$, etc., would be traced out in the case of zero conjectural variation: but this path converges to the equilibrium. Then the equilibrium is stable, according to the standard definition of stability.

Finally, the initial price $p_0{}^*$ set by the cheater is also lower than p_0, the joint profit maximizing price.[10] The cartel's response is independent of the cheater's first price, so p^* is set to maximize $p^*[\alpha - \beta p_0 + \gamma(p_0 - p^*)]$, the cheater's revenue rate during $(0, \tau)$, and

$$p_0^* = \alpha(\beta + \gamma)/4\beta\gamma \tag{9}$$

since $p_0 = \alpha/2\beta$.

This p_0^* will be smaller than p_0;

$$\alpha(\beta + \gamma)/4\beta\gamma < \alpha/2\beta$$

always follows from $\beta < \gamma$.

4. *The Profit-Maximizing Reaction as a Deterrent*

Assume that an interval of length τ elapses before the cartel reacts to one firm's price cut to p^*, and the cartel's strategy is the price p_0 in Equation 8, forcing the defector to p_0^{**}. Can the cartel prevent defection by adopting this profit-maximizing price P_0; if so, what is the maximum allowable lag for which this response is effective?

The defector will set p_0^*, given in Equation 9, during $(0, \tau)$, and will move to p_0^{**}, given in Equation 8, during (τ, ∞). With these prices, his profit stream over an infinite horizon has the present value

$$\frac{1}{\lambda}\left[\frac{\alpha^2(\beta+\gamma)^2}{16\beta^2\gamma}(1 - e^{-\lambda t}) + \frac{\alpha^2\gamma[n(\beta+\gamma)+2\delta]^2}{(4n\beta\gamma + \beta\delta + 3\gamma\delta)^2}e^{-\lambda t}\right] \tag{10}$$

which must be greater than the share of joint profits $\alpha^2/4\beta\lambda$ for defection to be profitable. For some values of α, β, γ, δ, λ, and n, there will be a value of τ for which Expression 10 equals $\alpha^2/4\beta\lambda$; in these cases, if expected τ is less than the breakeven value, no member of the cartel can profit from cheating. Table A.2 presents several such cases.[11]

TABLE A.2

Case	β	γ	δ	n	τ	*Case*	β	γ	δ	n	τ
1	1	2	$\frac{4}{3}$	10	*	9	.1	2	1.5	5	.38
2	1	5	4	10	*	10	.1	2	1.9	5	.78
3	1	5	3	3	.17	11	.1	2	1.3	5	.089
4	1	10	7	3	1.14	12	.1	2	1.2	5	*
5	1	10	7	10	*	13	.1	3	1.5	5	*
6	1	10	9	10	*	14	.1	2.4	1.5	5	.144
7	1	10	9	3	1.8	15	.1	1.6	1.5	5	.8
8	10	100	90	3	.18						

All cases use $\alpha = 100$, $\lambda = .09$.

* Denotes lack of effectiveness of optimal cartel counterstrategy as a deterrent; otherwise the value of τ indicates the *longest* time lag consistent with unprofitable cheating.

The examples indicate that the effectiveness of the cartel's strategy varies inversely with n, the number of firms in the cartel; directly with δ, the coefficient of the price differential in the cartel's demand relation; and inversely with γ, the coefficient of the price differential in the cheater's demand relation. These indications make sense on *a priori* grounds: the greater the impact of defection on the profits of each competitor, the more likely the competitors are to react with a large price cut and therefore the longer the interval preceding detection required for the cheater to break even.

5. *An "Arbitrary" Extension of the Profit-Maximizing Reaction*

A cartel strategy of maximizing joint profits of the loyal firms will make cheating unprofitable, in some cases. For other sets of parameters a price-cutter would profit, despite this "optimal" cartel counterstrategy. It will be seen that modification of the profit-maximizing rule reduces the set of cases for which cheating is profitable.

Consider cases 4 and 5 in Table A.2. The market demand relations differ in these cases; in 4, a one period reaction lag does not defeat the cartel strategy, while in 5, the maximizing strategy is always ineffective. In case 4, one-fourth of the firms cut price; in case 5, only one of eleven does. But if one of eleven cuts price and thereby profits, it may be taken as certain that others will follow as soon as the profitability of price-cutting is demonstrated. The idea that one successful cheater will be imitated by additional defectors suggests a policy to prevent price-cutting entirely.

Cheating may be profitable if the simple profit-maximizing strategy is used by the cartel, and if there are $n + 1$ firms in the cartel. However, identical market demand and response lag conditions might not yield profit to a price-cutter in the face of the same strategy if he were one of only $m + 1$ firms, where $n > m > 1$. *Then the group can deter price-cutting by responding to a nonadhering firm as though it were one of $m + 1$ firms, instead of one $n + 1$ firms.*

For example, assume that the demand parameters are $\alpha = 100$, $\beta = 1$, $\gamma = 10$, $\delta = 7$, and $\lambda = .09$; let the number of firms in the cartel be 12, and τ, the true reaction lag, be one period. We further assume that the sales loss to the loyal firms doubles with two price-cutters, triples with three, etc.; so 2γ and 2δ are the coefficients of the price differential when two firms cut price concurrently.[12] Given the assumed parameter values, it is profitable for one firm to cut price; in fact, the optimal cartel response also fails to protect against two defectors, as is seen in Table A.3. However, if three cheaters appear together, the cartel adopts a price that prevents gains from price-cutting. (These results, obtained for three cheaters in a twelve-firm cartel, are the same as would obtain in a cartel with the same

TABLE A.3

Number of Price-Cutters	*Cartel's Price*†	*Value of Future Earnings** *Cheating*‡	*Colluding*§
1	34.55	.4509	.25
2	25.65	.3153	.25
3	19.88	.2438	.25

* Both profit columns have been deflated by $\alpha^2/\lambda = 11{,}111$.
† Joint profit maximum price = 50.
‡ Profit figures are calculated from Expression 10.
§ Calculated from $\alpha^2/4\beta\lambda$, the collusive profit.

demand parameters, and one cheater among four firms.) The cartel can calculate $m + 1$ (in the example, a workable value of $m + 1$ is $\frac{12}{3}$ or 4), and respond to the first price cut with an immediate price reduction to the level consonant with one cheater in a cartel of that size. This policy would be sufficient to prevent cheating. However, the cartel will not be content just to find a value m that prevents gains from price-cutting; since cartel profits decline monotonically as the number of cheaters increases, it behooves the cartel to find the *largest* value of m consonant with profitless cheating (which we will call $\hat{m}$) and to react to the first cheater as though he were one of $\hat{m} + 1$ firms, instead of one among $n + 1$ firms.

In the unlikely event that cheaters emerge in an orderly succession, one at a time, the cartel would recalculate its optimal price as each new cheater emerged. It would eventually occur that the last firm to cut price (say, the kth),[13] would find cheating unprofitable. Perceiving this, the cartel makes known the policy of treating the first cheater as though he were the spokesman for a group of k cheaters. The first cheater will immediately find cheating unprofitable, and the possibility of mass defection based on the example of one successful cheater is averted. Knowledge that the cartel will calculate in the suggested way will lead to stability at the joint profit maximum price so long as the cartel's $\hat{m}$ calculation comes out greater than two.[14]

This alteration of the optimality calculations prescribes that the cartel should anticipate the set of reaction functions that are simultaneously inconsistent with profitable cheating, and maximum profit for the cartel: this supplants a price policy based upon an existing set of reaction functions.

Finally, a price policy successful in deterring, say, three or fewer cheaters in a cartel of twelve firms, may not prevent four or more firms from combining to cut price. Thus, the foregoing does not deny the analysis of coalitions as a valid treatment of the n-person nonzero sum game. The problems are different; formation of coalitions is not the usual cause of cartel breakdown.

6. *Summary, Extensions, and Conclusions*

In this paper it is alleged that in a number of cases a cartel can, by identifying and pursuing a consistent policy of profit maximization, effectively eliminate the profitability of price-cutting. In a number of additional cases the cartel can prevent cheating by treating one cheater as a harbinger of several future cheaters. These pricing policies may be supplemented by other techniques for reducing profits realized by a cheater (for example, shortening the reaction time τ).

Our analysis is shaped by three considerations: time lags in the flow of information, the linearity of all relations, and division of the market for a homogeneous good equally among the several sellers. The first consideration is treated very simply; we assume that the size of a price change does not affect the time required for the cartel to learn of its occurrence. However, the analysis can easily be extended to make the lags depend on the size of the price cut.

Suppose, for example, that a relation of the form

$$\tau \sim p^*K/(p - p^*) \tag{11}$$

holds; p^* is the cheater's price, p is the cartel price, and K is a constant related to the size of the market shift required to prove cheating. In Sections 3 to 5, p^* was chosen independently of τ (see Equation 9). However, with a relation like 11 in effect, the cheater's price and the reaction lag are not independent of each other. The relation 5 shows the dependence of the former on the latter: p^* will equal p for $\tau = 0$, and the price differential will increase with increases in τ until a maximum value of $\alpha(\gamma - \beta)/4\beta$ is reached at the limit. A plot of the relations 5 and 11 in the $(p - p^*, \tau)$ plane shows that the basic process of equilibration that held in Sections 3 to 5 is not upset by inclusion of the relation 11. See Diagram 2.

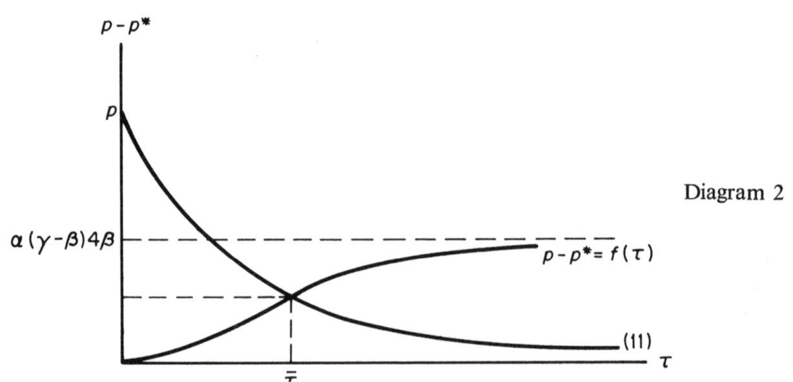

Diagram 2

The relation 11 begins at a positive level and is asymptotically zero, while the dependence of $p - p^*$ on τ (derived from the relation 5) begins at zero and has a positive asymptote. Since the relations are both continuous, the intersection will give a $p - p^*$ that has the same equilibrium properties as were found in the case discussed in Section 4.

The most novel aspects of this paper are the results of Sections 4 and 5. These results depend on the explicit nature of the impact of price-cutting on the cartel's profit. Such a representation necessarily has an *ad hoc* flavor; however, a tradition has been established for showing the effect of two prices in a market,[15] and our function 7 is consistent with this tradition. The precise assumptions about the effect of a price differential on demand emerge more clearly in a linear model than in a "more general" representation that deals with the slopes of functions in implicit form.[16]

The explicit introduction of location is a device calculated to avert ambush along the "Chamberlin vs. Chicago" route. The analysis relies on the assumption of a differentiated product with identical demand relations confronting each firm, and spatial separation seems the most plausible way to achieve this.

Two ways of eliminating the location assumption seem to be available. The first of these is to take into account the differences in cost and demand among the member firms. This, it turns out, is not viable. A joint profit maximum can only be effected by a side payment, and deviation from agreed prices may be interpreted only as evidence of dissatisfaction with the payment received. The center of attention should be the division of profits; this must be done equitably before any question of cheating is raised. A determinate one-price joint profit maximum seems to be necessary to a "general" analysis of the kind presented in this paper.

The second course is to focus directly on the rates of flow of price information within the market. A price cut by one firm would not immediately give him the entire market, since information reaches different customers at different times. The rate of information dissemination would depend on the size of the cut; the cartel would identify the cheater when his market share exceeded some threshold level. Such a model would study in isolation an extremely important consideration in conflict situations: the transmission of information concerning the actions of an opponent.

Notes

1. A survey of cartel experience in the United States shows several instances of successful market-sharing. Among these are the cast iron pipe cartel (1894 to 1898) which used a side payment mechanism to allocate sales; a contract was awarded to the firm which offered the largest bonus payment for division among

the other members [cf., *U.S.* vs. *Addyston Pipe and Steel Co. et al.*, 85 Fed. 271 (1898), p. 274]. The New York–Chicago trunk railways combined successfully *ca.* 1888 to 1895 by establishing freight quotas and allowing member lines to adjust their rates to assure that quotas were met. In the late 1920's, the Sugar Institute required member firms to divulge price and quantity information on all sales transactions, and boycotted brokers and warehousemen who would not sign a pledge to uphold the Institute's "code" [cf., *Sugar Institute* vs. *U.S.*, 297, U.S. 587 (1936)].

Also of interest is the earlier experience of the New York–Chicago railways. In 1885 the Western Traffic Association was established to administer rates; this effort was conspicuously unsuccessful because of rate-cutting by participating railroads. The response of the Association was suspension of the official rates, and all members were authorized to meet any cut in rates. [Cf., the *Railroad Gazette* (January 23, 1885), 56.]

2. Cf. footnote 2.

3. "Enforcement consists basically of detecting significant deviations from the agreed-upon prices. Once detected, the deviations will tend to disappear because they are no longer secret and will be matched by fellow conspirators if they are not withdrawn." [George J. Stigler, "A Theory of Oligopoly," *Journal of Political Economy*, *LXXII* (February, 1964), 46.]

4. As the model unfolds, it becomes clear why we insist upon differentiation solely on the basis of location: "quality" differentiation would almost certainly lead to different demand elasticities among firms, and the policy of joint profit maximization becomes enormously more difficult to exposit. With different demand elasticities confronting the individual firms, the cartel assigns different prices to its members and adjusts revenue differences by some sort of side payment mechanism, like the one used by the Southern Associated Pipe Producers (cf., footnote 1).

5. A price cut is effective as soon as customers are aware of its existence and have had time to purchase accordingly. This process of discovery and reaction on the part of customers is assumed to be sufficiently uniform to warrant treating the price cut as being effective at a single point in time.

The lag in the cartel's discovery of the price cut is caused by an asymmetry in the flow of market information. Certain customers of the cartel are aware of the lower price while competing sellers remain in ignorance. Presumably, the cartel becomes aware of the price cut from shifts in market share.

The customer, then, must conceal the presence of a price-cutter. At first this appears irrational: when other firms are made aware of the cut they will undercut, so the customer's interest is served by informing on the cheater. This overlooks the cost advantage over some or all competitors that can be maintained by protecting the cutter. Moreover, claims of preferred cuts in price are doubtful: customers may invent price cuts to induce another firm to "match." Most persuasive is the argument that the customers can calculate, too: they realize that the shorter the cartel's reaction lag is, the less likely it is that a price cut will be attempted again in the future.

6. With $\beta > \gamma$, a cheater would *raise* price on the cartel, instead of cutting price. The necessity for assuming different locations of the sellers becomes

evident here: otherwise, the cutter would corner the whole market as soon as all customers became aware of the price differential. Locational differentiation leads to buyer's expenditures on transit: a cheater may increase the radius of the area from which he draws customers by a cut in price, but a small cut does not assure that he will obtain the entire market.

7. One frequently encounters the argument that in a potentially unstable situation the greatest gain goes to the player that gets "first licks," and that the breakdown of the agreement is therefore inevitable, as in the "prisoner's dilemma" game [cf., R. D. Luce and H. Raiffa, *Games and Decisions* (New York: John Wiley and Sons, Inc., 1957), pp. 95 to 102]. Our objective is not to analyze behavior patterns arising from unstable situations; rather the discussion that follows is concerned with the market conditions that yield a *successful* cartel— or the market conditions that yield stability rather than some form of instability.

8. One should not encounter imaginary roots to this equation. Imaginary roots imply that the cheater never loses on *any* price cut.

9. The question of whether to cheat may be conceptualized as a nonzero sum game. If the cartel has no effective deterrent at its disposal, the game involves as many players as there are cartel members: there ensues the inadequately charted chaos of the n-person situation. However, an effective deterrent implies an equilibrium at the joint profit maximum, which can be illustrated in terms of a two-person game, cheater versus cartel. Of the innumerable possible alternative cartel policies, we defer consideration of those involving avoidable costs or losses, such as price wars. The threat of destructive reactions preserves an equilibrium until some pugnacious or overstocked challenger appears, as may happen at fairly regular intervals.

10. This condition was asserted in footnote 7; in the case discussed there, the cheater's price was set to maximize profits in the face of price-cut matching by the cartel. Here we show that if price is tampered with in the face of the optimal cartel response, it will be cut, not raised.

11. The intractability of the explicit form of breakeven $\tau = f(\alpha, \beta, \gamma, \delta, \lambda, n)$ precludes a more general solution.

12. For simplicity, we assume that the cartel regards groups of price-cutters as collusive. The cartel calculation for two cheaters assumes that each knows the other is cheating, and each selects a price appropriate to the situation in which two cheaters are present. This approach is appropriate for developing an effective deterrent, rather than for an accurate description of the time path of prices in a dissolving cartel.

13. Thus, k cheaters in a group of $n + 1$ firms are related to one cheater in a group of $m + 1$ firms by

$$\hat{m} + 1 = \left[\frac{n+1}{k}\right]$$

where $[(n + 1)/k]$ is the largest integer smaller than $(n + 1)/k$.

14. It is not at all clear that a group of less than half the cartel's membership could wage a successful price campaign against an equal or larger group in an effort to maintain stability at the joint profit maximum level. This is the implied sequence of events with $\hat{m}$ less than two.

15. See, for example, the "DD and dd curves" used profusely in E. H. Chamberlin, *The Theory of Monopolistic Competition* (6th ed., Cambridge and London, 1948): the diagram, p. 91, is one of many instances.

16. The results of Section 2 are readily obtained from an implicit form model. Let $\pi(p, P)$ represent the profit rate of a representative cartel member, where p is the member's price and P is the price of all other firms who act in unison. The conditions $\partial\pi/\partial p < 0$ and $\partial\pi/\partial P > 0$ are obvious. The cartel-established joint maximum profit is $\pi(P^*, P^*)$. Where the arguments p and P are equal, it follows that $\partial\pi/\partial p = -\partial\pi/\partial P$. When a member firm cuts price to p, the cartel price remains at P^* for τ time units, then is changed to P, a function of p. The cheater's present value then is (as in Expression 4),

$$\pi = \frac{1}{\lambda}[(1 - e^{-\lambda t})\pi(p, P^*) + e^{-\lambda t}\pi(p, P)]$$

The maximum condition, evaluated at $p = P^*$, is

$$\partial\pi/\partial p = \frac{1}{\lambda}\left\{(1 - e^{-\lambda t})\pi_1(P^*, P^*) + e^{-\lambda t}\left[\pi_1(P^*, P^*) + \pi_2(P^*, P^*)\frac{dP}{dp}\right]\right\}$$

But since $-\pi_1 = \pi_2$ at (P^*, P^*)

$$\partial\pi/\partial p = \frac{1}{\lambda}\left(1 - e^{-\lambda t}\right)\frac{dP}{dp}\pi_1(P^*, P^*) \tag{a}$$

If the cartel just matches the price p, $dP/dp = 1$, and the Expression (a) is always negative; it will always pay the individual firm to cut price if the cartel merely matches.

To obtain the results of Section 4 in this form, assumptions must be made about the relative magnitudes of the cartel response and the chiseler's response (similar to the $\gamma > \beta$ condition imposed in the linear model); to obtain the results of Section 5, the number of firms must be made an argument of the profit function. This all begins to look terribly *ad hoc* when cast in implicit form. Moreover, the "greater generality" of the implicit form representation is a delusion, since results like those obtained in this footnote are based on linear approximations of the functions and depend for validity on restrictive conditions of regularity.

Statistical Appendices

Statistical Appendix to Chapter 3

Grain Shipments from Chicago, 1874 to 1879

The trend of shares of each railroad $S = \alpha + \beta T + \gamma \overline{X} + \delta \overline{Y} + \nu$ has been estimated by the method of least squares, with S = percentage of total shipment from Chicago of wheat, corn, oats, and flour; T = month of shipment, numbered from 1 to 108 for the period 1871 to 1879; $\overline{X} = 0,1$ for the lakes closed to traffic or open to shipping; $\overline{Y}$ = the change in shares with the entry of the Baltimore and Ohio ($\overline{Y} = 1$ for all months after the Baltimore road's entry). The estimates of trend are:

The New York Central Railroad,

$$S = 60.271 + 0.0536T - 49.244\overline{X} - 2.861\overline{Y}$$
$$\hat{\rho}^2 = .917 \text{ and } Su = 10.646$$

The Pennsylvania Railroad,

$$S = 15.108 + 0.095T - 15.521\overline{X} - 2.211\overline{Y}$$
$$\hat{\rho}^2 = .798 \text{ and } Su = 5.917$$

The Baltimore and Ohio Railroad (1875 to 1879 only),

$$S = 2.613 - 0.022T - 3.451\overline{X} + 6.329\overline{Y}$$
$$\hat{\rho}^2 = .574 \text{ and } Su = 4.491$$

The "expected" share for any month is obtained by inserting the relevant values of T, $\overline{X}$, and $\overline{Y}$ in the correct equation. The actual shares are shown for 1874 to 1879, along with the difference between computed "expected" shares and the actual shares.

	The New York Central Railroad		The Pennsylvania Railroad		The Baltimore & Ohio Railroad	
	Percentage of Shipments	Excess or Deficit from "Expected"	Percentage of Shipments	Excess or Deficit from "Expected"	Percentage of Shipments	Excess or Deficit from "Expected"
November 1874	6.0	-7.5	1.1	-3.0	--	--
December 1874	68.6	+5.8	10.2	-9.5*	13.2	+5.7*
January 1875	77.1	+17.1*	10.8	-6.8*	1.2	-6.7*
February	71.1	+11.0*	20.5	+2.8	0.9	-7.0*
March	65.8	+5.7	16.6	-1.2	6.5	-1.3
April	54.4	-5.8	32.3	+14.5*	2.2	-5.6*
May	12.1	+1.1	10.0	+7.6*	1.9	-2.4
June	4.3	-6.8	8.7	+6.2*	1.9	-2.4
July	2.7	-8.4	0.5	-2.2	0.8	-3.5
August	4.6	-6.6	1.7	-2.0	2.5	-1.7
September	6.8	-4.4	3.6	+0.8	1.4	-2.8
October	7.1	-4.1	3.1	+0.2	0.6	-3.6
November	7.0	-4.3	6.7	+3.7	0.2	-4.0
December	43.9	-16.7*	32.0	+13.4*	6.3	-1.3
January 1876	45.1	-15.5*	9.3	-9.4*	19.6	+11.9*
February	39.1	-21.6*	16.1	-2.7	25.8	+18.1*
March	39.6	-21.2*	9.4	-9.5*	28.4	+20.8*
April	27.5	-33.3*	26.3	+7.3*	5.8	-1.7

	The New York Central Railroad		The Pennsylvania Railroad		The Baltimore & Ohio Railroad	
	Percentage of Shipments	Excess or Deficit from "Expected"	Percentage of Shipments	Excess or Deficit from "Expected"	Percentage of Shipments	Excess or Deficit from "Expected"
May	29.1	+17.4*	7.0	+3.4	7.3	+3.2
June	24.9	+13.2*	3.6	-0.1	6.8	+2.7
July	17.2	+5.4	2.6	-1.2	2.8	-1.2
August	9.3	-2.5	2.1	-1.8	3.8	-0.2
September	17.8	+5.9	4.4	+0.1	3.0	-1.0
October	17.9	+6.0	1.7	-2.3	2.3	-1.6
November	22.6	+10.6	3.0	-1.1	11.9	+7.9*
December	45.9	-15.3*	19.9	+0.0	27.8	+20.4*
January 1877	55.8	-5.5	21.4	+1.5	13.6	+6.2*
February	79.9	+18.5*	9.7	-10.2*	0.4	-6.9*
March	77.0	+10.5*	11.8	-8.2*	7.2	-0.1
April	29.0	+16.8*	5.8	+1.2	8.1	+4.2
May	18.1	+5.8	4.1	-0.6	2.6	-1.2
June	9.4	-3.0	2.3	-2.5	1.0	-2.8
July	7.1	-5.3	1.7	-3.2	0.5	-3.2
August	9.7	-2.8	1.7	-3.3	0.7	-3.0
September	12.9	+3.9	2.4	-2.7	1.6	-2.1

	The New York Central Railroad		The Pennsylvania Railroad		The Baltimore & Ohio Railroad	
	Percentage Of Shipments	Excess or Deficit from "Expected"	Percentage of Shipments	Excess or Deficit from "Expected"	Percentage of Shipments	Excess or Deficit from "Expected"
October	9.8	-2.8	2.7	-2.4	1.9	-1.8
November	8.4	-4.2	1.6	-3.6	1.1	-2.6
December	77.1	+15.1*	9.5	-11.4*	3.6	-3.5
January 1878	73.7	+11.7*	19.0	-2.0	5.5	-1.6
February	62.5	+0.4	23.2	+2.1	11.0	+3.9
March	65.1	+3.0	23.7	+2.5	9.4	+2.3
April	15.4	+2.5	9.2	+3.4	1.3	-2.2
May	28.0	+15.1*	9.2	+3.3	2.9	-0.6
June	16.0	+3.0	3.7	-2.2	1.0	-2.5
July	22.3	+9.3	3.2	-2.8	2.6	-0.9
August	14.5	+1.4	3.3	-2.8	0.9	-2.6
September	8.0	-5.2	4.1	-2.1	0.6	-2.8
October	8.5	-4.7	2.3	-4.0	0.9	-2.5
November	13.5	+0.2	2.8	-3.6	1.0	-2.4
December	60.8	-1.7	26.7	+4.7	2.8	-4.0
January 1879	58.1	-4.5	36.1	+13.9*	2.9	-4.0
February	56.8	-5.9	36.8	+14.6*	3.4	-3.4

	The New York Central Railroad		The Pennsylvania Railroad		The Baltimore & Ohio Railroad	
	Percentage of Shipments	Excess or Deficit from "Expected"	Percentage of Shipments	Excess or Deficit from "Expected"	Percentage of Shipments	Excess or Deficit from "Expected"
March	56.4	-6.3	36.2	+13.8*	6.5	-0.3
April	54.1	-8.7	25.5	+3.1*	3.4	-3.3
May	31.4	+17.8*	14.7	+7.6*	3.6	+0.2
June	30.5	+16.8*	10.5	+3.4	2.9	-0.4
July	13.7	+0.0	5.1	-2.0	1.1	-2.1
August	14.0	+0.3	4.7	-2.5	3.6	+3.5
September	13.4	-0.4	5.0	-2.4	3.6	+3.7
October	11.8	-2.0	4.7	-2.8	5.6	+2.4
November	12.3	-1.6	2.3	-5.2	2.9	-2.8
December	52.4	-10.8*	10.0	-13.1	13.9	+7.2*

* Denotes an excess or deficit from the trend line greater than the value of the standard error of estimate.

Source: From Shipment figures in the *Annual Reports* of the Chicago Board of Trade (Chicago: 1871-1879) or the *Daily Commercial Reports* of the Board of Trade as shown in the "Weekly Commercial Bulletins" (1871-1879).

Statistical Appendices to Chapter 4

Table A.1, Grain Shipments from Chicago, 1880 to 1886

The trend of the percentage share of grain tonnage of each railroad has been defined as

$$s = \alpha + \beta T + \gamma X + \nu$$

with s = a railroad's percentage of total tonnage of Chicago shipments; T = month of shipments, numbered from 1 to 84 for the period 1880 to 1886; X = "1" for the winter months in which the lakes were closed to traffic, "0" for the summer months. The estimates of trend, as calculated by the method of least squares, are:

The New York Central Railroad
$S = 30.056 + 0.045T + 15.451X; \quad \hat{\rho}^2 = .283$

The Pennsylvania Railroad
$S = 12.358 + 0.061T + 2.979X; \quad \hat{\rho}^2 = .074$

The Baltimore and Ohio Railroad
$S = 3.628 + 0.023T + 1.662X; \quad \hat{\rho}^2 = .125$

The Erie Railroad (from June, 1883 to December, 1886)
$S = 5.437 + 0.122T + 1.394X; \quad \hat{\rho}^2 = .124$

The Grand Trunk Railway
$S = 5.831 + 0.36T + 3.768X; \quad \hat{\rho}^2 = .165$

The "trend" share for any month is the value $\hat{S}$ from inserting the relevant values of T and X in the relevant computed equation. The actual shares are shown in the table, along with the difference between actual and "trend" shares for that month.

Grain 1880	The New York Central Railroad		The Pennsylvania Railroad		The Baltimore and Ohio Railroad		The Erie Railroad		The Grand Trunk Railway	
Month	Actual Share	Actual Share minus "Expected: Share	Actual Share	Actual Share minus "Expected" Share	Actual Share	Actual Share minus "Expected" Share	Actual Share	Actual Share minus "Expected" Share	Actual Share	Actual Share minus "Expected" Share
Jan.	65.4	19.8*	20.2	4.8	7.8	2.4			--	--
Feb.	63.9	18.4*	15.6	0.1	9.2	3.9*			7.6	-2.1
Mar.	54.6	9.0	22.1	6.5	12.7	7.3*			6.1	-3.6
Apr.	19.5	-26.2*	8.6	-7.0	4.2	-1.2			2.0	-7.7*
May	8.8	-21.5*	4.3	-8.4*	0.6	-3.1*			4.4	-1.6
June	15.2	-15.1*	4.2	-8.6*	1.2	-2.6*			3.3	-2.8
July	10.3	-20.1*	2.5	-10.3*	1.1	-2.7*			2.4	-3.7
Aug.	10.8	-19.6*	3.3	-9.5*	1.0	-2.8*			3.9	-2.2
Sept.	13.7	-16,8*	2.8	-10.1*	0.9	-3.0*			3.0	-3.1
Oct.	12.1	-18.4*	3.8	-9.2*	0.8	-3.0*			3.3	-2.9
Nov.	19.1	-11.5	5.6	-7.4*	2.7	-1.2			3.6	-2.7
Dec.	63.3	17.2*	19.2	3.1	4.1	-1.5			7.6	-2.4
1881										
Jan.	71.3	25.2*	8.7	-7.4*	6.3	0.7			10.8	0.7
Feb.	68.4	22.3*	12.7	-3.5	5.1	-0.5			10.6	0.5
Mar.	60.9	14.7*	14.6	-1.6	10.8	5.1			10.9	0.7
Apr.	50.9	4.7	22.6	6.3	5.8	0.1			12.1	1.9
May	11.0	-19.8*	8.5	-4.9	1.1	-2.9*			4.0	-2.4
June	16.5	-14.4*	5.5	-8.0*	1.7	-2.3			3.9	-2.5
July	19.1	-11.8	7.2	-6.3	2.1	-2.0			3.0	-3.5
Aug.	17.9	-13.1*	6.9	-6.7	2.0	-2.0			4.9	-1.7
Sept.	24.3	- 6.7	5.8	-7.8*	3.8	-0.3			4.6	-2.0
Oct.	23.5	- 7.6	5.4	-8.3*	2.7	-1.4			5.3	-1.3
Nov.	35.6	4.5	8.4	-5.3	1.0	-3.1*			6.7	0.0
Dec.	58.1	11.6	13.1	-3.7	3.0	-2.8*			15.8	5.3*

Grain 1882	The New York Central Railroad		The Pennsylvania Railroad		The Baltimore and Ohio Railroad		The Erie Railroad		The Grand Railway	
Month	Actual Share	Actual Share minus "Expected" Share	Actual Share	Actual Share minus "Expected" Share	Actual Share	Actual Share minus "Expected" Share	Actual Share	Actual Share minus "Expected" Share	Actual Share	Actual Share minus "Expected" Share
Jan.	65.3	16.7*	19.2	2.4	2.7	-3.2			9.3	-1.2
Feb.	53.0	6.3	23.5	6.5	4.9	-1.0			12.9	2.4
Mar.	50.8	4.1	25.4	8.4*	4.4	-1.5			13.7	3.2
Apr.	30.2	-16.5*	27.7	10.6*	7.5	1.6			23.9	13.3*
May	36.5	5.1	29.7	15.6*	9.3	5.0*			10.7	3.8
June	39.2	7.8	30.1	16.0*	11.5	7.1*			3.2	-3.7
July	34.3	2.8	25.0	10.8*	8.0	3.6*			11.0	4.1
Aug.	34.4	2.9	25.7	11.4*	6.7	2.3			14.1	7.1*
Sept.	36.6	5.0	28.0	13.6*	3.7	-0.6			18.2	11.2*
Oct.	37.8	6.2	31.3	16.9*	6.5	2.1			9.9	2.9
Nov.	41.3	9.7	25.8	11.3*	6.0	1.6			13.3	6.2*
Dec.	53.3	6.2	23.1	5.6	7.1	0.9			11.6	0.7
1883										
Jan.	51.7	4.5	20.7	3.1	9.2	3.1			9.7	-1.2
Feb.	n.a.		n.a.		n.a.				n.a.	
Mar.	n.a.		n.a.		n.a.				n.a.	
Apr.	n.a.		n.a.		n.a.				n.a.	
May	n.a.		n.a.		n.a.				n.a.	
June	33.5	1.5	28.1	13.1*	5.3	0.7	5.8	-4.8*	14.7	7.3*
July	33.3	1.3	20.2	5.3	3.0	-1.6	14.1	3.4	10.8	3.4
Aug.	41.2	9.2	20.5	5.5	5.4	0.8	14.0	3.2	10.1	2.6
Sept.	37.2	5.1	18.2	3.1	4.9	0.2	17.8	6.9*	16.2	8.8*
Oct.	45.6	13.5*	14.9	-0.3	6.5	1.9	14.7	3.7	10.6	3.1
Nov.	42.2	10.0	20.0	4.8	5.0	0.3	14.2	3.0	10.4	2.9
Dec.	45.7	-1.9	16.0	-2.2	2.6	-3.8*	13.5	3.5	13.5	2.2

Grain 1884	The New York Central Railroad		The Pennsylvania Railroad		The Baltimore and Ohio Railroad		The Erie Railroad		The Grand Trunk Railway	
Month	Actual Share	Actual Share minus "Expected" Share	Actual Share	Actual Share minus "Expected" Share	Actual Share	Actual Share minus "Expected" Share	Actual Share	Actual Share minus "Expected" Share	Actual Share	Actual Share minus "Expected" Share
Jan.	51.8	4.1	16.4	-1.9	3.4	-3.0*	9.5	-0.5	9.2	-2.2
Feb.	47.5	-0.2	7.5	-10.8*	3.1	-3.3*	6.4	-3.8	23.0	11.6*
Mar.	43.6	-4.2	10.8	-7.6*	6.1	-0.3	13.3	3.0	10.7	-0.7
Apr.	34.8	-13.0*	23.6	5.1	7.4	1.0	19.7	9.3*	9.9	-1.5
May	41.9	9.4	17.9	2.3	5.1	0.3	9.7	-2.2	18.3	10.5*
June	36.8	4.4	17.8	2.1	10.4	5.6*	10.7	-1.3	14.4	6.6*
July	38.1	5.5	18.9	3.2	8.6	3.7*	13.7	1.6	8.8	1.0
Aug.	44.2	11.7	18.9	3.1	7.8	2.9*	7.5	-4.7	8.3	0.4
Sept.	31.8	-0.8	18.1	2.2	7.3	2.3	13.4	1.0	4.7	-3.2
Oct.	49.0	16.3*	16.6	0.6	6.0	1.1	11.3	-1.2	2.8	-5.1*
Nov.	50.5	17.8*	9.0	-6.9	3.4	-1.6	17.1	4.4	3.9	-4.1
Dec.	44.3	-3.8	11.4	-7.6*	4.2	-2.4	12.4	1.1	12.1	0.3
1885										
Jan.	43.7	-4.5	17.4	-1.7	4.6	-2.1	8.6	-2.9	14.0	2.1
Feb.	43.9	-4.4	22.2	3.1	6.7	0.0	15.0	3.4	5.0	-6.8*
Mar.	38.9	-9.5	26.7	7.6*	7.5	0.7	11.3	-0.4	6.2	-5.7*
Apr.	32.1	-16.3*	32.4	13.1*	4.5	-2.2	6.8	-5.1*	13.4	-1.5
May	45.7	12.7*	20.0	3.7	8.1	3.0*	6.1	-7.2*	5.1	-3.1
June	39.0	5.9	17.7	1.3	5.0	-0.1	10.3	-3.2	7.2	-1.1
July	35.4	2.3	22.1	5.6	6.5	1.3	15.4	1.8	6.5	-1.7
Aug.	43.4	10.3	13.7	-2.8	5.9	0.7	13.4	-0.4	9.1	0.8
Sept.	48.4	15.3*	13.8	-2.7	5.7	0.5	11.8	-2.0	6.6	-1.7
Oct.	35.7	2.5	15.3	-1.3	5.2	-0.1	10.4	-3.6	6.5	-1.8
Nov.	37.7	4.4	18.5	1.9	3.8	-1.4	14.2	0.0	8.7	0.3
Dec.	38.5	-10.2	18.9	-0.8	6.9	0.0	12.2	-0.7	11.2	-1.0

Grain 1886	The New York Central Railroad		The Pennsylvania Railroad		The Baltimore and Ohio Railroad		The Erie Railroad		The Grand Trunk Railway	
Month	Actual Share	Actual Share minus "Expected" Share	Actual Share	Actual Share minus "Expected" Share	Actual Share	Actual Share minus "Expected" Share	Actual Share	Actual Share minus "Expected" Share	Actual Share	Actual Share minus "Expected" Share
Jan.	31.9	-16.9*	17.8	-2.0	10.2	3.2*	13.8	0.8	4.7	-7.6*
Feb.	43.9	-4.9	17.2	-2.7	3.8	-3.1	9.0	-4.0	16.8	4.5
Mar.	37.8	-11.1	10.1	-9.8*	7.7	0.7	10.8	-2.4	15.9	3.6
Apr.	28.2	-20.7*	11.0	-8.9*	8.5	1.5	9.2	-4.1	8.3	-4.0
May	25.8	-7.7	14.5	-2.5	6.0	0.6	15.9	1.0	4.0	-4.6*
June	30.3	-3.2	16.7	-0.4	4.9	-0.5	18.6	3.6	5.0	-3.6
July	40.3	6.6	16.8	-0.4	4.1	-1.3	11.7	-3.3	2.9	-5.8*
Aug.	43.2	9.5	10.8	-6.4	4.0	-1.4	13.5	-1.7	6.1	-2.6
Sept.	35.3	1.6	8.4	-8.8*	1.9	-3.6*	26.9	11.5*	5.4	-3.3
Oct.	26.9	-6.9	9.6	-7.8*	2.4	-3.1*	25.9	10.4*	5.1	-3.7
Nov.	27.9	-5.9	15.3	-2.1	5.6	0.1	16.7	1.2	7.3	-1.5
Dec.	28.8	-5.0	13.6	-3.8	6.8	1.3	15.0	-0.5	15.7	6.9*

* Greater than the standard error of estimate.

Source: Statistics on total monthly tonnage of wheat, corn, oats, and flour out of Chicago, and by railroad, as shown in the Annual Reports of the Chicago Board of Trade 1879 to 1887. Shares are calculated as percentages of total shipments, where the totals include short distance movement from Chicago storage to points South and West as well as eastbound tonnage over the trunk-line railroads and lake steamers.

Table A.2, Grain Shipments into the Eastern Seaboard, 1880 to 1886

The trend of shares of each railroad by $S = \alpha + \beta T + \gamma X + \delta Y + \nu$ has been estimated by the method of least squares, with S = percentage of total shipments from Chicago of wheat, corn, and oats into New York, Montreal, Boston, Baltimore, and Philadelphia; T = month of shipment, numbered from 1 to 84 for the period 1880 to 1886; X = 1,0 for the lakes closed to traffic or open to shipping; Y = the change in trend with the entry of the Lackawanna Railroad ($Y = 1$ for all months after this railroad's entry). The estimates of trend are:

The New York Central Railroad

$$S = 10.81 + 0.05T + 8.65X - 5.75Y; \quad \hat{\rho}^2 = .533; \quad S_u = 4.33$$

The Pennsylvania Railroad

$$S = 24.64 - 0.05T + 9.84X - 0.53Y; \quad \hat{\rho}^2 = .487; \quad S_u = 5.44$$

The Baltimore and Ohio Railroad

$$S = 6.93 + 0.00T + 6.91X + 1.50Y; \quad \hat{\rho}^2 = .429; \quad S_u = 4.15$$

The Erie Railroad

$$S = 9.37 + 0.00T + 3.45X - 3.19Y; \quad \hat{\rho}^2 = .275; \quad S_u = 3.68$$

The Delaware, Lackawanna and Western Railroad

$$S = 0.10 + 0.01T + 2.91X; \quad \hat{\rho}^2 = .354; \quad S_u = 2.02$$

The "trend" share for any month is the value $\hat{S}$ from inserting the relevant values of T, X, and Y in the correct equation. The actual shares are shown for 1880 to 1886, along with the difference between $\hat{S}$ and S for that month.

Grain	The New York Central Railroad		The Pennsylvania Railroad		The Baltimore and Ohio Railroad		The Erie Railroad		The Lackawanna Railroad	
1880										
Month	Actual Share	Actual Share minus "Expected" Share	Actual Share	Actual Share minus "Expected" Share	Actual Share	Actual Share minus "Expected" Share	Actual Share	Actual Share minus "Expected" Share	Actual Share	Actual Share minus "Expected" Share
Jan.	15.3	-4	29.3	-5	9.2	-5*	12.0			
Feb.	16.4	-3	39.9	+5	20.0	+6*	9.8	-3		
Mar.	24.0	+4	39.9	+5	15.9	+2	11.5	-1		
Apr.	22.4	+3	44.3	+10*	9.4	-4	13.6			
May	7.4	-4	26.5	+2	7.3		3.8	-5*		
June	11.0		27.6	+3	5.9	-1	7.7	-2		
July	10.8		23.5		13.4	+6*	7.4	-2		
Aug.	11.2		22.6	-1	10.4	+3	3.9	-5*		
Sept.	6.1	-5*	17.1	-7*	6.1		5.8	-4*		
Oct.	10.1	-1	25.0		9.3	+2	4.9	-4*		
Nov.	15.4	+4	24.9		12.1	+5*	8.3	-1		
Dec.	18.8	-1	35.1	+1	18.1	+4	13.2			
1881										
Jan.	13.5	-7*	40.5	+6*	16.8	+3	13.1			
Feb.	14.7	-5*	35.8	+2	17.9	+4	12.7			
Mar.			32.8		19.7	+6*	13.4			
Apr.	22.5	+2	26.4	-7*	9.9	-4	20.6	+7*		
May	8.2	-3	29.8	+6*	7.0		16.5	+7*		
June	11.7		27.1	+3	6.2		6.4	-3		
July	16.9	+5*	25.9	+2	6.1		13.1	+4*		
Aug.	16.4	+5*	34.3	+10*	10.1	+3	11.0	+2		
Sept.	18.6	+7*	22.3	-1	6.5		11.8	+2		
Oct.	13.1	+1	21.9	-2	7.6		15.3	+6*		
Nov.	16.2	+4	18.6	-5	3.9	-3	7.7	-2		
Dec.	19.3	-1	24.8	-8*	7.6	-6*	9.9	-3		

Grain 1882	The New York Central Railroad		The Pennsylvania Railroad		The Baltimore and Ohio Railroad		The Erie Railroad		The Lackawanna Railroad	
Month	Actual Share	Actual Share minus "Trend" Share	Actual Share	Actual Share minus "Trend" Share	Actual Share	Actual Share minus "Trend" Share	Actual Share	Actual Share minus "Trend" Share	Actual Share	Actual Share minus "Trend" Share
Jan.	28.9	+ 8*	27.6	- 6*	6.5	- 7*	20.0	+ 7*		
Feb.	26.4	+ 6*	29.2	- 4	5.1	- 9*	17.3	+ 4*		
Mar.	27.1	+ 6*	25.5	- 8*	8.6	- 5*	15.1	+ 2		
Apr.	20.4		34.8	+ 2	10.3	- 4	11.1	- 2		
May	9.2	- 3	18.6	- 5	2.9	- 4	6.8	- 3		
June	16.8	+ 4	18.4	- 5	4.0	- 3	10.3	+ 1		
July	10.0	- 2	28.5	+ 5	14.9	+ 8*	9.5			
Aug.	13.3		34.5	+11*	11.7	+ 5*	6.7	- 3		
Sept.	15.9	+ 3	25.2	+ 2	10.0	+ 3	9.1			
Oct.	11.8		14.2	- 9*	7.6		6.2	- 3		
Nov.	15.6	+ 3	17.9	- 5	8.3	+ 1	12.3	+ 3		
Dec.	23.0	+ 2	29.2	- 3	12.3	- 2	14.7	+ 2		
1883										
Jan.	21.3	+ 6*	37.0	+ 5	13.4	- 2	10.8	+ 1	1.2	- 2
Feb.	17.2	+ 2	42.8	+11*	14.5		10.1		3.3	
Mar.	19.4	+ 4	24.4	- 8*	16.9	+ 2	9.0		5.4	+ 2
Apr.	13.4	- 2	33.2	+ 1	10.5	- 5*	9.8		7.7	+ 4*
May	8.2	+ 1	27.4	+ 5	5.4	- 3	6.1		4.0	+ 3*
June	4.8	- 2	16.2	- 6*	5.7	- 3	5.7		0.7	
July	3.7	- 4	15.9	- 6*	8.5		6.1		0.5	
Aug.	6.6		28.8	+ 7*	13.5	+ 5*	9.0	+ 3	0.2	
Sept.	7.3		18.1	- 4	8.3		12.9	+ 7*	1.2	
Oct.	6.3	- 1	17.9	- 4	6.0	- 2	10.2	+ 4	1.2	
Nov.	9.6	+ 2	14.6	- 7*	9.7	+ 1	7.4	+ 1	0.3	
Dec.	13.7	- 2	26.0	- 6*	10.7	- 5*	9.6		1.5	- 2

Grain 1884	The New York Central Railroad		The Pennsylvania Railroad		The Baltimore and Ohio Railroad		The Erie Railroad		The Lackawanna Railroad		The New York West Shore and Buffalo Railway
Month	Actual Share	Actual Share minus "Trend" Share	Actual Share	Actual Share minus "Trend" Share	Actual Share	Actual Share minus "Trend" Share	Actual Share	Actual Share minus "Trend" Share	Actual Share	Actual Share minus "Trend" Share	Actual Share
Jan.	26.9	+11*	30.7		10.8	- 5*	8.6	- 1	5.0	+ 2	
Feb.	20.2	+ 4	24.0	- 7*	15.2		3.9	- 6*	2.7		
Mar.	22.2	+ 6*	23.4	- 8*	27.2	+12*	6.9	- 3	1.0	- 3*	
Apr.	13.1	- 3	27.2	- 4	17.2	+ 2	16.9	+ 7*	3.7		
May	9.3	+ 2	21.4		11.3	+ 3	8.6	+ 3	0.3		
June	9.8	+ 2	21.3		7.9		14.0	+ 8*	1.5		
July	7.4		25.9	+ 4	10.2	+ 2	7.3	+ 1	1.0		
Aug.	10.4	+ 3	26.1	+ 5	12.8	+ 4	4.7	- 2	1.3		
Sept.	7.3		20.0		8.9		7.4	+ 1	0.8		
Oct.	7.3		16.1	- 5	6.3	- 2	3.5	- 3	0.5		
Nov.	10.7	+ 3	14.8	- 6*	7.6		4.8	- 1	0.8		
Dec.	19.5	+ 3	31.9		15.6		3.6	- 6*	4.4		
1885											
Jan.	13.7	- 3	29.6	- 1	20.6	+ 5*	6.9	- 2	5.1	+ 1	4.8
Feb.	15.7	- 1	36.9	+ 6*	27.1	+12*	8.4	- 1	2.8		2.4
Mar.	21.4	+ 5*	31.7		14.7		12.6	+ 3	3.2		4.4
Apr.	17.8		39.0	+ 8*	11.8	- 4	9.6		2.2	- 1	5.9
May	11.5	+ 3	29.5	+ 9*	9.2		6.5		0.8		6.5
June	6.9	- 1	20.6		4.0	- 4	6.1		0.2		2.3
July	8.9		16.6	- 4	4.1	- 4	6.7		0.2		1.1
Aug.	9.8	+ 1	17.7	- 3	5.5	- 3	6.4		0.4		1.0
Sept.	10.2	+ 2	18.5	- 2	3.9	- 5*	10.6	+ 4*	0.4		5.4
Oct.	8.0		24.2	+ 4	5.4	- 3	4.2		0.0		3.6
Nov.	10.2	+ 2	16.0	- 4	6.7	- 2	8.5	+ 2	0.2		0.7
Dec.	21.5	+ 4	24.8	- 5	13.9	- 1	15.5	+ 6*	1.9	- 2	1.5

Grain 1886	The New York Central Railroad (including the West Shore)		The Pennsylvania Railroad		The Baltimore and Ohio Railroad		The Erie Railroad		The Lackawanna Railroad	
Month	Actual Share	Actual Share minus "Expected" Share	Actual Share	Actual Share minus "Expected" Share	Actual Share	Actual Share minus "Expected" Share	Actual Share	Actual Share minus "Expected" Share	Actual Share	Actual Share minus "Expected" Share
Jan.	15.9	-2	40.6	+10*	25.2	+9*	10.4		1.5	-2
Feb.	21.3	+4	25.2	-5	15.0		12.9	+3	5.7	+2
Mar.	15.8	-2	25.2	-5	18.2	+3	9.1		13.6	+9*
Apr.	5.0	-13*	34.4	+4	23.3	+8*	6.9	-3	2.5	-1
May	5.5	-3	18.5	-2	10.6	+2	13.0	+7*	2.8	+2
June	44.4	-5*	16.8	-3	6.3	-2	4.8		0.5	
July	9.7		26.0	+6	12.0	+4	2.5	-4*	0.4	
Aug.	8.6		25.3	+5	12.4	+4	2.6	-4*	0.9	
Sept.	11.2	+2	19.2		8.4		8.0	+2	2.4	+1
Oct.	8.9		18.3		7.8		8.3	+2	2.8	+2
Nov.	7.9	-1	21.6		9.2		9.8	+4*	2.5	+2
Dec.	22.0	+4	38.1	+8*	16.8	+1	6.5	-3	3.3	

* Greater than the standard error of estimate.

Source: Total receipts into the east coast seaports each month for 1880 to 1886 are given in the *Annual Statistical Reports* of the New York Produce Exchange 1880 to 1884 and the *Annual Reports of the Baltimore Corn and Flour Exchange*, 1880 to 1886. Receipts into New York each month for each railroad are given in the *Annual Statistical Reports* of the New York Produce Exchange, 1880 to 1884; these statistics were compiled for 1885 to 1886 from the daily reports of receipts, by shipper and by railroad, in the *New York Journal of Commerce* for these years. Receipts into Baltimore by railroad are given in the *Annual Reports* of the Baltimore Corn and Flour Exchange 1880 to 1886; Receipts into Philadelphia by line are not available in this or Philadelphia sources, so that the Pennsylvania Railroad is given credit for all receipts into this city. Receipts into Boston and Montreal are not given by railroad, and no railroad is given credit for the volumes. Percentages for each railroad equal $\sum_{i=1}^{9} R_{ji} / \sum_{i=1}^{9} C_i$, where R is railroad "j"'s receipts into city i, and C_i is total receipts of city i. Thus for Boston and Montreal, $R_j = 9$, but $C > 0$.

Table A.3. Railroad Rates and Lard Price Differences: 1884 to 1886

Month	Official Rate Cents per 100 lb	Board of Trade Rate Cents per 100 lb	Chicago-New York Western Lard Price Mean Difference Cents per 100 lb
January 1884	33	33	36
February	35	33	32
March	28	25	27
April	25	25	32
May	25	25	24
June	26	26	24
July	30	30	33
August	35	35	40
September	35	30	33
October	35	30	32
November	30	30	28
December	30	30	45
January 1885	30	27	34
February	30	28	32
March	30	25	33
April	29	24	32
May	25	17	--
June	25	13	20
July	25	22	28
August	25	25	33
September	25	13	23
October	25	25	31
November	27	27	38
December	30	30	40
January 1886	30	30	35
February	30	30	30
March	30	30	33
April	30	28	32
May	30	29	28
June	30	28	21
July	30	29	22
August	30	29	15
September	30	29	--
October	30	28	29
November	30	30	30
December	31	31	33

Source: The official rates were the Central Traffic Association seventh-class rate in 1884 and twelfth-class rate in 1885 and 1886 as listed (with the change in classification) in the Interstate Commerce Commission, *Railways in the United States in 1902*, Vol. II, p. 78. The Board of Trade rate is from the "Weekly Review of Freights," the *Chicago Daily Commercial Bulletin*, 1884 to 1886. The western lard price differences are the averages of the differences in the daily spot prices shown in the *Chicago Daily Commercial Bulletin* and the *Annual Report of the New York Produce Exchange* for 1884; for 1885 and 1886 New York spot prices were those shown in the *Chicago Daily Commercial Bulletin*.

Table A.4, Lard Shipments from Chicago, 1884 to 1886

The trend of the percentage share of lard tonnage of each railroad has been defined as

$$s = \alpha + \beta T + \gamma X + \nu$$

with s = a railroad's percentage of total tonnage of Chicago shipments; T = month of shipments, numbered from 1 to 84 for the period 1880 to 1896; X = "1" for the winter months in which the lakes were closed to traffic, "0" for the summer months. The estimates of trend, as calculated by the method of least squares, are:

The New York Central Railroad

$$S = 24.845 + .020T + 6.119X; \quad \hat{\rho}^2 = .092$$

The Pennsylvania Railroad

$$S = 48.028 - .234T + 0.457X; \quad \hat{\rho}^2 = .114$$

The Baltimore and Ohio Railroad

$$S = -4.552 + .155T + 2.447X; \quad \hat{\rho}^2 = .461$$

The Erie Railroad (from June 1883, to December, 1886)

$$S = 12.407 - .078T + 0.393X; \quad \hat{\rho}^2 = .019$$

The Grand Trunk Railway

$$S = 12.779 - .024T + 1.027X; \quad \hat{\rho}^2 = .009$$

The "trend" share for any month is calculated from inserting the relevant values of T and X in the computed equation. The actual share is shown in the table, along with the difference between actual and trend shares for that month.

Lard 1884	The New York Central Railroad		The Pennsylvania Railroad		The Baltimore and Ohio Railroad		The Erie Railroad		The Grand Trunk Railway	
Month	Actual Share	Actual Share minus "Trend" Share	Actual Share	Actual Share minus "Trend" Share	Actual Share	Actual Share minus "Trend" Share	Actual Share	Actual Share minus "Trend" Share	Actual Share	Actual Share minus "Trend" Share
Jan.	30.7	- 1.2	51.9	14.9*	4.8	- 0.7	12.4	3.4	16.5	3.8
Feb.	40.9	- 9.0	36.4	- 0.4*	3.5	- 2.2	0.0	- 8.9*	21.1	8.5*
Mar.	29.9	- 2.1	14.7	-21.9*	5.4	- 0.4	6.9	- 1.9	21.7	9.1*
Apr.	27.4	- 4.5	24.2	-12.2	3.2	- 7.8	5.2	- 3.5	27.7	15.1*
May	16.6	- 9.3	34.8	- 0.9	2.6	- 1.1	14.8	6.5	25.5	14.0*
June	14.0	-11.9*	27.7	- 7.7	9.9	6.1*	20.0	11.8*	20.9	9.4*
July	16.8	- 9.1	17.0	-18.2*	2.5	- 1.5	19.6	11.4*	32.0	20.6*
Aug.	15.0	-11.0*	12.8	-22.2*	2.0	- 2.1	12.9	4.9	12.8	1.4
Sept.	12.7	-13.3*	60.0	25.3*	2.6	- 1.7	4.5	- 3.5	7.1	- 4.3
Oct.	26.9	0.9	48.2	13.7*	2.2	- 2.2	5.4	- 2.5	3.0	- 8.3*
Nov.	37.5	11.5*	34.3	--	1.3	- 3.3*	3.3	- 4.6	4.7	- 6.6*
Dec.	24.4	- 7.7	32.7	- 1.7	5.2	- 2.0	1.8	- 6.3	24.0	11.6*
1885										
Jan.	40.3	8.2	24.4	- 9.9	3.9	- 3.5*	5.2	- 2.9	16.2	3.8
Feb.	39.6	7.4	21.5	-12.5*	5.5	- 2.0	20.0	12.1*	3.3	- 9.0
Mar.	36.8	4.6	29.6	- 4.2	10.3	2.6	5.7	- 2.2	5.3	- 7.0*
Apr.	32.6	0.4	38.8	5.3	8.6	0.7	3.7	- 4.1	5.6	- 6.6
May	36.9	10.8*	31.4	- 1.5	5.2	- 0.4	4.1	- 3.3	5.6	- 5.6
June	24.8	- 1.4	41.7	9.0	7.6	2.0	1.6	- 5.7	8.7	- 2.5
July	21.3	- 4.8	40.6	8.2	1.6	- 4.3*	9.7	2.5	9.1	- 2.0
Aug.	21.4	- 4.8	39.6	7.4	2.0	- 4.0*	12.9	5.7	12.3	1.1
Sept.	30.4	4.2	47.0	15.1*	1.6	- 4.5*	7.1	0.1	6.3	- 4.8
Oct.	27.0	0.7	40.9	9.2	3.7	- 2.6	1.1	- 5.9	11.6	0.5
Nov.	23.9	- 2.4	41.6	10.1	16.2	9.7*	1.1	- 5.8	8.1	- 2.9
Dec.	34.9	2.5	34.8	3.1	11.6	2.5	1.1	- 6.1	11.4	- 0.7

Lard 1886	The New York Central Railroad		The Pennsylvania Railroad		The Baltimore and Ohio Railroad		The Erie Railroad		The Grand Trunk Railway	
Month	Actual Share	Actual Share minus "Expected" Share	Actual Share	Actual Share minus "Expected" Share	Actual Share	Actual Share minus "Expected" Share	Actual Share	Actual Share minus "Expected" Share	Actual Share	Actual Share minus "Expected" Share
Jan.	32.7	0.4	42.0	10.6	8.8	-0.4	.9	-6.2	7.5	-4.5
Feb.	20.5	-11.9*	40.8	9.6	14.9	5.5*	3.8	-3.3	11.8	+0.2
Mar.	25.2	-7.2	39.1	8.1	12.3	2.7	5.5	-1.5	9.4	-2.5
Apr.	16.4	-16.0*	26.8	-4.0	10.2	0.5	23.6	16.8*	13.3	+1.3
May	22.1	-4.2	29.6	-0.5	9.8	2.4	12.2	5.7	10.4	-0.5
June	14.9	-11.5*	38.5	8.7	8.8	1.2	6.8	0.4	13.6	+2.7
July	14.0	-12.4*	35.4	5.8	8.1	0.4	4.3	-2.0	9.1	-1.8
Aug.	55.1	28.8*	15.4	-13.9*	1.8	-6.0*	1.5	-4.7	7.9	-2.9
Sept.	40.9	14.5*	15.7	-13.5*	2.0	-6.0*	19.0	12.9*	3.0	-7.8
Oct.	40.0	13.5*	26.1	-2.7	5.9	-2.3	1.9	-4.1	15.2	+4.4
Nov.	44.7	18.2*	22.7	-5.9	12.3	4.0*	1.5	-4.5	5.8	-5.0
Dec.	31.5	-1.1	18.2	-10.6	18.4	7.4*	11.7	5.4	8.8	-3.0

* Greater than the standard error of estimate.

Source: The total tonnages of monthly shipments from Chicago, and by railroad, as shown in the Annual Reports of the Chicago Board of Trade, 1884 to 1886.

Statistical Appendices to Chapter 5

Table A.1, Grain Shipments from Chicago, 1887 to 1899

The trend of the percentage share of grain tonnage of each railroad has been defined as

$$s = \alpha + \beta T + \gamma X + \nu$$

with s = a railroad's percentage of total tonnage of Chicago shipments; T = month of shipments, numbered from 1 to 156 for the period 1887 to 1899; X = "1" for the winter months in which the lakes were closed to traffic, "0" for the summer months. The estimates of trend, as calculated by the method of least squares, are:

The New York Central Railroad

$$S = 31.428 - .014T + 2.050X; \quad \hat{\rho}^2 = .036; \quad S_u = 6.33$$

The Pennsylvania Railroad

$$S = 13.557 + .033T + 1.717X; \quad \hat{\rho}^2 = .083; \quad S_u = 5.66$$

The Baltimore and Ohio Railroad

$$S = 5.031 + .014T + 0.318X; \quad \hat{\rho}^2 = .036; \quad S_u = 3.38$$

The Erie Railroad

$$S = 11.837 + .010T - 1.583X; \quad \hat{\rho}^2 = .057; \quad S_u = 3.72$$

The Grand Trunk Railway

$$S = 11.065 - .038T + 4.395X; \quad \hat{\rho}^2 = .291; \quad S_u = 4.45$$

The "trend" share for any month is determined by inserting the relevant values of T and X in the computed equations. The actual shares are shown in the table, along with the difference between actual and trend shares for that month.

Grain 1887	The New York Central Railroad		The Pennsylvania Railroad		The Baltimore and Ohio Railroad		The Erie Railroad		The Grand Trunk Railway	
Month	Actual Share	Actual Share minus "Trend" Share	Actual Share	Actual Share minus "Trend" Share	Actual Share	Actual Share minus "Trend" Share	Actual Share	Actual Share minus "Trend" Share	Actual Share	Actual Share minus "Trend" Share
Jan.	28.5	- 5.0	10.2	- 5.1	2.7	- 2.7	15.4	5.1*	15.6	0.2
Feb.	30.5	- 2.9	19.5	4.2	3.7	- 1.7	18.3	8.0*	8.4	- 7.0*
Mar.	19.0	-14.5*	32.4	17.0*	9.4	4.0	9.9	- 0.4	11.4	- 3.9
Apr.	32.4	- 1.0	22.6	7.2*	4.1	- 1.3	11.6	1.3	8.5	- 6.8*
May	29.6	- 1.8	14.7	0.9	8.8	3.7*	9.7	- 2.2	9.3	- 1.6
June	26.5	- 4.8	19.2	5.4	11.1	6.0*	7.4	- 4.5*	6.8	- 4.0
July	25.6	- 5.7	31.5	17.7*	8.4	3.3	6.3	- 5.6*	6.9	- 3.9
Aug.	33.5	2.2	23.0	9.1*	8.5	3.4*	10.5	- 1.4	6.6	- 4.2
Sept.	33.5	2.2	14.3	0.4	7.7	2.6	5.7	- 6.3*	15.7	5.0*
Oct.	24.7	- 6.6*	19.4	5.5	6.2	1.0	11.2	- 0.7	12.7	2.1
Nov.	37.3	6.0	21.3	7.3*	4.9	- 0.2	11.2	- 0.8	7.5	- 3.1
Dec.	39.0	5.7	16.3	0.6	4.0	- 1.5	8.3	- 2.0	12.3	- 2.7
1888										
Jan.	30.7	- 2.6	13.8	- 1.9	6.9	1.4	7.8	- 2.6	20.1	5.1*
Feb.	26.2	- 7.0*	17.6	1.8	15.4	9.9*	6.3	- 4.1	15.7	0.7
Mar.	23.4	- 9.9*	23.0	7.3*	8.7	3.1	4.4	- 6.0*	20.0	5.1*
Apr.	27.7	- 5.6	17.6	1.8	6.5	0.9	9.5	- 0.9	23.2	8.3*
May	27.9	- 3.3	22.4	8.2*	6.8	1.5	7.6	- 4.4*	6.5	- 3.9
June	25.0	- 6.2	16.4	2.2	4.9	- 0.4	10.5	- 1.6	4.2	- 6.1*
July	25.7	- 5.6	14.7	0.5	3.0	- 2.3	9.5	- 2.5	5.4	- 5.0*
Aug.	24.6	- 6.6*	21.3	7.1*	5.0	- 0.4	14.4	2.3	6.4	- 3.9
Sept.	22.6	- 8.6*	15.6	1.4	7.7	2.4	12.4	0.4	8.6	- 1.6
Oct.	30.7	- 0.4	15.8	1.5	4.0	- 1.3	13.9	1.9	14.2	3.9
Nov.	29.7	- 1.4	12.8	- 1.5	3.7	- 1.7	14.1	2.1	19.9	9.7*
Dec.	35.7	2.6	12.5	- 3.6	10.6	5.0*	12.6	2.2	18.2	3.7

Grain 1889	The New York Central Railroad		The Pennsylvania Railroad		The Baltimore and Ohio Railroad		The Erie Railroad		The Grand Trunk Railway	
Month	Actual Share	Actual Share minus "Trend" Share	Actual Share	Actual Share minus "Trend" Share	Actual Share	Actual Share minus "Trend" Share	Actual Share	Actual Share minus "Trend" Share	Actual Share	Actual Share minus "Trend" Share
Jan.	32.6	- 0.5	16.9	0.8	12.2	6.5*	12.0	1.5	17.9	3.4
Feb.	27.2	- 5.9	17.7	1.5	4.3	- 1.5	8.0	- 2.5	25.0	10.5*
Mar.	26.0	- 7.2*	15.4	- 0.8	2.7	- 3.0	14.2	3.7*	28.8	14.4*
Apr.	31.1	- 1.9	16.4	0.2	4.6	- 1.1	12.8	2.3	16.6	2.2
May	31.5	0.4	18.3	3.8	6.9	1.5	12.5	0.4	10.3	0.3
June	32.2	1.2	15.6	1.0	7.0	1.6	12.2	0.0	6.8	- 3.1
July	32.4	1.4	14.8	0.2	7.0	1.5	5.8	- 6.4*	10.1	0.2
Aug.	39.1	8.1*	11.5	- 3.2	3.8	- 1.7	13.4	1.2	12.7	2.9
Sept.	37.1	6.2	11.5	- 3.1	2.3	- 3.2	17.8	5.6*	8.0	- 1.8
Oct.	32.2	1.3	14.6	- 0.1	7.8	2.3	19.3	7.1*	6.8	- 2.9
Nov.	33.1	2.1	15.0	- 0.2	13.4	7.9*	13.3	1.1	8.8	- 0.9
Dec.	34.0	1.0	18.2	- 0.2	16.4	10.5*	7.8	- 2.7	15.0	0.9
1890										
Jan.	39.5	6.6*	13.0	- 3.5	11.8	5.9*	11.1	0.5	14.2	0.2
Feb.	36.1	3.1	17.3	0.8	7.2	1.3	13.0	2.4	9.9	- 4.1
Mar.	40.2	7.3*	15.4	- 1.2	12.1	6.2*	8.2	- 2.5	9.3	- 4.7*
Apr.	37.5	4.6	15.8	- 0.8	6.2	0.3	13.5	2.8	11.8	- 2.1
May	38.4	7.6*	12.8	- 2.1	8.1	2.5	12.9	0.7	8.5	- 1.0
June	36.7	5.8	11.7	- 3.3	6.5	0.8	12.6	0.3	9.8	0.4
July	35.8	5.0	14.4	- 0.5	3.3	- 2.3	16.9	4.7*	9.6	0.2
Aug.	32.9	2.0	14.4	- 0.6	4.1	- 1.6	12.3	0.1	10.9	1.5
Sept.	30.4	- 0.3	11.9	- 3.1	5.0	- 0.6	16.8	4.5*	13.5	4.1
Oct.	34.8	4.0	10.8	- 4.2	3.1	- 2.6	14.7	2.4	16.8	7.5*
Nov.	36.3	5.5	13.0	- 2.1	2.9	- 2.8	12.4	0.1	13.3	4.0
Dec.	38.3	5.5	14.6	- 2.3	5.3	- 0.7	12.4	- 0.3	8.8	- 4.8*

Grain 1891	The New York Central Railroad		The Pennsylvania Railroad		The Baltimore and Ohio Railroad		The Erie Railroad		The Grand Trunk Railway	
Month	Actual Share	Actual Share minus "Trend" Share	Actual Share	Actual Share minus "Trend" Share	Actual Share	Actual Share minus "Trend" Share	Actual Share	Actual Share minus "Trend" Share	Actual Share	Actual Share minus "Trend" Share
Jan.	35.6	2.8	11.2	- 5.7*	5.2	- 0.8	15.7	4.9*	18.5	4.9*
Feb.	38.2	5.4	10.3	- 6.6*	5.0	- 1.1	14.2	3.5	14.3	0.7
Mar.	36.9	4.2	9.6	- 7.3*	4.4	- 1.7	13.8	3.1	15.1	1.6
Apr.	35.8	3.0	10.5	- 6.4*	4.2	- 1.8	13.1	2.3	14.8	1.4
May	36.1	5.4	14.1	- 1.2	5.1	- 0.7	12.0	- 0.4	7.0	- 2.0
June	28.6	- 2.0	14.9	- 0.4	6.0	0.2	18.0	5.7*	3.6	- 5.4*
July	34.0	3.3	15.8	0.4	5.4	- 0.4	15.2	2.8	5.1	- 3.9
Aug.	36.7	6.0	11.7	- 3.7	4.7	- 1.1	18.1	5.7*	7.0	- 1.9
Sept.	44.5	13.9*	10.4	- 5.1	3.9	- 2.0	13.6	1.2	5.8	- 3.1
Oct.	35.1	4.5	15.2	- 0.3	4.8	- 1.0	11.7	- 0.6	5.0	- 3.8
Nov.	40.2	9.6*	10.6	- 4.9	2.8	- 3.0	12.2	- 0.2	14.4	5.6*
Dec.	38.2	5.5	20.9	3.7	4.6	- 1.6	9.4	- 1.4	15.9	2.7
1892										
Jan.	43.6	11.0*	21.1	3.8	6.6	0.4	6.5	- 4.3	9.3	- 3.9
Feb.	40.1	7.5*	18.8	1.5	6.5	0.2	7.4	- 3.4	8.3	- 4.8*
Mar.	32.9	0.3	12.4	- 5.0	5.2	- 1.1	11.3	0.5	11.6	- 1.5
Apr.	29.6	- 3.0	9.0	- 8.4*	3.5	- 2.8	14.5	3.6	14.7	1.7
May	29.5	- 1.0	16.0	0.3	3.6	- 2.4	11.2	- 1.2	2.8	- 5.8*
June	25.3	- 5.3	10.7	- 5.0	3.4	- 2.6	9.0	- 3.5	7.8	- 0.7
July	30.8	0.3	11.2	- 4.5	3.6	- 2.4	10.9	- 1.5	3.8	- 4.7*
Aug.	28.1	- 2.4	11.3	- 4.5	4.1	- 1.9	10.7	- 1.8	4.2	- 4.3
Sept.	39.3	8.8*	11.6	- 4.2	2.9	- 3.1	11.3	- 1.2	5.3	- 3.1
Oct.	44.6	14.1*	11.3	- 4.6	2.8	- 3.2	10.1	- 2.5	6.4	- 2.0
Nov.	43.5	13.1*	12.7	- 3.2	3.8	- 2.2	7.6	- 4.9*	7.6	- 0.8
Dec.	44.0	11.6*	13.1	- 4.5	2.6	- 3.8*	7.1	- 3.8*	12.0	- 0.7

Grain 1893	The New York Central Railroad		The Pennsylvania Railroad		The Baltimore and Ohio Railroad		The Erie Railroad		The Grand Trunk Railway	
Month	Actual Share	Actual Share minus "Trend" Share	Actual Share	Actual Share minus "Trend" Share	Actual Share	Actual Share minus "Trend" Share	Actual Share	Actual Share minus "Trend" Share	Actual Share	Actual Share minus "Trend" Share
Jan.	36.7	4.3	13.2	- 4.5	3.8	- 2.6	7.7	- 3.3	16.8	4.1
Feb.	35.9	3.4	14.6	- 3.1	6.7	0.3	9.5	- 1.4	10.9	- 1.7
Mar.	32.9	0.5	15.0	- 2.8	4.4	- 2.0	13.0	2.0	12.4	- 0.2
Apr.	35.0	2.6	14.2	- 3.6	4.6	- 1.8	8.2	- 2.7	14.6	2.0
May	35.5	5.1	16.1	- 0.1	4.1	- 2.0	8.6	- 4.0*	10.7	2.6
June	40.1	9.8*	17.5	1.4	4.2	- 1.9	12.1	- 0.5	5.2	- 2.9
July	34.2	3.9	16.1	- 0.1	3.2	- 3.0	14.3	1.7	4.3	- 3.7
Aug.	30.9	0.6	16.4	0.2	2.6	- 3.6*	15.3	2.7	5.6	- 2.4
Sept.	35.3	5.0	11.5	- 4.8	4.2	- 2.0	15.2	2.5	4.8	- 3.1
Oct.	32.7	2.5	16.3	0.0	3.0	- 3.2	17.2	4.5*	2.5	- 5.5*
Nov.	31.4	1.2	17.0	0.7	1.8	- 4.4*	15.4	2.8	5.0	- 2.9
Dec.	40.3	8.0*	22.2	4.1	1.8	- 4.8*	7.8	- 3.3	11.5	- 0.7
1894										
Jan.	27.7	- 4.6	33.0	15.0*	2.8	- 3.7*	10.0	- 1.1	9.7	- 2.5
Feb.	32.9	0.6	14.7	- 3.5	4.9	- 2.4	11.2	0.2	10.6	- 1.5
Mar.	36.0	3.7	16.0	- 2.2	9.2	2.6	7.9	- 3.2	11.6	- 0.5
Apr.	25.8	- 6.5*	20.5	2.3	5.2	- 1.4	8.0	- 3.1	10.2	- 1.9
May	27.5	- 2.6	12.6	- 4.0	3.9	- 2.4	9.1	- 3.6	14.4	11.7*
June	24.8	- 5.4	11.9	- 4.7	4.1	- 2.2	11.4	- 1.3	12.1	4.5*
July	29.6	- 0.5	15.3	- 1.3	5.5	- 0.8	10.4	- 2.3	9.4	1.8
Aug.	17.0	-13.1*	11.7	- 4.9	3.3	- 3.0	16.2	3.5	9.5	1.9
Sept.	18.3	-11.8*	8.3	- 8.3*	1.6	- 4.8*	15.5	2.8	14.3	6.8*
Oct.	17.6	-12.5*	10.7	- 6.0*	6.9	0.5	11.5	- 1.2	11.6	4.1
Nov.	18.3	-11.8*	8.4	- 8.3*	4.1	- 2.2	16.3	3.6	15.3	7.8*
Dec.	16.4	-15.7*	12.9	- 5.5	4.9	- 1.8	15.2	4.0*	14.6	2.8

Grain 1895	The New York Central Railroad		The Pennsylvania Railroad		The Baltimore and Ohio Railroad		The Erie Railroad		The Grand Trunk Railway	
Month	Actual Share	Actual Share minus "Trend" Share	Actual Share	Actual Share minus "Trend" Share	Actual Share	Actual Share minus "Trend" Share	Actual Share	Actual Share minus "Trend" Share	Actual Share	Actual Share minus "Trend" Share
Jan.	21.9	-10.2*	14.3	- 4.2	11.3	4.6*	9.8	- 1.4	6.2	- 5.6*
Feb.	28.9	- 3.2	29.5	11.0*	6.9	0.1	5.2	- 6.0*	5.9	- 5.8*
Mar.	27.9	- 4.2	8.3	-10.3*	1.7	- 5.1*	14.2	3.0	13.8	2.2
Apr.	28.7	- 3.4	15.3	- 3.2	2.9	- 3.9*	11.3	0.1	14.5	2.9
May	23.3	- 6.8*	26.8	9.9*	1.4	- 5.1*	15.5	2.7	3.4	- 3.7
June	19.0	-11.0*	17.6	0.7	6.7	0.3	8.8	- 4.0*	5.7	- 1.5
July	33.5	3.5	13.5	- 3.4	8.6	2.1	5.8	- 7.0*	6.6	- 0.5
Aug.	14.2	-15.8*	11.8	- 5.6*	4.5	- 2.0	11.3	- 1.5	17.1	10.0*
Sept.	17.7	-12.3*	16.1	- 0.9	4.9	- 1.6	10.1	- 2.8	24.3	17.3*
Oct.	27.2	- 2.8	13.8	- 3.3	5.3	- 1.2	12.5	- 0.3	14.0	7.0*
Nov.	32.8	2.9	11.1	- 6.0*	5.1	- 1.4	10.7	- 2.2	12.3	5.3*
Dec.	32.9	0.9	11.3	- 7.0*	3.9	- 2.3	5.9	- 5.4*	15.6	4.3
1896										
Jan.	28.9	- 3.1	10.8	- 8.1*	3.9	- 3.0	16.5	5.2*	18.2	6.9*
Feb.	25.6	- 6.4*	24.6	5.7*	5.2	- 1.7	18.7	7.4*	12.2	1.0
Mar.	25.7	- 6.3*	10.7	- 8.2*	5.7	- 1.2	22.1	10.8*	14.3	3.2
Apr.	25.8	- 6.1	15.1	- 3.8	6.7	- 0.2	17.2	5.9*	8.1	- 3.1
May	26.8	- 3.0	15.9	- 1.4	7.1	0.5	14.7	1.7	5.0	- 1.7
June	28.3	- 1.6	11.7	- 5.6*	5.7	- 0.9	14.3	1.4	7.0	0.3
July	22.2	- 7.6*	13.3	- 4.0	9.1	2.5	14.0	1.1	7.0	0.3
Aug.	23.2	- 6.7*	14.8	- 2.6	11.1	4.5*	11.6	- 1.4	2.3	- 4.3
Sept.	34.9	- 5.1	13.2	- 4.2	5.3	- 1.4	9.3	- 3.7*	8.0	1.5
Oct.	27.4	- 2.3	13.7	- 3.8	10.2	3.5*	13.9	0.9	10.2	3.7
Nov.	30.6	0.8	11.6	- 5.9*	14.9	8.2*	8.9	- 4.0*	9.0	2.5
Dec.	29.5	- 2.4	12.3	- 6.9*	9.6	2.6	11.7	0.3	6.6	- 4.3

Grain 1897	The New York Central Railroad		The Pennsylvania Railroad		The Baltimore and Ohio Railroad		The Erie Railroad		The Grand Trunk Railway	
Month	Actual Share	Actual Share minus "Trend" Share	Actual Share	Actual Share minus "Trend" Share	Actual Share	Actual Share minus "Trend" Share	Actual Share	Actual Share minus "Trend" Share	Actual Share	Actual Share minus "Trend" Share
Jan.	33.5	1.7	24.7	5.5	3.0	- 4.0*	7.9	- 3.5	11.6	0.8
Feb.	30.4	- 1.4	27.4	8.0*	7.1	0.0	13.1	1.7	7.8	- 3.0
Mar.	30.5	- 1.3	22.6	3.3	5.0	- 2.1	7.1	- 4.3*	4.3	- 6.4*
Apr.	29.4	- 2.4	16.1	- 3.3	3.9	- 3.2	6.9	- 4.6*	2.6	- 8.1*
May	28.9	- 0.8	17.4	- 0.4	3.8	- 2.9	5.5	- 7.5*	3.1	- 3.1
June	19.5	-10.2*	24.3	6.6*	5.4	- 1.4	9.2	- 3.8*	8.0	1.7
July	19.1	-10.5*	26.3	8.5*	4.4	- 2.4	11.1	- 2.0	11.3	5.1*
Aug.	41.8	12.1*	10.9	- 6.9*	2.1	- 4.7*	9.0	- 4.0*	10.7	4.6*
Sept.	35.4	5.7	19.2	1.4	2.5	- 4.3*	15.2	2.2	6.6	- 0.5
Oct.	27.5	- 2.1	18.9	1.0	5.6	- 1.3	15.7	2.6	5.2	- 0.8
Nov.	17.2	-12.4*	33.5	15.6*	3.1	- 3.7*	11.1	- 2.0	0.2	- 5.8*
Dec.	30.7	- 0.9	27.1	7.5*	5.4	- 1.8	8.2	- 3.3	1.0	- 9.4*
1898										
Jan.	31.2	- 0.4	29.7	10.0*	5.1	- 2.1	11.9	0.3	10.2	- 0.2
Feb.	35.8	4.2	24.2	4.5	4.7	- 2.5	12.6	1.0	12.3	2.0
Mar.	39.4	7.8*	16.5	- 3.3	8.0	0.7	10.9	- 0.6	10.4	0.1
Apr.	43.1	11.6*	12.4	- 7.4*	15.8	8.6*	12.3	0.7	4.7	- 5.6*
May	28.4	- 1.1	12.9	- 5.2	10.0	3.1	15.2	2.1	0.3	- 5.5*
June	38.9	9.4*	14.2	- 3.9	9.2	2.2	13.8	0.6	0.5	- 5.3*
July	24.0	- 5.5	21.4	3.2	6.4	- 0.6	16.0	2.8	0.9	- 4.9*
Aug.	31.8	2.3	11.8	- 6.4*	4.5	- 2.5	26.3	13.1*	0.5	- 5.2*
Sept.	28.8	- 0.7	17.4	- 0.9	4.6	- 2.4	26.1	12.9*	3.3	- 2.4
Oct.	26.7	- 2.7	24.6	6.4*	7.4	0.3	24.0	10.8*	3.6	- 2.1
Nov.	24.6	- 4.8	33.7	15.4*	14.4	7.4*	12.5	- 0.7	6.6	1.0
Dec.	33.6	2.2	26.0	6.0*	7.7	0.4	12.5	0.9	9.2	- 0.7

Grain 1899	The New York Central Railroad		The Pennsylvania Railroad		The Baltimore and Ohio Railroad		The Erie Railroad		The Grand Trunk Railway	
Month	Actual Share	Actual Share minus "Trend" Share	Actual Share	Actual Share minus "Trend" Share	Actual Share	Actual Share minus "Trend" Share	Actual Share	Actual Share minus "Trend" Share	Actual Share	Actual Share minus "Trend" Share
Jan.	36.0	4.5	25.9	5.8*	6.1	- 2.3	8.6	- 3.1	11.6	1.7
Feb.	26.6	- 4.9	24.6	4.5	9.7	2.3	12.0	0.3	15.3	5.4*
Mar.	33.5	2.0	19.9	- 0.3	10.2	2.8	10.5	- 1.1	15.0	5.2*
Apr.	37.0	5.6	23.8	3.6	10.8	3.4*	8.4	- 3.2	8.3	- 1.5
May	29.9	0.5	29.4	10.9*	11.4	4.2*	13.4	0.2	4.2	- 1.2
June	40.5	11.2*	10.7	- 7.8*	13.6	6.5*	12.6	- 0.6	6.8	1.5
July	30.7	1.4	31.2	12.6*	14.6	7.4*	8.2	- 5.1*	4.5	- 0.7
Aug.	28.9	0.4	27.7	9.1*	15.0	7.8*	12.0	- 1.2	6.0	0.7
Sept.	36.6	7.3*	20.0	1.4	14.6	7.4*	10.0	- 3.3	6.8	1.6
Oct.	33.5	4.3	24.5	5.9*	13.4	6.2*	9.4	- 3.9*	8.4	3.2
Nov.	35.0	5.8	20.1	1.4	13.3	6.1*	11.9	- 1.4	6.7	1.6
Dec.	30.2	- 1.1	26.9	6.5*	7.7	0.1	12.1	0.4	6.3	- 3.1

* Greater than the standard error of estimate.

Source: Weekly shipments of wheat, corn, oats, and flour, both by railroad and as a total, as listed in the *Chicago Daily Commercial Bulletin*, 1880 to 1899.

Table A.2, Grain Shipments to the Eastern Seaboard, 1887 to 1899

The trend of the percentage share of grain tonnage of each railroad has been defined as

$$s = \alpha + \beta T + \gamma X_1 + \delta X_2 + \nu$$

with s = a railroad's percentage of total tonnage into Montreal, Boston, New York, Philadelphia, and Baltimore; T = month of shipments, numbered from 1 to 156 for the period 1887 to 1899; X_1 = "1" for the winter months in which the lakes were closed to traffic, "0" for the summer months, $X_2 = 0$ for the months before the establishment of the Lehigh Valley as a lake-rail shipper, "1," thereafter. The estimates of trend, as calculated by the method of least squares, are:

The New York Central Railroad

$$S = 13.82 + 0.02T + 3.26X_1 + 0.49X_2; \quad \hat{\rho}^2 = .092; \quad S_u = 6.17$$

The Pennsylvania Railroad

$$S = 21.29 + 0.00T + 7.56X_1 - 5.74X_2; \quad \hat{\rho}^2 = .438; \quad S_u = 5.11$$

The Baltimore and Ohio Railroad

$$S = 7.73 + 0.08T + 3.82X_1 - 4.24X_2; \quad \hat{\rho}^2 = .209; \quad S_u = 4.19$$

The Erie Railroad

$$S = 8.05 - 0.02T + 0.64X_1 + 2.16X_2; \quad \hat{\rho}^2 = .097; \quad S_u = 2.97$$

The Lehigh Valley Railway

$$S = 5.00 + 0.02T + 0.95X_1; \quad \hat{\rho}^2 = .318; \quad S_u = 3.05$$

The Lackawanna Railroad

$$S = 1.64 + 0.00T + 0.71X_1 + 0.52X_2; \quad \hat{\rho}^2 = .091; \quad S_u = 1.43$$

The "trend" share for any month is obtained from inserting the relevant values of T and X in the computed equations. The actual shares are shown in the table, along with the difference between actual and trend shares for that month.

Grain 1887	The New York Central Railroad		The Pennsylvania Railroad		The Baltimore and Ohio Railroad		The Erie Railroad		The Lackawanna Railroad	
Month	Actual Share	Actual Share minus "Trend" Share	Actual Share	Actual Share minus "Trend" Share	Actual Share	Actual Share minus "Trend" Share	Actual Share	Actual Share minus "Trend" Share	Actual Share	Actual Share minus "Trend" Share
Jan.	36.0	18*	29.9	1	13.2	1	11.1	2	0.0	- 2
Feb.	15.4	- 1	31.6	2	19.3	7	9.2		2.6	
Mar.	15.9	- 1	34.9	6*	11.7		8.0		2.0	
Apr.	18.0		30.8	1	4.7	- 7*	11.0	2	3.4	1
May	12.4	- 1	20.1	- 1	7.9		13.5	5*	1.1	
June	9.8	- 4	18.2	- 3	9.1		7.1		1.1	
July	7.2	- 6*	22.9	1	10.3	2	8.8		0.9	
Aug.	8.2	- 5	23.0	1	8.2		4.8	- 3*	1.2	
Sept.	11.3	- 2	19.1	- 2	6.8	- 1	4.8	- 3*	1.5	
Oct.	13.2		18.3	- 2	6.4	- 2	5.3	- 2	1.5	
Nov.	14.6		17.0	- 4	5.0	- 3	5.8	- 2	1.7	
Dec.	23.2	5	25.5	- 3	7.7	- 4*	9.3		2.4	
1888										
Jan.	30.0	12	30.9	2	10.5	- 2	8.3		2.4	
Feb.	23.2	5	24.2	- 4	23.6	10*	6.0	- 2	2.6	
Mar.	20.6	3	28.9		10.2	- 2	6.4	- 1	2.8	
Apr.	19.5	2	34.8	5*	10.5	- 2	7.3	- 1	4.5	2*
May	12.0	- 2	21.2		10.7	1	15.1	7*	2.0	
June	8.8	- 5	15.4	- 5*	8.5		7.7		1.2	
July	9.0	- 5	24.6	3	11.2	1	8.4		2.4	
Aug.	10.2	- 4	19.3	- 1	10.0		4.8	- 2	1.2	
Sept.	12.1	- 2	16.2	- 5*	6.7	- 2	5.5	- 2	1.4	
Oct.	10.4	- 3	21.2		4.7	- 4*	5.5	- 2	1.3	
Nov.	14.0		21.0		7.3	- 2	7.7		1.5	
Dec.	16.1	- 1	35.2	6*	24.1	10*	10.8	2	1.6	

Grain 1889	The New York Central Railroad		The Pennsylvania Railroad		The Baltimore and Ohio Railroad		The Erie Railroad		The Lehigh Valley Railroad		The Lackawanna Railroad	
Month	Actual Share	Actual Share minus "Trend" Share	Actual Share	Actual Share minus "Trend" Share	Actual Share	Actual Share minus "Trend" Share	Actual Share	Actual Share minus "Trend" Share	Actual Share	Actual Share minus "Trend" Share	Actual Share	Actual Share minus "Trend" Share
Jan.	21.2	3	16.8	- 6*	10.5	1	8.0	- 2	3.6	- 2	2.7	
Feb.	16.4	- 1	20.1	- 3	11.0	1	6.2	- 4*	1.1	- 5*	2.5	
Mar.	12.2	- 5	20.5	- 2	3.6	- 5*	9.9		0.8	- 5*	12.8	9*
Apr.	21.2	3	16.6	- 6*	6.2	- 3	12.9	2	1.8	- 4*	5.5	2*
May	15.9	1	11.9	- 3	3.0	- 2	8.7		1.7	- 3*	2.0	
June	17.1	2	12.1	- 3	3.9	- 1	5.9	- 3*	7.5	1	2.6	
July	9.2	- 5	13.1	- 2	11.1	5*	5.8	- 3*	5.1		1.6	
Aug.	10.1	- 4	14.7		9.8	3	6.3	- 3*	1.7	- 3*	2.3	
Sept.	13.8	- 1	11.0	- 4	4.3	- 1	5.3	- 4*	4.6	- 1	1.7	
Oct.	14.9		9.8	- 5*	7.9	1	9.7		3.3	- 2	2.1	
Nov.	11.9	- 3	9.9	- 5*	10.7	4*	8.2	- 1	2.7	- 3*	2.2	
Dec.	16.5	- 1	14.3	- 8*	19.3	9*	12.4	2	5.6	- 1	3.1	
1890												
Jan.	14.4	- 3	28.8	5*	15.0	4*	10.0		5.6	1	4.2	- 1
Feb.	10.3	- 8*	35.4	12*	11.2		12.5	2	6.4		2.5	
Mar.	7.9	-10*	29.2	6*	5.7	- 4*	8.8	- 1	3.6	- 3*	1.5	- 1
Apr.	9.1	- 9*	31.5	8*	7.0	- 3	9.9		7.0		2.3	
May	13.2	- 1	21.9	6*	5.5	- 1	7.4	- 1	5.2		1.7	
June	9.9	- 5	20.8	5*	1.6	- 5*	7.5	- 1	6.3		1.9	
July	4.6	-10*	26.1	10*	9.6	2	5.5	- 3*	3.2	- 2	0.5	- 1
Aug.	8.5	- 6*	24.9	9*	8.9	1	6.5	- 2	5.2		1.4	
Sept.	10.7	- 4	20.6	5*	6.3		7.4	- 1	3.9	- 2	1.9	
Oct.	15.4		16.6	1	5.4	- 1	9.2		4.5	- 1	2.2	
Nov.	16.9	1	17.6	2	5.7	- 1	8.4		5.0		1.3	
Dec.	18.3		23.0		7.6	- 3	8.7	- 1	7.6		2.2	

Grain 1891	The New York Central Railroad		The Pennsylvania Railroad		The Baltimore and Ohio Railroad		The Erie Railroad		The Lehigh Valley Railroad		The Lackawanna Railroad	
Month	Actual Share	Actual Share minus "Trend" Share	Actual Share	Actual Share minus "Trend" Share	Actual Share	Actual Share minus "Trend" Share	Actual Share	Actual Share minus "Trend" Share	Actual Share	Actual Share minus "Trend" Share	Actual Share	Actual Share minus "Trend" Share
Jan.	23.3	4	17.2	- 5*	10.1	- 1	10.1		8.0	1	2.5	
Feb.	21.5	2	19.2	- 3	11.8		15.7	5*	7.1		2.0	
Mar.	22.7	4	17.3	- 5*	10.0	- 1	14.6	4*	8.2	1	2.2	
Apr.	24.1	5	17.4	- 5*	8.5	- 2	14.2	4*	6.6		1.4	- 1
May	22.0	6*	15.1		7.5		12.7	3*	8.0	1	1.5	
June	14.5		11.3	- 4	5.9	- 1	11.0	1	4.5	- 1	0.8	- 1
July	21.8	6*	13.0	- 2	20.0	12*	7.5	- 1	3.9	- 2	0.6	- 1
Aug.	17.5	2	15.9		13.3	5*	8.9		5.7		0.4	- 1
Sept.	18.6	3	11.8	- 3	5.6	- 2	12.8	3*	6.6		0.8	- 1
Oct.	18.4	2	12.5	- 3	5.3	- 2	13.1	4*	6.1		1.1	- 1
Nov.	20.0	4	11.7	- 3	8.6		11.6	2	8.8	2	1.6	
Dec.	20.8	2	19.7	- 3	10.2	- 1	11.3	1	7.4		2.9	
1892												
Jan.	24.5	5	23.3				8.5	- 1	5.2	- 1	2.0	
Feb.	22.4	3	23.9				7.7	- 1	5.5	- 1	1.1	- 1
Mar.	16.0	- 2	28.2	5*			11.1	1	3.2	- 4*	0.9	- 1
Apr.	22.0	3	28.3	5*			12.6	3*	7.3		0.9	- 1
May	20.7	5	20.3	4			12.1	3*	7.9	1	0.6	- 1
June	20.2	4	17.6	2			7.9		6.4		0.5	- 1
July	22.5	6*	17.5	1			10.5	1	5.0	- 1	0.7	- 1
Aug.	16.5		20.1	4			9.2		6.4		0.8	- 1
Sept.	24.5	8*	14.0	- 1			11.8	2	8.8	2	1.0	- 1
Oct.	23.9	8*	14.1	- 1			7.6	- 1	7.7	1	1.2	
Nov.	25.5	9*	16.0				7.5	- 1	4.8	- 1	1.3	
Dec.	20.9	1	18.6	- 4			10.6	1	6.8		1.7	- 1

Grain 1893	The New York Central Railroad		The Pennsylvania Railroad		The Erie Railroad		The Lehigh Valley Railroad		The Lackawanna Railroad	
Month	Actual Share	Actual Share minus "Trend" Share	Actual Share	Actual Share minus "Trend" Share	Actual Share	Actual Share minus "Trend" Share	Actual Share	Actual Share minus "Trend" Share	Actual Share	Actual Share minus "Trend" Share
Jan.	23.9	4	21.4	- 1	8.2	- 1	7.0		3.1	
Feb.	14.6	- 4	20.9	- 2	6.3	- 3*	7.0		3.8	
Mar.	17.8	- 1	24.6	1	6.7	- 2	7.6		4.6	1
Apr.	20.9	1	25.6	2	9.4		6.3	- 1	5.4	2*
May	20.2	4	14.6		8.8		6.3		3.2	1
June	13.5	- 2	11.9	- 3	7.4	- 1	5.2	- 1	2.7	
July	11.2	- 4	11.7	- 3	5.1	- 3*	3.8	- 2	2.7	
Aug.	13.7	- 2	12.0	- 3	4.9	- 3*	3.6	- 3*	1.7	
Sept.	11.5	- 4	12.8	- 2	6.4	- 2	2.6	- 4*	1.6	
Oct.	11.8	- 4	11.0	- 4	5.7	- 2	4.6	- 2	1.8	
Nov.	14.6	- 1	12.5	- 3	5.7	- 2	5.3	- 1	3.2	1
Dec.	19.6		16.3	- 6*	5.6	- 3*	5.5	- 2	3.6	
1894										
Jan.	23.3	4	22.8		6.1	- 3*	4.1	- 3*	1.2	- 1
Feb.	18.9		23.0		3.9	- 5*	4.6	- 3*	1.4	- 1
Mar.	17.9	- 1	25.3	2	6.3	- 2	3.3	- 4*	1.2	- 1
Apr.	22.7	3	22.4		11.6	2	4.0	- 3*	3.2	
May	12.4	- 3	14.7		6.5	- 1	3.8	- 2	1.8	
June	11.0	- 5	13.5	- 2	6.0	- 2	6.0		1.3	
July	14.7	- 1	20.3	4	2.9	- 5*	2.7	- 4*	0.8	- 1
Aug.	19.4	5	18.8	3	4.1	- 4*	4.6	- 2	1.4	
Sept.	19.8	4	19.8	5*	5.3	- 3*	2.9	- 3*	1.0	- 1
Oct.	19.9	4	15.3		5.4	- 2	2.7	- 4*	1.4	
Nov.	11.7	- 4	16.6	1	4.3	- 4*	4.2	- 2	2.5	
Dec.	22.2	2	17.8	- 5*	5.9	- 3*	3.0	- 4*	3.5	

Grain 1895	The New York Central Railroad		The Pennsylvania Railroad		The Erie Railroad		The Lehigh Valley Railroad		The Lackawanna Railroad	
Month	Actual Share	Actual Share minus "Trend" Share	Actual Share	Actual Share minus "Trend" Share	Actual Share	Actual Share minus "Trend" Share	Actual Share	Actual Share minus "Trend" Share	Actual Share	Actual Share minus "Trend" Share
Jan.	14.7	- 4	28.2	5*	7.8	- 1	5.1	- 2	2.0	
Feb.	15.1	- 4	27.5	4	5.3	- 3	5.0	- 2	1.0	- 1
Mar.	21.1	1	21.4	- 1	7.7	- 1	6.7	- 1	0.5	- 2*
Apr.	23.0	3	19.3	- 3	10.6	1	8.2		1.8	- 1
May	23.2	6*	12.9	- 2	13.9	5*	4.7	- 2	0.3	- 1
June	17.5	1	16.1		17.9	9*	4.3	- 2	0.6	- 1
July	15.5		15.2		13.9	5*	2.3	- 4*	0.5	- 1
Aug.	26.9	10*	13.9	- 1	15.5	7*	2.3	- 4*	0.4	- 1
Sept.	27.9	11*	10.2	- 5*	16.4	8*	2.8	- 4*	0.6	- 1
Oct.	24.8	8*	12.4	- 3	14.2	6*	5.4	- 1	0.9	- 1
Nov.	28.0	11*	10.8	- 4	13.1	5*	5.8	- 1	0.8	- 1
Dec.	27.5	7*	12.5	-10*	11.4	2	7.6		0.8	- 2*
1896										
Jan.	15.3	- 4	22.8		5.0	- 3*	5.9	- 2	0.5	- 2*
Feb.	14.4	- 5	24.8	1	6.9	- 1	5.6	- 2	1.2	- 1
Mar.	20.7		16.1	- 7*	9.9	1	4.9	- 3*	0.6	- 2*
Apr.	22.7	2	16.5	- 6*	10.0	1	6.2	- 1	0.5	- 2*
May	15.8		9.1	- 6*	8.6		5.3	- 1	0.2	- 1
June	13.3	- 3	12.0	- 3	6.5	- 1	6.0	- 1	0.5	- 1
July	9.7	- 6*	12.4	- 3	4.8	- 3*	4.9	- 2	0.5	- 1
Aug.	7.9	- 8*	11.0	- 4	4.7	- 3*	10.9	3*	0.5	- 1
Sept.	7.6	- 9*	14.6		6.1	- 1	9.7	2	0.3	- 1
Oct.	16.7		12.8	- 2	6.0	- 1	6.0	- 1	0.4	- 1
Nov.	21.7	5	14.2	- 1	8.9	1	7.2		0.6	- 1
Dec.	19.0		18.3	- 4	8.9		6.5	- 1	0.8	- 2*

Grain 1897[1]	The New York Central Railroad		The Pennsylvania Railroad		The Erie Railroad		The Lehigh Valley Railroad		The Lackawanna Railroad	
Month	Actual Share	Actual Share minus "Trend" Share	Actual Share	Actual Share minus "Trend" Share	Actual Share	Actual Share minus "Trend" Share	Actual Share	Actual Share minus "Trend" Share	Actual Share	Actual Share minus "Trend" Share
Jan.	10.8	- 8*	20.4	- 2	6.1	- 2			0.5	- 2*
Feb.	14.0	- 5	25.4	1	7.5		5.6	- 3*	0.2	- 2*
Mar.	14.0	- 6*	25.3	2	7.1	- 1	4.0	- 3*	0.2	- 2*
Apr.	16.8	- 3	27.8	5*	8.3		3.4	- 4*	0.3	- 2*
May	12.7	- 4	12.1	- 3	3.8	- 5*	4.6	- 2	0.3	- 1
June	19.1	3	12.2	- 3	13.3	5*	4.1	- 3*	0.3	- 1
July	18.4	1	13.3	- 2	9.8	2	2.8	- 4*	0.5	- 1
Aug.	14.3	- 2	14.3	- 1	12.1	4*	2.8	- 4*	0.4	- 1
Sept.	21.5	3	15.8	1	8.9	1	6.2	- 1	0.4	- 1
Oct.	20.6	3	14.2	- 1	8.6	1	7.4		0.5	- 1
Nov.	24.7	7*	14.2	- 1	5.5	2	9.5	2	1.0	- 1
Dec.	16.6	- 3	18.2	- 4	6.7	- 2	10.5	2	1.9	
1898										
Jan.	15.5	- 4	23.2		8.7		7.7		0.7	- 2*
Feb.	15.2	- 5	28.9	5*	4.5	- 3*	7.9		0.8	- 2*
Mar.	15.0	- 5	22.1	- 1	8.5		10.7	2	0.9	- 1
Apr.	30.5	10*	13.0	-10*	7.0	- 1	12.0	3*	3.3	
May	23.3	6*	19.4	3	7.7		10.1	2	3.3	1
June	16.6		13.6	- 1	9.5	2	18.4	10*	3.6	1
July	19.4	2	10.2	- 5*	10.1	2	3.2	- 4*	1.0	- 1
Aug.	19.7	2	17.4	1	5.9	- 1	2.6	- 5*	2.6	
Sept.	25.8	8*	14.7		7.5		2.7	- 5*	2.1	
Oct.	19.3	2	23.6	8*	8.1		2.9	- 4*	5.7	3*
Nov.	25.0	7*	19.2	3	7.2		6.2	- 1	1.9	
Dec.	15.8	- 4	20.3	- 2	5.2	- 2	13.2	4*	0.9	- 1

Grain 1899	The New York Central Railroad		The Pennsylvania Railroad		The Erie Railroad		The Lehigh Valley Railroad		The Lackawanna Railroad	
Month	Actual Share	Actual Share minus "Trend" Share	Actual Share	Actual Share minus "Trend" Share	Actual Share	Actual Share minus "Trend" Share	Actual Share	Actual Share minus "Trend" Share	Actual Share	Actual Share minus "Trend" Share
Jan.	15.0	- 5	23.9		4.1	- 3*	12.2	3*	0.5	- 2
Feb.	12.8	- 7*	23.9		4.0	- 3*	5.9	- 2	0.9	- 1
Mar.	15.9	- 4	20.1	- 3	6.7	- 1	7.9		0.8	- 2
Apr.	12.7	- 7*	15.9	- 7*	8.3		6.9	- 2	1.1	- 1
May	19.8[1]	3[1]	14.1	- 1	7.1		6.6	- 1	0.7	- 1
June	16.6		15.6		5.5	- 1	6.7	- 1	0.4	- 1
July	18.3		15.8		4.0	- 3	8.4		1.0	- 1
Aug.	20.0	2	15.4		3.6	- 3	7.1		0.5	- 1
Sept.	18.8	1	19.6	4	4.2	- 2	6.1	- 1	2.7	
Oct.	21.5	4	17.2	1	2.5	- 4*	6.5	- 1	4.2	2
Nov.	13.0	- 4	22.6	7*	3.9	- 3*	10.2	2	4.8	2
Dec.	14.8	- 5	20.9	- 2	7.0		10.2	1	3.4	

* Greater than the standard error of estimate.

[1] Additional data: because of absence of information when the trend lines were calculated, the percentage was not included but was added at a later time. As a consequence, the sum of the residuals does not equal zero.

Note: Receipts into Baltimore are not available from January 1892 on.

Source: Shipments into Montreal, Boston, New York, Philadelphia and Baltimore of wheat, corn, oats, and flour as listed in the Annual Reports of the New York Produce Exchange, 1880 to 1899. Shipments into these locations by railroad are from the same source, or from the Annual Reports of the Baltimore Corn and Flour Exchange, 1880 to 1891. **Shipments into three cities by the Pennsylvania Railroad are "forced," given that they** are the sum of listed figures into New York and Baltimore, and of all of the shipments into Philadelphia.

Table A.3, Lard Shipments from Chicago, 1887 to 1896

The trend of the percentage share of lard tonnage of each railroad has been defined as

$$s = \alpha + \beta T + \gamma X + \nu$$

with s = a railroad's percentage of total tonnage of Chicago shipments; T = month of shipments, numbered from 1 to 156 for the period 1887 to 1899; X = "1" for the winter months in which the lakes were closed to traffic, "0" for the summer months. The estimates of trend, as calculated by the method of least squares, are:

The New York Central Railroad

$$S = 35.295 - 0.074T + 2.651X; \quad \hat{\rho}^2 = .179; \quad S_u = 7.832$$

The Pennsylvania Railroad

$$S = 16.122 + 0.084T - 0.504X; \quad \hat{\rho}^2 = .189; \quad S_u = 7.903$$

The Baltimore and Ohio Railroad

$$S = 8.555 + 0.051T + 3.453X; \quad \hat{\rho}^2 = .170; \quad S_u = 6.247$$

The Erie Railroad

$$S = 10.120 - 0.009T - 2.312X; \quad \hat{\rho}^2 = .033; \quad S_u = 6.602$$

The Grand Trunk Railway

$$S = 11.671 - 0.051T + 0.227X; \quad \hat{\rho}^2 = .175; \quad S_u = 5.118$$

The "trend" share for any month is obtained from inserting the relevant values of T and X in the computed equation. The actual shares are shown in the table, along with the difference between actual and trend shares for that month.

Lard 1887	The New York Central Railroad		The Pennsylvania Railroad		The Baltimore and Ohio Railroad		The Erie Railroad		The Grand Trunk Railway	
Month	Actual Share	Actual Share minus "Trend" Share	Actual Share	Actual Share minus "Trend" Share	Actual Share	Actual Share minus "Trend" Share	Actual Share	Actual Share minus "Trend" Share	Actual Share	Actual Share minus "Trend" Share
Jan.	24.1	-13.8*	20.1	4.4	3.9	- 8.1*	5.1	- 2.7	32.2	20.3*
Feb.	34.8	- 3.0	29.0	13.2*	4.3	- 7.8*	5.5	- 2.3	11.9	0.0
Mar.	29.8	- 8.0*	27.9	12.1*	8.0	- 4.2	11.3	3.4	7.8	- 4.0
Apr.	45.9	8.2*	16.7	0.8	6.7	- 5.6	8.7	0.9	7.1	- 4.6
May	18.2	-16.7*	22.4	5.9	16.7	7.8*	19.0	8.9*	10.1	- 1.3
June	20.0	-14.8*	30.9	14.2*	10.5	1.6	12.1	2.0	14.2	2.8
July	24.9	- 9.9*	29.7	13.0*	2.8	- 6.1	10.2	0.1	13.4	2.1
Aug.	30.8	- 3.9	23.8	7.0	4.6	- 4.4	9.1	- 0.9	10.4	- 0.8
Sept.	31.1	- 3.5	20.9	4.0	9.0	- 0.1	12.2	2.1	10.8	- 0.4
Oct.	23.6	-10.9*	15.7	- 1.3	9.1	0.0	15.8	5.8	9.9	- 1.2
Nov.	29.0	- 5.5	16.6	- 0.4	4.2	- 4.9	11.1	1.1	21.6	10.5*
Dec.	30.9	- 6.2	19.9	3.3	14.9	2.3	7.7	0.0	14.0	2.7
1888										
Jan.	25.7	-11.3*	16.9	0.2	15.4	2.7	6.2	- 1.5	18.3	7.0*
Feb.	36.5	- 0.5	19.9	3.2	8.4	- 4.3	8.3	0.6	11.4	0.2
Mar.	37.6	0.8	18.6	1.7	11.1	- 1.7	5.9	- 1.8	8.8	- 2.4
Apr.	35.6	- 1.2	16.3	- 0.7	8.3	- 4.6	7.8	0.1	10.9	- 0.2
May	31.6	- 2.4	12.8	- 4.7	4.0	- 5.4	6.7	- 3.3	12.8	2.0
June	25.8	- 8.2	17.6	- 0.0	4.0	- 5.5	5.2	- 4.7	15.5	4.8
July	30.6	- 3.3	16.5	- 1.3	4.0	- 5.6	18.7	8.7*	11.4	0.7
Aug.	40.8	7.0	17.8	- 0.0	3.5	- 6.1	1.8	- 8.1*	14.1	3.5
Sept.	31.7	- 2.0	18.0	0.1	8.0	- 1.6	4.0	- 5.9	12.5	1.9
Oct.	40.2	6.5	23.0	5.0	3.8	- 5.9	1.4	- 8.5*	15.6	5.1
Nov.	34.9	1.3	16.5	- 1.6	13.9	4.2	3.1	- 6.8*	9.6	- 0.9
Dec.	38.4	2.2	17.9	0.3	16.6	3.4	3.8	- 3.8	6.8	- 3.8

Lard 1889	The New York Central Railroad		The Pennsylvania Railroad		The Baltimore and Ohio Railroad		The Erie Railroad		The Grand Trunk Railway	
Month	Actual Share	Actual Share minus "Trend" Share	Actual Share	Actual Share minus "Trend" Share	Actual Share	Actual Share minus "Trend" Share	Actual Share	Actual Share minus "Trend" Share	Actual Share	Actual Share minus "Trend" Share
Jan.	31.4	- 4.7	15.5	- 2.2	16.7	3.4	4.2	- 3.4	10.9	0.4
Feb.	30.1	- 6.0	19.2	1.4	13.5	0.2	1.4	- 6.1	9.1	- 1.5
Mar.	33.2	- 2.8	19.1	1.2	8.3	- 5.1	2.3	- 5.3	8.9	- 1.6
Apr.	38.7	2.8	23.7	5.8	6.8	- 6.6*	1.6	- 5.9	9.6	- 0.8
May	50.2	17.0*	13.0	- 5.6	8.1	- 1.9	5.0	- 4.8	11.5	1.3
June	39.2	6.2	8.0	-10.6*	11.4	1.4	10.3	0.4	15.6	5.4*
July	37.4	4.4	9.1	- 9.6*	10.7	0.6	12.7	2.8	15.0	4.9
Aug.	34.0	1.1	14.4	- 4.4	10.3	0.1	2.6	- 7.3*	16.3	6.3*
Sept.	31.8	- 1.0	14.1	- 4.8	14.5	4.2	0.4	- 9.5*	13.6	3.6
Oct.	37.8	5.1	10.1	- 8.9*	18.6	8.3*	0.5	- 9.3*	14.8	4.9
Nov.	49.9	17.2*	18.9	- 0.2	10.7	0.3	0.2	- 9.6*	7.4	- 2.5
Dec.	56.8	21.5*	14.1	- 4.5	13.6	- 0.3	4.4	- 3.1	5.0	- 5.0
1890										
Jan.	59.0	23.8*	14.6	- 4.0	13.6	- 0.3	2.1	- 5.4	3.5	- 6.5*
Feb.	48.7	13.6*	14.9	- 3.9	12.2	- 1.7	5.5	- 2.0	4.6	- 5.3*
Mar.	38.8	3.7	14.0	- 4.8	15.6	1.6	10.9	3.5	4.6	- 5.3*
Apr.	39.5	4.5	16.5	- 2.4	17.2	3.2	6.1	- 1.4	4.7	- 5.2*
May	36.1	3.8	16.6	- 2.9	16.9	6.2	6.8	- 3.0	5.5	- 4.1
June	30.6	- 1.6	14.5	- 5.1	9.4	- 1.3	7.1	- 2.6	9.2	- 0.3
July	38.3	6.1	14.1	- 5.7	20.1	9.4*	5.5	- 4.2	9.4	- 0.0
Aug.	31.3	- 0.8	20.3	0.5	13.3	2.5	0.7	- 9.0*	17.6	8.2*
Sept.	24.0	- 7.9*	23.7	3.8	15.0	4.1	2.8	- 7.0*	18.4	9.1*
Oct.	41.1	9.2*	23.9	3.9	10.3	- 0.6	2.2	- 7.5*	11.0	1.7
Nov.	50.6	18.7*	12.0	- 8.0*	18.0	7.0*	0.1	- 9.6*	7.5	- 1.7
Dec.	49.4	15.0*	9.7	- 9.9*	19.0	4.5	0.3	- 7.1	11.2	1.8

Lard 1891	The New York Central Railroad		The Pennsylvania Railroad		The Baltimore and Ohio Railroad		The Erie Railroad		The Grand Trunk Railway	
Month	Actual Share	Actual Share minus "Trend" Share	Actual Share	Actual Share minus "Trend" Share	Actual Share	Actual Share minus "Trend" Share	Actual Share	Actual Share minus "Trend" Share	Actual Share	Actual Share minus "Trend" Share
Jan.	48.2	13.9*	11.4	- 8.3*	13.4	- 1.1	2.4	- 4.9	8.3	- 1.1
Feb.	54.4	20.1*	10.8	- 9.0*	10.1	- 4.4	2.1	- 5.3	10.5	1.2
Mar.	46.0	11.8*	11.3	- 8.6*	10.7	- 3.9	4.4	- 3.0	14.4	5.2
Apr.	42.3	8.1	12.6	- 7.3	9.4	- 5.2	12.6	5.2	5.1	- 4.1
May	36.6	5.2	13.5	- 7.0	8.5	- 2.7	17.7	8.1	2.1	- 6.8*
June	32.8	1.6	12.2	- 8.4*	15.5	4.2	14.2	4.6	6.0	- 2.8
July	30.5	- 0.8	7.6	-13.1*	10.4	- 1.0	14.3	4.7	10.6	1.8
Aug.	31.5	0.4	7.9	-12.9*	15.5	4.1	15.9	6.3	5.6	- 3.1
Sept.	29.4	- 1.7	11.9	- 9.0*	15.8	4.3	27.9	18.3*	1.2	- 7.5*
Oct.	28.4	- 2.6	14.0	- 7.0	18.7	7.2	27.1	17.5*	2.1	- 6.6*
Nov.	35.2	4.3	9.8	-11.2*	15.8	4.2	16.8	7.2*	11.9	3.3
Dec.	31.6	- 1.9	6.6	-14.0*	22.7	7.6*	18.4	11.2*	10.6	1.8
1892										
Jan.	32.6	- 0.8	6.3	-14.5*	24.0	8.9*	18.9	11.6*	7.4	- 1.3
Feb.	37.9	4.6	9.3	-11.5*	16.3	1.2	11.2	3.9	11.4	2.7
Mar.	46.8	13.7	10.9	- 9.9*	17.6	2.4	15.7	8.4*	6.6	- 2.0
Apr.	29.1	- 4.2	14.3	- 6.6	13.2	- 2.0	22.7	15.5*	5.2	- 3.4
May	26.4	- 4.1	25.3	3.7	8.6	- 3.2	15.7	6.2	3.3	- 5.0
June	28.1	- 2.3	16.4	- 5.2	12.4	0.5	19.0	9.5*	4.6	- 3.7
July	22.1	- 8.2	20.0	- 1.7	10.6	- 1.3	18.7	9.2*	8.2	- 0.0
Aug.	17.0	-13.3*	28.5	6.7	9.2	- 2.8	15.2	5.8	18.3	-10.1*
Sept.	17.9	-12.2*	17.3	- 4.5	10.3	- 1.8	41.9	32.4*	5.8	- 2.3
Oct.	24.7	- 5.4	23.8	1.8	19.9	7.8*	14.4	4.9	6.6	- 1.4
Nov.	41.4	11.4*	13.7	- 8.4*	11.1	- 1.1	23.9	14.5*	3.4	- 4.6
Dec.	29.8	- 2.7	22.3	0.7	10.8	- 4.9	16.6	9.4*	14.4	6.2*

Lard 1893	The New York Central Railroad		The Pennsylvania Railroad		The Baltimore and Ohio Railroad		The Erie Railroad		The Grand Trunk Railway	
Month	Actual Share	Actual Share minus "Trend" Share	Actual Share	Actual Share minus "Trend" Share	Actual Share	Actual Share minus "Trend" Share	Actual Share	Actual Share minus "Trend" Share	Actual Share	Actual Share minus "Trend" Share
Jan.	24.8	- 7.7	23.9	2.2	7.2	- 8.5*	5.8	- 1.3	28.8	20.7*
Feb.	31.1	- 1.4	20.7	- 1.1	12.9	- 2.9	12.9	5.8	8.9	0.8
Mar.	27.5	- 4.9	20.6	- 1.3	30.7	14.9*	10.4	3.3	3.4	- 4.6
Apr.	34.7	2.4	19.4	- 2.6	17.9	2.0	9.4	2.3	3.8	- 4.2
May	31.2	1.6	23.2	0.6	20.6	8.2*	6.7	- 2.7	4.4	- 3.3
June	32.1	2.6	27.3	4.7	18.2	5.6	9.4	0.0	2.8	- 4.8
July	41.8	12.4*	20.8	- 1.9	16.3	3.7	10.9	1.6	2.3	- 5.3*
Aug.	19.1	-10.3*	32.6	9.8*	8.2	- 4.4	23.7	14.3*	2.3	- 5.2*
Sept.	21.6	- 7.7	36.6	13.7*	9.6	- 3.0	15.4	6.0	3.3	- 4.2
Oct.	29.1	- 0.1	30.6	7.7*	14.0	1.2	11.4	2.0	2.4	- 5.1*
Nov.	29.6	0.5	32.5	9.4*	12.0	- 0.8	10.1	0.7	1.2	- 6.2*
Dec.	30.4	- 1.4	35.6	13.0*	14.5	- 1.8	9.7	2.7	0.8	- 6.8*
1894										
Jan.	19.3	-12.4*	47.8	25.1*	9.0	- 7.4*	7.3	0.3	6.8	- 0.7
Feb.	32.7	1.1	32.3	9.5*	9.9	- 6.5*	5.6	- 1.4	6.3	- 1.2
Mar.	27.3	- 4.2	29.1	6.2	19.6	3.2	9.9	2.9	0.2	- 7.2*
Apr.	26.9	- 4.5	35.7	12.8*	13.0	- 3.5	8.7	1.8	1.5	- 5.8*
May	14.2	-14.5*	21.0	- 2.6	16.2	3.1	6.0	- 3.3	5.7	- 1.4
June	22.8	- 5.9	29.8	6.1	10.8	- 2.3	9.4	0.1	7.0	- 0.1
July	31.9	3.4	30.2	6.5	4.1	- 9.0*	12.4	3.1	12.0	5.1*
Aug.	24.7	- 3.8	30.2	6.4	8.9	- 4.3	8.3	- 1.0	11.5	4.6
Sept.	22.9	- 5.5	26.5	2.6	17.1	3.8	5.4	- 3.8	13.3	6.4*
Oct.	14.3	-14.0*	39.5	15.5*	15.7	2.4	2.8	- 6.5	4.3	- 2.5
Nov.	19.8	- 8.5*	45.7	21.6*	16.6	3.2	0.9	- 8.4*	3.5	- 3.3
Dec.	19.9	-10.9*	36.9	13.3*	16.0	- 0.8	1.2	- 5.8	5.2	- 1.8

Lard 1895	The New York Central Railroad		The Pennsylvania Railroad		The Baltimore and Ohio Railroad		The Erie Railroad		The Grand Trunk Railway	
Month	Actual Share	Actual Share minus "Trend" Share	Actual Share	Actual Share minus "Trend" Share	Actual Share	Actual Share minus "Trend" Share	Actual Share	Actual Share minus "Trend" Share	Actual Share	Actual Share minus "Trend" Share
Jan.	18.6	-12.2*	22.0	- 1.7	36.5	19.5*	2.9	- 4.0	5.8	- 1.1
Feb.	25.6	- 5.1	31.1	7.3	28.3	11.3*	2.0	- 4.9	3.0	- 3.8
Mar.	35.2	4.6	22.3	- 1.6	8.5	- 8.6*	11.4	4.5	3.1	- 3.7
Apr.	33.4	2.9	23.9	- 0.1	13.0	- 4.0	12.5	5.6	2.0	- 4.8
May	30.8	2.9	44.6	20.1*	5.8	- 8.0*	5.7	- 3.5	0.0	- 6.4*
June	32.8	5.0	32.6	8.0*	16.9	3.2	2.4	- 6.7*	1.8	- 4.6
July	31.3	3.6	35.3	10.6*	16.0	2.2	2.9	- 6.3	3.4	- 3.0
Aug.	31.6	4.0	33.3	8.5*	10.4	- 3.4	0.2	- 9.0*	4.8	- 1.5
Sept.	28.0	0.4	32.4	7.5	3.9	-10.0*	0.7	- 8.5*	11.5	5.2*
Oct.	29.8	2.3	23.5	- 1.5	5.8	- 8.1*	15.4	6.3	4.5	- 1.7
Nov.	35.0	7.6	24.1	- 1.0	12.9	- 1.1	8.8	- 4.0	0.2	- 6.0*
Dec.	35.1	5.1	19.2	- 5.5	28.8	11.3*	0.2	- 6.6*	0.9	- 5.4*
1896										
Jan.	28.5	- 1.4	24.5	0.2	24.6	7.1*	0.1	- 6.7*	3.1	- 3.1
Feb.	18.7	-11.1*	20.9	- 3.9	37.5	19.9*	1.1	- 5.7	1.9	- 4.3
Mar.	21.2	- 8.5*	15.8	- 9.1*	27.2	9.5*	0.2	- 6.6*	2.1	- 4.1
Apr.	22.4	- 7.3	23.3	- 1.7	23.9	6.2*	5.2	- 1.6	4.6	- 1.5
May	20.6	- 6.3	25.7	0.2	19.9	5.6	6.0	- 3.1	2.9	- 3.0
June	24.4	- 2.4	28.0	2.3	24.6	10.3*	0.6	- 8.4*	2.8	- 2.9
July	22.1	- 4.7	19.4	- 6.3	31.5	17.1*	1.3	- 7.8*	10.3	4.6
Aug.	39.6	12.8*	12.7	-13.1*	23.1	8.6*	3.5	- 5.6	8.6	2.9
Sept.	31.2	4.6	22.2	- 3.7	23.7	9.2*	2.2	- 6.8*	3.7	- 1.9
Oct.	30.3	3.8	18.6	- 7.4	18.6	4.1	4.6	- 4.4	4.0	- 1.6
Nov.	32.1	5.6	19.8	- 6.2	23.0	8.4*	4.1	- 5.0	2.9	- 2.6
Dec.	30.0	0.9	22.9	- 2.7	29.5	11.4*	4.0	- 2.7	1.4	- 4.3

* Greater than the relevant standard error of estimate S_u.

Source: Shipments of "western barreled lard" by railroad, and total shipments, as listed in the weekly market summaries of the Chicago Daily Commercial Bulletin, 1883-1897.

Table A.4. Railroad Rates and Lard Price Differences: 1891 to 1896

Month	Official Rate Cents per 100 lb	Board of Trade Rate Cents per 100 lb	Chicago-New York Western Lard Price Mean Difference Cents per 100 lb	Number of Days in Month in Which There Were Both Chicago and New York Prices
January 1891	30	30	36	26
February	30	30	36	23
March	30	30	33	25
April	30	30	25	25
May	30	28	28	25
June	30	25	28	25
July	30	30	24	25
August	30	30	32	26
September	30	30	34	25
October	30	30	33	26
November	30	30	34	23
December	30	30*	41	25
January 1892	30	30*	37	25
February	30	30*	34	23
March	30	30	33	27
April	30	30	33	24
May	30	30	29	23
June	30	27	29	26
July	25	25	29	24
August	25	25	31	27
September	25	25	30	24
October	30	30	34	23
November	30	30*	56	24
December	30	30	--	--

Month	Official Rate Cents per 100 lb	Board of Trade Rate Cents per 100 lb	Chicago-New York Western Lard Price Mean Difference Cents per 100 lb	Number of Days in Month in Which There Were Both Chicago and New York Prices
January 1893	30	30	36	25
February	30	30	32	22
March	30	30	29	26
April	30	30	30	23
May	30	30	32	25
June	30	30	28	26
July	30	30	30	24
August	50**	50**	50	27
September	50**	50**	47	24
October	50**	50**	50	24
November	50**	49	52	23
December	50**	40	46	24
January 1894	30	30	38	26
February	30	30	44	22
March	30	28*	40	26
April	30	30	41	24
May	30	30	36	26
June	30	30*	36	26
July	30	30*	49	24
August	30	30	37	27
September	30	25	30	24
October	30	24	28	27
November	30	28	36	24
December	30	30	35	24

Month	Official Rate Cents per 100 lb	Board of Trade Rate Cents per 100 lb	Chicago-New York Western Lard Price Mean Difference Cents per 100 lb	Number of Days in Month in Which There Were Both Chicago and New York Prices
January 1895	30	26	29	26
February	30	25	31	22
March	30	30	31	26
April	30	27	27	24
May	30	30	30	26
June	30	20	19	25
July	30	30	***	
August	30	30	38	27
September	30	30	39	23
October	30	30	35	27
November	30	30	32	24
December	30	24	15	24
January 1896	30	30	27	26
February	30	30	30	23
March	30	30	30	26
April	30	30	29	24
May	30	30	31	25
June	30	30	29	26
July	30	28	25	25
August	30	30	40	26
September	30	30	***	
October	30	30	40	26
November	30	30*	40	23
December	30	30	35	25

* Blockade of transport, as noted in the Chicago Daily Commercial Bulletin.

** Suspected shift of lard to third class.

*** "No shipping demand" according to the Chicago Daily Commercial Bulletins for that month.

Source: As in Table 5.3.

Statistical Appendix to Chapter 6

Exports of Grain: 1871 to 1899

	Total Exports from the United States	Total Value of These Exports	
Fiscal Year July 1-June 30	Millions of Pounds of Wheat and Corn	Millions of Dollars	Average Value, Cents per Pound
1871-1872	433.3	81.9	18.9
1872-1873	537.1	177.9	33.1
1873-1874	750.0	157.0	20.9
1874-1875*	605.4	109.0	18.0
1875-1876	733.2	127.4	17.4
1876-1877*	748.6	111.9	15.0
1877-1878*	1040.9	171.4	16.5
1878-1879*	1378.4	202.0	14.7
1879-1880	1639.6	277.2	16.9
1880-1881	1641.9	264.6	16.1
1881-1882*	962.7	179.1	18.6
1882-1883	1120.3	203.4	18.2
1883-1884*	928.3	154.6	16.7
1884-1885*	1091.3	153.9	14.1
1885-1886*	930.5	121.3	13.0
1886-1887*	1154.6	162.6	14.1
1887-1888	859.3	125.1	14.6
1888-1889	928.1	120.7	13.0
1889-1890	1235.4	145.8	11.8
1890-1891	817.5	124.7	15.3
1891-1892	1783.2	279.3	15.7
1892-1893	1415.7	194.5	13.7
1893-1894*	1357.6	159.6	11.8
1894-1895*	1029.0	110.8	10.8
1895-1896	1324.6	130.2	9.8
1896-1897*	1871.9	170.7	9.1
1897-1898*	2491.0	290.8	11.7

* Fiscal years in which there were extended breakdowns in rates.

Source: Annual Reports of the New York Produce Exchange, 1888 and 1899.

Selected Bibliography

Books

Aldrich, W. R. "Wholesale prices, wages and transportation," in United States Congress, *Report of the Senate Committee on Finance* (March 3, 1893).

The Annual Report of the Interstate Commerce Commission, December 1887–December 1899. Washington: U.S. Government Printing Office.

Annual Report of the Massachusetts Board of Railroad Commissioners. Boston: Wright and Patton, January, 1870–1880.

The Articles of Organization of the Joint Traffic Association. Washington, D.C.: Bureau of Railway Economics, Association of American Railroads, November 19, 1895.

Baltimore Corn and Flour Exchange. *Annual Report*. 1880–1899.

Burgess, G. H., and M. C. Kennedy. *Centennial History of the Pennsylvania*. Philadelphia: Pennsylvania Railroad Company, 1949.

Chicago Board of Trade. *Annual Report*. Chicago: 1870–1900.

———. *Report of the Commission on Railway Discriminations*. Chicago: 1876.

———. *Underbilling*. Chicago: 1872.

Chicago and Grand Trunk Railway. *Arguments on the Apportionment of East Bound Traffic*. 1880.

Clark, V. S. *History of Manufactures in the United States*. Vol. II. New York: 1929.

Division of Dead Freight in Livestock 1882. Abstract of Proceeding of Chicago Committee. New York: Russell Brothers, Printers, 1882.

Duke, B. W. *Transportation Tariffs*. Louisville: 1886.

Financial Chronicle. Chicago, Illinois: 1870–1899.

Fink, Albert. *Bill of Lading Exposed*. New York: 1886.

———. *Cost of Railway Transportation*. Louisville: 1875.

———. *The Railroad Problem and Its Solution, 1880*. N.p.

———. *Report on Adjustment of Railway Transportation Rates, 1882*. N.p.

———. *Statistics Covering the Movement of East Bound and West Bound Traffic over the Trunk Lines and Connecting Roads*. N.p., n.d.

———. *Why Railway Tariffs are not Maintained*. N.p.: 1881.

The Great Railway Conflict: Remarks of John W. Garrett, President, Made on April 14, 1875 at the Regular Monthly Meeting of the Board of Directors of the Baltimore and Ohio Railroad Co. Baltimore: The Sun Book and Job Printing Co., 1875.

Grodinsky, J. *The Iowa Pool: A Study in Railroad Competition, 1870–1884.* Chicago: The University of Chicago Press, 1950.

Illinois Railroad and Warehouse Commission. *Annual Report.* 1870–1900.

Interstate Commerce Commission. *Investigation of the Joint Traffic Association.* 55th Congress, 2d Session, Senate Document No. 133, n.d.

———. *Railways in the United States in 1902.* Washington: U.S. Government Printing Office, 1903.

Johnson, G. R., *et al. History of Domestic Trade and Foreign Commerce.* Washington: 1915.

Lansing, G. L. *Natural Principles Regulating Railroad Rates.* Chicago: Railway Age Publishing Company, 1887.

Little, I. M. D. *A Critique of Welfare Economics.* Oxford Paperbacks, 1960.

Luce, R. D., and H. Raiffa. *Games and Decisions.* New York: Wiley and Sons, 1957.

Manifest Book: Cesna, Pennsylvania. Archives of the Pennsylvania Railroad.

The Manual of Statistics: Railroads, Grain, and Produce Cotton, Petroleum, mining dividends and production. N.p.: 1885.

New York Journal of Commerce. 1885–1886.

New York Produce Exchange. *Annual Report.* 1870–1899.

———. *Annual Statistical Report.* 1882–1884.

New York State Legislative Assembly, Special Committee on Railroads. *Railroad Investigation, 1879.* New York: Evening Post Steam Presses, 1879.

Pittsburgh Chamber of Commerce. *Discriminatory Freight Rates on Pittsburgh Coal.* 1911.

Poor, Henry V. *Manual of Railroads of the United States.* New York (70 Wall Street): H. V. and H. W. Poor.

Proceedings and Circulars of the Joint Executive Committee. New York: Russell Brothers Printers, 1880.

Proceedings and Circulars of the Joint Rate Committee of the Trunk Line, Central Traffic, and Western Freight Associations. Chicago: Henry O. Sheppard Company, 1890+.

Proceedings and Circulars of the Trunk Line Committee. New York: Russell Brothers, Printers, 1887+.

Proceedings and Circulars of the Trunkline Committees (freight department), 1884. Washington, D.C.: Bureau of Railway Economics, Archives of the Association of American Railroads, January 1884.

The Proceedings and Circulars of the Trunkline Executive Committee, 1880–1881. Washington, D.C.: Bureau of Railway Economics, Archives of the Association of American Railroads.

Proceedings of the Joint Traffic Association, 1895–1896. N.p., n.d.

Railroad Commission of the State of Ohio. *Annual Report.* 1870–1899.

Report of the Commissioner of Agriculture (for the years 1870 to 1872). Washington: U.S. Government Printing Office.

Ringwalt, J. L. *Development of Transportation Systems in the United States.* Philadelphia: 1888.

St. Louis Union Merchants' Exchange. *Annual Report*. St. Louis: 1871.

Snyder, C. *Business Cycles and Business Measurements.* New York: 1929.

Taylor, C. H. *History of the Board of Trade of Chicago*. Chicago: 1917.

Taylor, G. R. *The Transportation Revolution 1815–1860.* New York: Rinehart and Company, 1951.

Messrs. Thurman, Washburn, and Cooley. *Advisory Committee on Differential Rates.* New York: 1882.

Trunkline Executive Committee. *Report on Relative Costs of Transporting Livestock*. 1883.

U.S. Industrial Commission. *Report of the U.S. Industrial Commission*, Volumes 4, 9, 19. 57th Congress, 1st Session, House Document No. 178. Washington 1902.

U.S. Senate Committee on Interstate Commerce. *Hearings Before the Senate Committee on Interstate Commerce*, Volumes 1, 2, and 3. 59th Congress, 1st Session, Document No. 244. Washington: U.S. Government Printing Office, 1905.

———. *Regulation of Railway Rates*. Washington: U.S. Government Printing Office, 1905.

———. *Report on Transportation Routes to the Seaboard.* 43rd Congress, 1st Session, Senate Report No. 307, two parts. Washington: U.S. Government Printing Office, 1877–1878.

Watkins, J. E. *History of the Pennsylvania Railroad, 1846–1896.* Unpublished manuscript in the Smithsonian Institute Library.

Williamson, H. F. (ed.). *The Growth of the American Economy*. New York: 1951.

Articles and Notes

"Awards, 1879–1880," *Proceedings of the Joint Executive Committee, 1880*. New York: Russell Brothers, Printers, 1880.

Blanchard, G. R. Testimony in *Railroad Investigation 1879*. New York State Legislature, Special Committee on Railroads.

Chicago Board of Trade. "Weekly Commerical Bulletins," *Daily Commercial Report*. 1871–1879.

Chicago Daily Commercial Bulletin. April 4, 1881–December 30, 1899.

Chicago Times, December 22, 1886.

Chicago Tribune. 1875–1876.

"The Division of Trade from New York," *Annual Report of the New York Produce Exchange*. New York: 1877.

Fink, Albert. Letter of March 11, 1878, to John W. Garrett. Archives of the Baltimore and Ohio Railroad.

———. Speech reported in *The Twenty-Second Annual Report of the Trade and Commerce of Chicago of the Board of Trade*. Chicago: 1880.

Fishlow, A., and P. A. David. "Optimum Resource Allocation in an Imperfect Market Setting," *Journal of Political Economy*, *LXIX* (1961), 529–546.

Garrett, J. W. Letter of February 15, 1875, to T. A. Scott, as shown in unpublished manuscript entitled "The Pennsylvania Railroad Company" in the Archives of the Smithsonian Institution.

———. Letter of July 13, 1877, to T. A. Scott. Baltimore and Ohio Railroad Archives.

Gilchrist, D. T. "Albert Fink and the Pooling System," *Business History Review*, *XXXIV* (1960), 25–49.

Gray, G. M. Report contained in *Argument Regarding the Division of East Bound Freight from Chicago*. New York: Russell Brothers, Printers, 1879.

Hill, H. C. "The Development of Chicago as a Center of the Meat Packing Industry," *Mississippi Valley Historical Review*, *X* (1923), 253–273.

King, John. Letter of November 25, 1874, to president of the Baltimore and Ohio Railroad. Archives of the Baltimore and Ohio Railroad.

———. Letter of July 12, 1877, to J. H. Rutter. Archives of the Baltimore and Ohio Railroad.

———. Letter of July 3, 1879, to J. H. Rutter. Archives of the Baltimore and Ohio Railroad.

———. Atlantic and Pacific Telegram to John W. Garrett, February 7, 1879. Archives of the Baltimore and Ohio Railroad.

———. Baltimore and Ohio Telegram to John W. Garrett, February 22, 1879. Archives of the Baltimore and Ohio Railroad.

Mills, Arthur. Letter of November 14, 1887, to Mr. N. Guilford. Cleveland, Ohio: Archives of the New York Central Railroad.

The New York Times. November 29, 1878–March 25, 1897.

"Percentages for the Division of Eastbound Dead Freight from Chicago," File 50114, June 20, 1891, the Archives of the Pennsylvania Railroad System, Philadelphia, Pennsylvania.

Perkins, C. E. Letter of September 23, 1887, to Mr. E. P. Ripley. Archives of the Chicago, Burlington and Quincy Railroad.

The Railroad Gazette. January 6, 1872–October 28, 1898.

"The Railroads: Important Meeting of Eastern and Western Managers," *Chicago Tribune*, December 19, 1878.

"Report of the Chicago Committee," *Proceedings Western Railroads, 1877–1880*. New York: Russell Brothers, Printers, 1880.

Report of the Senate Select Committee on Interstate Commerce. Statement of John W. Grubbs, 49th Congress, 1st Session, Report 46, Part 1. Washington: U.S. Government Printing Office, 1886.

"A Review of the Operation of the East Bound Pool," *Tenth Annual Report of the Massachusetts Railroad Commissioners*. Boston: 1879.

Ripley, E. P. Letter of September 22, 1887, to Mr. C. E. Perkins, Archives of the Chicago, Burlington and Quincy Railroad Company.

Rose, A. C. Letter of September 17, 1877, to J. W. Garrett. Archives of the Baltimore and Ohio Railroad.

Rutter, J. H. Dispatch to John King, Jr., July 10, 1877 (7:43 P.M.). Archives of the Baltimore and Ohio Railroad.

———. Dispatch to John King, Jr., July 11, 1877 (12:20 P.M.). Archives of the Baltimore and Ohio Railroad.

Scott, T. A. Letter of February 15, 1875, to J. W. Garrett, as shown in unpublished manuscript entitled "The Pennsylvania Railroad Company" in the Archives of the Smithsonian Institution.

"Statement of Tons of Each Class of East Bound Freight from Chicago from March 11th to April 30th, 1878 . . . ," *Proceedings Western Railroads, 1877–1880*. New York: Russell Brothers, Printers, 1880.

Tullis, E. A. *Historical Memorandum Concerning Class Rates in Central Freight Association Territory*. Chicago: Pennsylvania Railroad Company (historical files), 1918.

Yance, J. V. "A Model of Price Flexibility," *American Economic Review, 50* (June, 1960), 401–418.

Cases and Legal Materials

H. Bates, et al. vs. *The Pennsylvania Railroad Company, ICC Reports IV* (1890), 281.

The Board of Trade Union of Farmington et al. vs. *The Chicago, Milwaukee and St. Paul Railroad Company, ICC Reports, I* (1887), 215.

The Boston Chamber of Commerce vs. *The Lake Shore and Michigan Southern Railway Company, et al., ICC Reports, I* (1887), 436.

Brown vs. *Walker*, 161 U.S. 591 (1896).

Cincinnati, New Orleans and Texas Pacific Railway Company vs. *Interstate Commerce Commission et al.*, 162 U.S. 184 (1896).

Compulsory Testimony Act of February 11, 1893, 27 Stat. 443.

Counselman vs. *Hitchcock*, 142 U.S. 547 (1892).

Coxe Brothers and Company vs. *Lehigh Valley Railroad Company, ICC Reports, IV* (1891), 536.

Foster vs. *C. C. C. and St. L. Railway Company*, 56 Fed. 434 (1893).

In the Matter of Alleged Excessive Freight Rates and Charges on Food Products, ICC Reports, IV (1890), 48.

In Re Louisville & Nashville Railroad Company, ICC Reports, I (1887), 31.

Independent Refiners Association of Titusville et al. vs. *Western New York and Pennsylvania Railroad Company et al., ICC Reports, V* (1892), 415.

Interstate Commerce Commission vs. *Alabama Midland Railway Company*, 168 U.S. 144 (1897).

Interstate Commerce Commission vs. *Baltimore and Ohio Railway Company*, 43 Fed. 37 (1890).

Interstate Commerce Commission vs. *Cincinnati, New Orleans, and Texas Pacific Railway Company*, 167 U.S. 479 (1897).

P. N. C. Logan, F. D. Babcock and E. M. Parsons, et al. vs. *Chicago & Northwestern Railway*, Vol. III, *The Annual Report of the Interstate Commerce Commission* (1889).

The Manufacturers' and Jobbers' Union of Mankato, Minnesota vs. *The Minneapolis and St. Louis Railway Company et al., ICC Reports, IV* (1890), 79.

The New York Board of Trade and Transportation et al. vs. *The Pennsylvania Railroad Company*, *ICC Reports*, *IV* (1891), 447.

New York Produce Exchange vs. *Baltimore and Ohio Railroad*, *ICC Reports*, *VIII* (1898), 614.

The New York Produce Exchange vs. *The New York Central and Hudson River Railroad Company, et al.*, *ICC Reports*, *III* (1889), 137.

Osborne vs. *Chicago and Northwestern Railroad Company*, 52 Fed. 912 (1892).

E. B. Raymond vs. *Chicago, Milwaukee & St. Paul Railway Company*, *ICC Reports*, *I* (1887), 231.

John P. Squire and Company vs. *Michigan Central Railroad Company et al.*, *ICC Reports*, *IV* (1891), 611.

Union Pacific Railway Company vs. *Goodrich*, 149 U.S. 680 (1893).

United States vs. *The Joint Traffic Association*, 19 Sup. Ct. 506, 171 U.S. 505 (1898).

United States vs. *Michigan Central Railroad Company*, 43 Fed. 26 (1890).

U.S. vs. *Trans-Missouri Freight Association*, 166 U.S. 290 (1897).

Index

Act to Regulate Commerce: definition of legal and illegal rate setting, 111; standards for just and reasonable rates, 113–115; standards for long- and short-haul rate differences, 116–118

Alleged Excessive Freight Rates and Charges on Food Products IV *I.C.C.R.* 48 (1890), 146n

Boston and Albany Railroad Company vs. *Boston and Lowell Railroad Company* I *I.C.C.R.* 158 (1887), 115

Baltimore and Ohio: entry into Chicago, 42; effect on grain price differential, 42–45; effect on railroad rates, 42–44; effect on Western Railroad Bureau, 45–49; success of independent rate cutting, 47–49

Bates, H., et al. vs. *The Pennsylvania Railroad Company* IV *I.C.C.R.* 281 (1890), 146n

Blanchard, G. R., 29n

Board of Trade Union of Farmington et al. vs. *The Chicago, Milwaukee and St. Paul Railroad Company* I *I.C.C.R.* 215 (1887), 117n

Boston Chamber of Commerce vs. *The Lake Shore and Michigan Southern Railway Company et al.* I *I.C.C.R.* 451 (1888), 117

Brown vs. *Walker* 161 U.S. 591 (1896), 159n

Cartel: prerequisites for, 15–20; results of successful, 21–23

Cartel of 1879: effect on shippers, 90–91, 93; effect on commodity prices, 80–82, 93, 98–99, 102; effect on variation between long- and short-distance rates, 90–91, 103, 104, 107; effect on loyal railroads, 106–109; effect on disloyal railroads, 102–103, 104

Cartel of 1886–1888: procedures to comply with *Act to Regulate Commerce,* 123; effect on rates, 129–135, 193–195, 201; effect on provision prices, 136–140; effect on grain prices, 137–140; effect on railroad profits, 142–143, 196–201

Central Traffic Association of 1891: scale of rates, 149–151; effect on grain prices, 151–152; effect on provision prices, 151–153; rate war of 1893–1894, 159–161

Chicago Daily Commercial Bulletin, consumers reports of rates paid, 27n

Cincinnati, New Orleans and Texas Pacific Railroad Company vs. *Interstate Commerce Commission* 162 U.S. 184 (1896), 156n, 164, 183–185

Clark, V. S., 1

Collusive agreements of early 1870's, 26–30; effect on grain markets, 32–39

Collusive agreements of 1876–1878, 50–59; effects of rate cutting on, 56–58, 66–71; effects of breakdown on disparity between long- and short-distance rates, 71–73; effects of breakdown on profits of individual railroads, 71–77

Compulsory Testimony Act, 164n

Cooley, T. M., 114

Counselman vs. *Hitchcock* 142 U.S. 547 (1892), 158n, 159

Cowles, Alfred, 200n

Coxe Brothers and Company vs. *Lehigh Railroad Company* IV *I.C.C.R.* 536 (1891), 145n

David, P. A., 204n

Depew, Chauncey, 165

Field, Marshall, 127

Fink, Colonel Albert, 52n, 55n

Fishlow, A., 204n

Fog, B., 32n

Foster vs. *C.C.C. and St. Louis Railway Company* 56 Fed. 434 (1893), 145n

Garrett, John W., 40n, 41, 51n, 52n, 53n, 55, 71n

Gilchrist, D. T., 52
Gray, G. M., 72
Great Lakes transportation, in 1870's, 6
Grodinsky, J., 120n
Grubbs, John W., 110
Guilford, John, 53

Hill, H. C., 1
Hill, J. B., 149n

Independent Refiners Association of Titusville et al. vs. *Western New York and Pennsylvania Railroad Company et al.* V *I.C.C.R.* 415 (1892), 145n
Interstate Commerce Commission: regulation of rate wars, 118; regulation of pooling, 120; and freight classification, 121; interpretation of "just and reasonable," 146–148; standards for rate changes 1891–1893, 147–148; and Lake rail rates, 157
Interstate Commerce Commission vs. *Alabama Midland Railway Company* 168 U.S. 144 (1897), 185n
Interstate Commerce Commission vs. *Baltimore and Ohio Railway Company* 43 Fed. 37 (1890), 133n
Interstate Commerce Commission vs. *Cincinnati, New Orleans and Texas Pacific Railway Company* 167 U.S. 479 (1897), 184

Jewett, John, 25
Johnson, G. R., 1
Joint Traffic Association: organization in 1895, 177; effect on grain prices and shipment shares, 179–182, 189–191; and the Sherman Act, 187–191, 193
Joyce, W. H., 132

King, John, 10n, 25n, 51n, 53n, 57, 58

Lancaster, Kelvin, 204n
Lipsey, R. G., 204n
Little, I. M. D., 201n
Logan, P. N. C., F. D. Babcock and E. M. Parsons, et al. vs. *Chicago & Northwestern Railway* III *I.C.C.R. Ann. Rep.* 127 (1889), 119n
Louisville and Nashville Railroad Company I *I.C.C.R.* 31 (1887), 114

McCabe, D. T., 129n
Manufacturers' and Jobbers' Union of Mankato, Minnesota vs. *The Minnesota and St. Louis Railway Company* IV *I.C.C.R.* 79 (1890), 145n
Markets for transport services in the Midwest, 1, 2, 6
Maximum Freight Rates Case, see *Interstate Commerce Commission* vs. *Cincinnati, New Orleans and Texas Pacific Railway Company*
Mayer Buggy Company, see *Cincinnati, New Orleans and Texas Pacific Railroad Company* vs. *Interstate Commerce Commission*
Morgan, J. P., 171n

New York Board of Trade and Transportation et al. vs. *The Pennsylvania Railroad Company* IV *I.C.C.R.* 447 (1891), 146n
New York Central, independent rate setting with Wabash Railroad, 52–54
New York Produce Exchange vs. *Baltimore and Ohio Railroad* VIII *I.C.C.R.* 614 (1898), 142n
New York Produce Exchange vs. *The New York Central and Hudson River Railroad Company et al.* 3 *I.C.C.R.* 137 (1889), 117n

Osborne vs. *Chicago and Northwestern Railroad Company* 52 Fed. 912 (1892), 155–156

Perkins, C. E., 120n, 127n
Pooling of revenues, 92
Poor, Henry V., 103

Raymond, E. B., vs. *The Chicago, Milwaukee and St. Paul Railway Company* I *I.C.C.R.* 231 (1887), 117n
Ripley, E. P., 120n, 127n
Rose, A. C., 53n
Rutter, J. H., 53n, 57n

Scott, T. A., 25n, 40n, 41n, 52n
Snyder, C., 200n
Social Circle Case, see *Cincinnati, New Orleans and Texas Pacific Railroad* vs. *Interstate Commerce Commission*
Squire, J. P., and Company vs. *Michigan Central Railroad et al.* IV *I.C.C.R.* 611 (1891), 147n

Taft, Justice W. H., 17
Taylor, C. H., 1
Taylor, G. R., 1
Tiller, Sir Henry, 88
Trans-Missouri Freight Rate Association, 187–188
Troy Case, see *Interstate Commerce Commission* vs. *Alabama Midland Railway Company*
Tullis, E. A., 148–150

Union Pacific Railway Company vs. *Goodrich* 149 U.S. 680 (1893), 145n
United States vs. *The Joint Traffic Association* 19 Sup. Ct. 506 (1898), 187n, 204n

United States vs. *Michigan Central Railroad Company* 43 Fed. 26 (1890), 133n

Vanderbilt, Alfred, 53, 88
Veblan, T. B., 7
Vermont Central Case, see *Boston and Albany Railroad Company* vs. *Boston and Lowell Railroad Company*

Western Railroad Bureau, establishment of, 39
Williamson, H. F., 1

Yance, J. V., 34n